CONTENTS.

4 CONTENTS.

ILLUSTRATIONS.

THE GOLD PLACERS OF PARTS OF SEWARD PENINSULA, ALASKA.

By Arthur J. Collier, Frank L. Hess, Philip S. Smith, and Alfred H. Brooks.

PREFACE.

By Alfred H. Brooks.

This paper is devoted to a discussion of the auriferous gravels of the southern and northwestern parts of Seward Peninsula. Its purpose is to meet the wants of the miner and prospector by presenting the results of the reconnaissance surveys made from 1899 to 1903, so far as they bear on the distribution and origin of the placer gold. As a necessary preliminary the geology of the peninsula will be presented in outline, but as this volume must be kept to a reasonable size, much matter not directly pertinent to the discussion of the mineral wealth will perforce have to be omitted. More extended accounts of the geology will find place in future publications, to be prepared when the more detailed surveys, already begun, permit the presentation of conclusions with more confidence.

Geologic and topographic surveys of the peninsula were begun in the fall of 1899 by F. C. Schrader[a] and the writer, who spent a few weeks in examining the auriferous gravels in the immediate vicinity of Nome, while D. C. Witherspoon made a topographic survey of the same area. The increase of public interest led to the dispatching of three parties to this field in 1900. One of these, led by W. J. Peters and W. C. Mendenhall,[b] made reconnaissance surveys of the southeastern part of the peninsula. Another party under charge of E. C.

---

[a] Schrader, F. C., and Brooks, A. H., Preliminary report on the Cape Nome gold region, a special publication of the U. S. Geol. Survey, 1900.

[b] Mendenhall, W. C., Reconnaissances in the Cape Nome and Norton Bay regions, Alaska, in 1900, a special publication of the U. S. Geol. Survey, 1901, pp. 181–218.

7

Barnard, assisted by D. L. Reaburn, H. G. Hefty, and R. B. Robertson, made a topographic survey of the southwestern part of the peninsula, including Port Clarence to the west and Ophir Creek to the northeast. This work was materially aided by the cordial cooperation of J. F. Pratt and J. J. Gilbert, of the United States Coast and Geodetic Survey, who commanded a large force of men engaged during the same summer in charting the southern coast line of the peninsula. At the same time the geologic work in this region was intrusted to the writer,[a] who, assisted by G. B. Richardson and A. J. Collier, carried a reconnaissance over much of the southern half of the peninsula and extended it into the then newly discovered Kougarok district. In the fall of 1900 the writer also made a geologic and topographic reconnaissance survey of the York district, at the west end of the peninsula. In 1901 W. C. Mendenhall [b] and D. L. Reaburn made a hasty examination of part of the northern coast of the peninsula, after the completion of an exploration extending from the Yukon to Kotzebue Sound. The growth of the mining interests led to the extension of the reconnaissance survey over the northwestern part of the peninsula in 1901 by T. G. Gerdine and A. J. Collier.[c] The northeastern quarter of the peninsula still remained to be surveyed, and this was done in 1903 by D. C. Witherspoon and F. H. Moffit.[d]

The reports on these surveys included an account of practically every gold-producing creek in the peninsula, but the rapid progress of mining not only led to the discovery of many new placers, but also revealed many new facts bearing on the occurrence and distribution of the auriferous gravels. This made it desirable to supplement these investigations, and to this task A. J. Collier, assisted by F. L. Hess, was assigned in 1903. In the fall of 1904 Mr. Collier was able to spend a few weeks in the peninsula to collect some additional data. The writer's own investigations in 1899 and 1900 have been referred to; in addition to these he has made brief examinations of some of the mining districts along the southern coast in 1903, 1904, and 1905.

The results here set forth are based (1) on the studies of Mr. Collier and Mr. Hess in 1903, and (2) on the earlier work in this province. In some cases this fact is plainly brought out by direct quotations from the published reports, but in many others there has been so gradual an evolution of opinion as to make it impossible to give the proper credit. Although not all the subjects treated in the

---

[a] Brooks, A. H., Reconnaissances in the Cape Nome and Norton Bay regions, Alaska, in 1900, a special publication of the U. S. Geol. Survey, 1901, pp. 1–180.

[b] Mendenhall, W. C., Reconnaissance from Fort Hamlin to Kotzebue Sound by way of Dall, Kanuti, Allen, and Kowak rivers: Prof. Paper U. S. Geol. Survey No. 10, 1902.

[c] Collier, A. J., A reconnaissance of the northwestern portion of Seward Peninsula, Alaska: Prof. Paper U. S. Geol. Survey No. 2, 1902.

[d] Moffit, F. H., The Fairhaven gold placers, Seward Peninsula, Alaska: Bull. U. S. Geol. Survey No. 247, 1905.

earlier publications are here again considered, yet in a large measure this volume is in fact to be regarded as a revised edition of the report of 1900.

As this manuscript goes to press it has been possible to incorporate in it a part of the results of the surveys of 1906. P. S. Smith, who made a hasty study of some of the placers of the Solomon, Casadepaga, and Council regions, has incorporated some of his results with Mr. Collier's and Mr. Hess's descriptions of these districts. Mr. Smith has also prepared the account of the Iron Creek basin here included. The description of the Nome region has had the critical reading of F. H. Moffit, who has been for two years engaged in a detailed geologic study of the areas covered by the Nome and Grand Central special maps. It is not intended, however, to set forth here anything but a general account of this district. For the details the reader is referred to Mr. Moffit's report [a] now in preparation.

In 1906 the writer made a reexamination of a part of the Kougarok district, and his report is included in this volume.

The composite authorship of this report is indicated in above account of its preparation. It should be added that the manuscript submitted by Mr. Collier was revised by Mr. Hess, who utilized as far as possible the data collected by the later investigations. No attempt has been made to revise the geologic descriptions except so far as it has been necessary to make them consistent with the results of the later investigations.

The unfortunate delay in the completion of this volume has been in a large measure due to the fact that the time of the authors has been perforce devoted to other work, but also because each season's field work accumulated many new facts, which it seemed desirable to incorporate.

It is a matter of deep gratification to the writer that these investigations have received much cordial support and help on the part of the residents of the peninsula. If this work serves to aid those who are engaged in exploiting the mineral resources of the district, the purpose of its publication will be accomplished.

---

[a] Moffit, F. H., Hess, F. L., and Smith, P. S., Geology of the Nome and Grand Central areas, Seward Peninsula, Alaska.

THE DEVELOPMENT OF THE MINING INDUSTRY.

By Alfred H. Brooks.

FOREWORD.

The large peninsula thrust out from northwestern Alaska, dividing the Bering Sea from the Polar Sea, was, up to a few years ago, unnamed [a] and almost unknown. This peninsula, bordered by the Arctic Circle, affords a striking example of the rapid industrial changes that may be wrought in an isolated province by the exploitation of rich gold placers. A decade ago Seward Peninsula was little more than a barren waste, unpeopled except for a few hundred Eskimos and a score of white men, whereas it is now the scene of intense commercial activity, supporting a permanent population of 3,000 or 4,000 people, which in summer is more than doubled. Then the igloo of the Eskimos and a mission were the only permanent habitations; now a well-built town with all the adjuncts of civilization looks out on Bering Sea, and a dozen smaller settlements are scattered through the peninsula. This region, which then produced nothing except a few furs, now increases the wealth of the world annually by nearly $8,000,000. A decade ago the only communication with the civilized world was through the annual visit of the Arctic whaling fleet and the revenue cutter; now a score of ocean liners ply between Nome and Puget Sound during the summer months, and even in winter a weekly mail service is maintained by dog teams. Moreover, military telegraph lines, cables, and wireless systems, and a private telephone system keep all parts of the peninsula in close touch with the outer world. Railways connecting some of the inland mining centers with tide water traverse regions which a few years ago were almost unknown to white men.

This industrial improvement is, as has been stated, the result of the discovery and exploitation of the gold deposits. As there has been

---

[a] The name Seward Peninsula, given in honor of William H. Seward, was first published in "Preliminary report on the Cape Nome gold region," by F. C. Schrader and A. H. Brooks, U. S. Geol. Survey, 1900. W. H. Dall had as long ago as 1870 suggested the name "Kaviak Peninsula," but it did not find acceptance in any except his own publications and has never appeared on any map.

10

only one successful attempt at auriferous lode mining, practically all of the gold production has been taken from the placers. Lode and placer tin have been found in the peninsula, but their exploitation has not yet been brought to the profitable stage, though a few tons of tin are annually exported. The tin deposits have, however, added to the commercial prosperity of the peninsula by helping to focus public attention on its mineral resources.

The following figures of the output of the placer mines are based on the best estimates obtainable, but it should be borne in mind that there are no accurate statistics to be had. The limit of error in the total production since 1898, as shown in the table, is believed to be about $500,000.

Approximate value of placer-gold production of Seward Peninsula, Alaska.[a]

1898	$75,000	1904	$4,164,600
1899	2,800,000	1905	4,800,000
1900	4,750,000	1906	7,500,000
1901	4,130,000		
1902	4,561,800		
1903	4,465,600		37,247,000

The total production of $37,000,000 for nine years is small compared with the output from the California placers, which in two years (1851 to 1853) are estimated to have yielded $62,000,000.[b] It is also small compared with the production of the Klondike placers, whose output in the first decade is valued at $118,000,000.[c]

Fig. 1 illustrates approximately the relative size of the gold-bearing regions of Seward Peninsula, California, and the Klondike. The auriferous gravels of California, roughly outlined, probably cover an area about equal to that occupied by the auriferous gravels of the Seward Peninsula, but the Klondike gold field is probably less than one-tenth as large. The California placers are not only ideally located for economic exploitation, but their gold content averaged no less than that of the Seward Peninsula gravels. Moreover, the high gravels of California are far more extensive than those of the Alaska

[a] Brooks, A. H., Report on progress of investigation of mineral resources of Alaska: Bull. U. S. Geol. Survey No. 314, 1907, p. 21.

[b] Whitney, J. D., The auriferous gravels of the Sierra Nevada of California: Mem. Mus. Comp. Zoology, Cambridge, Mass., vol. 6, No. 1, 1880, p. 367.

[c] The estimates of the annual output of placer gold in the Klondike since its discovery are as follows:

1896	$300,000	1903	$12,250,000
1897	2,500,000	1904	10,000,000
1898	10,000,000	1905	7,300,000
1899	16,000,000	1906	5,600,000
1900	22,275,000		
1901	18,000,000		
1902	14,500,000		118,725,000

field. With abundant water supply, steep stream gradients, heavy gravel deposits, accessibility, and salubrious climate, it is no wonder that the California placers far outstripped the northern field in the first years of production.

The Klondike, on the other hand, is less favorably situated than Seward Peninsula, and its water supply available for mining is much less. It appears, however, that the placers of such creeks as Eldorado and Bonanza in the Klondike averaged richer than any deposits of similar extent yet found in the peninsula. It was the exploitation of these almost fabulously rich and relatively shallow gravels that brought the Klondike gold output up with a bound, and it is their

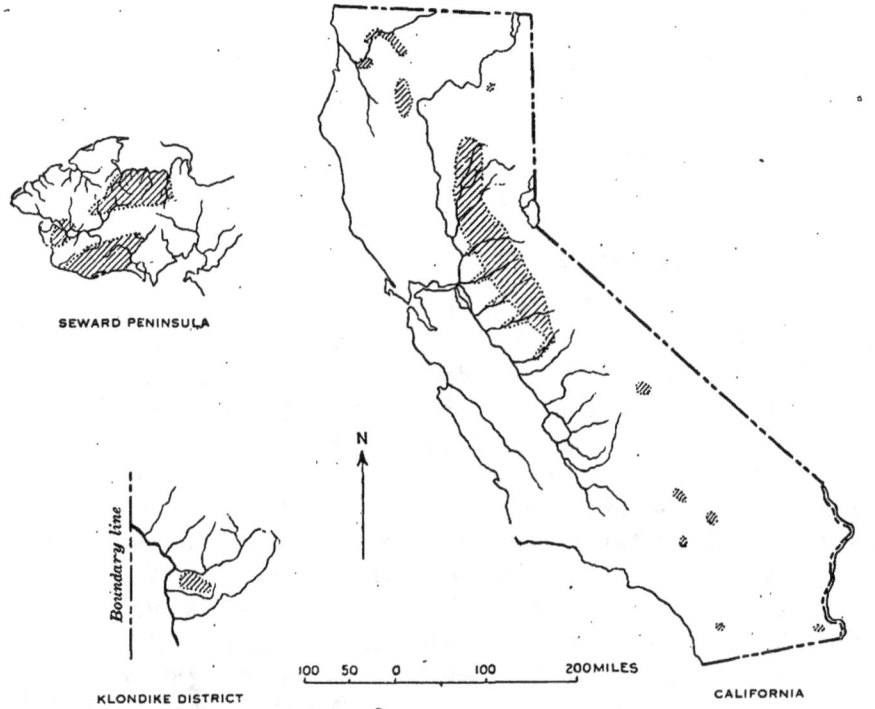

SEWARD PENINSULA

Boundary line

N

100 50 0   100   200 MILES

KLONDIKE DISTRICT   CALIFORNIA

FIG. 1.—Sketch maps showing relative gold-bearing areas in Seward Peninsula, California, and the Klondike region, drawn to the same scale.

quick exhaustion that has caused an almost equally rapid decline of the annual yield. There are still extensive bodies of lower-grade gravels to mine in the Klondike, but these can be developed only by means of extensive water conduits or by dredging. Mining in the Klondike has passed its zenith, whereas in Seward Peninsula the maximum yearly output is still to be reached.

In the comparison of the Seward Peninsula placer fields with others, it must be borne in mind that probably three-fourths of its entire production has been drawn from the region adjacent to Nome and from Ophir Creek and its tributaries. Therefore, though the gold-bearing area is large, yet only a few square miles have been extensively exploited. In other words, only a few of the creeks (but probably the richest ones of Seward Peninsula) have been exploited, and it is probable that it will be some time before the maximum annual yield will be attained. The future of the placers receives further consideration in the section devoted to economic geology (pp. 135–139), and the annual gold product is discussed in detail where the growth of the mining industry is sketched.

DISCOVERY OF GOLD.

The position of Seward Peninsula, whose western point is less than 60[a] miles from the Siberian coast, made it the first part of Alaska to be known to the Russians. The continent lying beyond Bering Strait was reported by the natives to the Russians as early as 1709; but so far as known the first landing on its borders was not until 1731, and then from a vessel which had been blown out of her course. Certain it is that the first survey of the coast line was made by Capt. James Cook in 1778. During the succeeding century of the Russian occupation, though the entire coast line was explored,[b] but little was learned of the interior of the peninsula. In 1822 the Russian Captain Kromchenko, commanding the *Golofnin*, surveyed Norton Sound and Golofnin Bay. It was not until 1835 that a trading post was established on St. Michael Island, the first Russian settlement north of the Aleutian Islands. From this point, trade was carried on with the natives to the north. The stories current that the Russians mined placer gold near Nome are utterly without foundation in fact, for the Russians had absolutely no knowledge of the mineral resources of the peninsula and never had a permanent settlement within its borders.

Probably the first important inland exploration was that made in 1865–66 by a party led by Baron Otto von Bendeleben, who, while seeking a route for a telegraph line, ascended the Niukluk and crossed the portage to the Kruzgamepa and thence continued to Port Clarence. This survey was made in the interests of the great transcontinental telegraph route which was planned to connect Europe and America. Von Bendeleben's party actually constructed a few miles of telegraph line, and thus were the first to start a commercial enterprise on Seward Peninsula. Of greater interest is the fact that Von Bendeleben is said to have found alluvial gold on Niukluk River.[c] If such was the case placer gold was found here earlier than in any other part of Alaska except at Cook Inlet, where Doroshin, a

[a] Chart T, U. S. Coast and Geodetic Survey.
[b] For an account of the Russian explorations, see Brooks, A. H., Geography and geology of Alaska: Prof. Paper U. S. Geol. Survey No. 45, 1906, pp. 104–132.
[c] W. H. Libby, a member of Von Bendeleben's party, is authority for this statement.

mining engineer sent out by the Russian Government, discovered gold in 1854. The telegraphic construction was discontinued after the successful laying of the Atlantic cable, Von Bendeleben's party disbanded, and a third of a century passed before any of its members found their way back to the scene of the gold discovery. It seems probable, therefore, that no one attached any importance to this occurrence of placer gold.

The peninsula continued to be regarded as a barren waste for many years after Alaska passed into the hands of the United States. The whalers had a rendezvous at Port Clarence, where they anchored in the early summer while waiting for the polar ice cap to break. Occasionally a trader came from the nearest permanent posts, Unalaklik or St. Michael, and bartered with the natives, but otherwise conditions were as primitive as during the Russian occupation.

Galena was discovered in the eastern part of the peninsula by the natives, who used it for making bullets, and brought it to the attention of traders at least as early as 1880. In 1881 a party under the leadership of John C. Green, with the help of the natives, located the source of this galena on what he named Fish River, tributary to Golofnin Bay. This was one of the first lode locations made in Alaska, being antedated only by those made near Sitka in 1877 and possibly by some of those at Juneau, where gold placers were found in 1880. Green and his party organized the Fish River mining district in July, 1881, and began the exploitation of this deposit under the corporate title, The Alaska Gold and Silver Mining, Milling, and Trading Company.

It was undoubtedly some of those connected with this company who first made public the occurrence of alluvial gold in Seward Peninsula.[a] It has been elsewhere suggested [b] that the discovery and first mining of alluvial gold is to be credited to the natives. This now seems to be an error, for the natives appear never to have had any knowledge of gold or methods of extracting it.

The operations of the Alaska Gold and Silver Mining, Milling, and Trading Company, later called the Russian-American Mining and Exploration Company, were continued up to about 1891 at Omalik, but though some ore was shipped, the mine never actually reached a paying stage. This enterprise, however, led to a better knowledge of the region and to the establishment of a trading post on Golofnin Bay by John Dexter, an employee of the company.

Dexter's post was long the only white settlement, but about 1890 a Congregational mission was established at Cape Prince of Wales, and later a Swedish Evangelical mission was founded on Port Clarence,

---

[a] An employee of this company named Sanderson is said to have found alluvial gold on the Niukluk about 1892.

[b] Schrader, F. C., and Brooks, A. H., Preliminary report on the Cape Nome gold region, a special publication of the U. S. Geol. Survey, 1900, p. 31.

and another on Golofnin Bay. About the same time the Government began the importation of reindeer from Siberia, and stations were established at various points in the peninsula.

Previous to 1897, though no doubt a few prospectors [a] had roamed over parts of the peninsula, no mineral wealth had been found, save the Omalik [b] galena deposit and a little alluvial gold in the Fish River basin. A little placer mining is said to have been done at this time. As operations at the silver mine had practically ceased, the whole population included only a few missionaries, traders, and Government reindeer herders. This region then like most of Alaska, except the Pacific seaboard, had made almost no industrial advance since the Russian occupation.

While Seward Peninsula was neglected up to 1897, the ever active prospector had penetrated the Yukon basin and discovered placers which had been productive in a small way for upward of a decade. This activity was suddenly augmented in 1896, when the Klondike placers were found, and in less than a year this isolated region became a focal point of public interest throughout the civilized world. Then the wild excitement of the California rush of half a century before was duplicated, and probably 50,000 gold seekers turned their faces northward. The popular belief that Alaska was a barren field for all except the fur hunters quickly disappeared and was replaced by the equally absurd conclusion that the entire territory was gold bearing. As a result of this erroneous opinion, expeditions were equipped to search for gold in all parts of the territory. Most of these, however, were bent on reaching the interior, and Seward Peninsula was practically overlooked.

The Kotzebue Sound region attracted about 1,500 people, and some of them, becoming discouraged at the outlook, made their way southward to Golofnin Bay. To the same point came other groups of prospectors from St. Michael who despaired of being able to ascend the Yukon. It was some of these men, whom chance rather than design had brought to the district, that made the third [c] discovery of alluvial gold in the peninsula, and they were the first to prove its commercial importance. The party which found these workable placers included Daniel B. Libby, a member of the Von Bendeleben expedition of 1866, L. S. Melsing, A. P. Mordaunt, and H. L. Blake. The first gold was discovered in the gravels of what they called Melsing Creek, in March, 1898, and a little later placers were found on an adjacent stream named Ophir Creek. In accordance with the United States mining laws and in keeping with a long-established

---

[a] There is a story current that a Swede named Johansen found placer gold and did a little sluicing at the mouth of the Casadepaga in 1894.

[b] Omalik is said to be the Eskimo word for high chief.

[c] As has been set forth, the first was made by the telegraph expedition, and the second probably by some of those connected with the silver-mining enterprise on Fish River.

custom, a "miners' meeting" was called, and the "Discovery district" was organized April 25, and a recorder elected. As soon as the snow had disappeared, mining was begun as systematically as the means at hand would allow. The equipment of these pioneer placer miners of the peninsula was very meager, and consequently the efforts at exploitation were crude. Fortunately the region afforded some spruce, and sluice boxes could therefore be' constructed from whipsawed lumber. Such methods yielded a return during the summer variously estimated at $30,000 to $100,000.

In the summer of 1898 there were probably several hundred men in the new district, but the interest in it hardly extended beyond its borders. Even at St. Michael, only 100 miles away, which the writer visited in September of that year, hardly any attention was paid to this new camp. The reason for this apathy was twofold—first, the Alaskan public had become tired of unfounded rumors of rich discoveries, and, second, the excavations on Ophir Creek had not by any means gone far enough to prove the great richness of its gravels.

There was at least one little group of men among those in the new district who had the enterprise to extend their field of prospecting beyond the narrow limits in which the placers had been found. These men are said to have been influenced in directing their journey by the rumor of a discovery of coarse gold on Sinuk River by a Government reindeer herder. Be that as it may, a party said to have included N. C. Hultberg, J. J. Brynteson, H. L. Blake, and J. L. Haggalin sailed westward from Golofnin Bay in a small boat in July, 1898, and became stormbound near the present site of Nome. Prospecting the Snake River bars revealed some fine gold colors, and coarser gold is said to have been found, July 26 or 27, on what was afterwards called Anvil Creek. Most of the party, however, considered the prospect not encouraging and decided to continue the journey to Sinuk. As nothing was found there, the party returned to Golofnin Bay. The gold found on Snake River was, however, regarded by some as sufficiently encouraging to warrant further investigation, so Brynteson, with two companions, [a] Jafet Lindeberg and Erik O. Lindblom, returned to the scene of this discovery about the middle of September, and on the 20th they found, on what they named Anvil Creek, the rich placers which were destined to render this region famous throughout the world.[b] Though none of the party were ex-

---

[a] According to some accounts A. N. Kittleson, G. W. Price, and a Laplander named John Tornensis were also in this party.

[b] Since the above was written a letter has been received by Frank L. Hess from Jafet Lindeberg, which, from its interest, is quoted at length:

"DEAR SIR: As per your request, I herewith make a report concerning the discovery of gold and its development on the Seward Peninsula, which dates from 1897, and will say that it is correct, as I am one of the early prospectors of the mining camps that now figure so prominently in that part of Alaska.

"The discovery of gold on the Seward Peninsula dates back several years before the location of mining claims on Ophir Creek, in the spring of 1898. John Dexter, the

perienced gold prospectors, yet they could not but realize the importance of this discovery. The season was far advanced, but two crude rockers were constructed, and by using heated water about $1,800 worth of gold was extracted from the gravels of Anvil Creek and

---

well-known trader at Golofnin Bay, supplied several natives with gold pans and taught them how to wash out a pan of dirt. For that purpose he obtained some gold dust from St. Michael, and would sprinkle a little in a pan of beach sand at Golofnin Bay. These natives would carry the gold pans with them on their fishing and squirrel-hunting trips, and it was while out on one of these trips that Tom Guarick, a native Esquimaux, made the discovery of gold on Ophir Creek, in the Council City district, which at that time had, in all likelihood, never been visited by white man. Tom, on his return to Dexter's, reported his discovery and brought back about one-half ounce of gold dust. This was in August, 1897. One month later Daniel B. Libby, A. P. Mordaunt, L. S. Melsing, and Harry Blake, who were sent by San Francisco capitalists in the spring of 1897 to roam around that part of Alaska under a grub-stake contract, landed at Golofnin Bay, and on reaching Dexter's house saw this gold. On inquiry they were told when and how it was found, and a few days later, with Tom, the native, as a guide, they were taken to Ophir Creek, which was later named, and, on prospecting there, found that native Tom had made a discovery of gold on that creek. All through the winter of 1897 and 1898 the party, with other residents of the vicinity, prospected the different creeks, and finally in April, 1898, staked mining claims on Ophir Creek and Melsing Creek. They then organized the El Dorado mining district, elected a mining recorder, and formulated a set of mining rules. The first stakers and organizers of the district were Daniel B. Libby, A. P. Mordaunt, L. S. Melsing, H. L. Blake, Dr. A. N. Kittleson, who had been in charge of the Government reindeer station at Port Clarence; Rev. N. C. Hultberg, a missionary at Golofnin Bay; P. H. Anderson, a missionary teacher at Golofnin Bay, and John A. Dexter. This discovery was soon noised abroad, and a stampede to the new diggings followed from near-by points.

"I, Jafet Lindeberg, a native of Norway, came to this country in the spring of 1898 with Sheldon Jackson, superintendent of the Government reindeer enterprise, for the express purpose of going to Plover Bay, in eastern Siberia, to relieve Captain Kelly, who was trading at that place for reindeer for the Government. In furtherance of this agreement, I left Seattle on the steamer *Del Norte* early in 1898, taking with me a stock of provisions, and not interfering with the Government business for which I was to be assigned. On arrival at St. Michael, news was brought to Doctor Jackson that Captain Kelly had been driven away from Plover Bay by hostile natives. It was then decided that it would be unwise to send me over there, and, being left without a suitable position, Doctor Jackson gave me permission to leave the Government employ. This I did, and, taking my outfit, made for the new diggings at Council City, which had been located on the banks of the Niukluk River, near Ophir Creek.

"John Brynteson, a native of Sweden and an experienced coal and iron miner, who for seven years had worked in the mines in Michigan, determined to go to Alaska and look for coal. Arriving at St. Michael and hearing of the discoveries on Ophir Creek, he promptly left St. Michael for Council City, arriving there early in the summer of 1898.

"Erik O. Lindblom, a native of Sweden, by profession a tailor, and for years following his trade in San Francisco, while there, hearing of the fabulous reports from Kotzebue Sound, joined that mad stampede, going north on the bark *Alaska*. Arriving at Port Clarence on his way to Kotzebue Sound and hearing of the gold discovery on Ophir Creek, he left the ship and proceeded to Golofnin Bay, thence to Council City.

"We three men met by chance at Council City in August, 1898; after prospecting around in that district for some time and staking claims, formed a prospecting companionship and decided to prospect over a wider range of territory. Even at this early date the Council City district was overrun by stampeders, and staked to the mountain tops; so we proceeded to Golofnin Bay, and taking a large open boat and an outfit of provisions, on September 11, 1898, started up the coast toward Port Clarence, stopping at the various rivers to prospect on the way, in which we found signs of gold, but not in paying quantities, and finally arriving at what is now known as the town of Nome. From there we proceeded up Snake River, which we named, and camped at the mouth of Glacier Creek, prospecting as we went along. The first encouraging signs of gold we found on the banks of Snake River was at about the place where Lane's pumping plant is now located. After locating our camp as before mentioned, we proceeded to prospect along the tributaries of Snake River, which tributaries we named

Snow Gulch. Claims were also staked on Anvil and Glacier creeks, as well as on Snow Gulch and other small creeks.

On the return of this party to Golofnin Bay early in October the news quickly spread, and soon all within its reach started for Anvil Creek. On October 18, at a miners' meeting, the Cape Nome mining district was formed, and A. N. Kittleson was elected recorder.

---

as follows: Anvil Creek (taking the name from an anvil-shaped rock which stands on the mountain on the east side of the creek), Snow Gulch, Glacier Creek, Rock Creek, and Dry Creek, in all of which we found gold in paying quantities, and proceeded to locate claims, first on Anvil Creek, because we found better prospects in that creek than in the others, and where we located the "discovery claim" in the name of us three jointly. In addition to this, each man staked a separate claim in his own name on the creek. This was the universal custom in Alaska, as it was conceded that the discoverer was entitled to a discovery claim and one other. After locating on Anvil Creek, claims were staked on Snow Gulch, Dry Creek, and Rock Creek, after which we returned to Golofnin Bay and reported the discovery.

"It was then decided to form a mining district, so we three original discoverers organized a party, taking with us Dr. A. N. Kittleson, G. W. Price, P. H. Anderson, and a few others, again proceeded to Nome in a small schooner which we chartered at Golofnin Bay, purchasing as many provisions as we could carry on the boat, and on our arrival the Cape Nome mining district was organized, and Dr. A. N. Kittleson elected the first recorder. Rules were formulated, after which the party prospected and staked claims, finally returning to Golofnin Bay for winter quarters. The news spread like wildfire, and soon a wild stampede was made to the new diggings from Council City, St. Michael, and the far-off Yukon.

"At this period very few mining men were in the country, the newcomers in many instances being from every trade known. The consequence of this was soon well known; a few men with a smattering of education gave their own interpretation to the mining laws, hence jumping mining claims soon became an active industry. Especially from Council City came the jumpers, who were the original men John Dexter, by an Esquimaux, had guided to the first discovery of gold on the Seward Peninsula. They were angry to think that they had not been taken in at the beginning, so a few of them promptly jumped nearly every claim on Anvil Creek, although there was an abundance of vacant and unlocated ground left which has since proved to be more valuable than the original claims located by us and our second party who helped us to form the district. This jumping, or relocating of claims by the parties above named, poisoned the minds of all the newcomers against every original locator of mining claims, and as a consequence every original claim was relocated by from one to a dozen different parties.

"At that time L. B. Shepard was United States commissioner at St. Michael, and in no case did a jumper have a chance to profit by his villainy, if Judge Shepard could prevent it. Another strong factor for good government at St. Michael and vicinity was Capt. E. S. Walker, of the United States Army. With exceptionally good judgment and a fearless attitude he held the lawless element in check, and great credit should be given him.

"In the early months of 1899 we hauled supplies to the creeks, and as soon as the thaw came began active mining on Snow Gulch and on Anvil Creek. Soon a large crowd flocked to Nome, which was then known as Anvil City. Among this crowd was a large element of lawless men who soon joined forces with the Council City jumpers, and every effort was made by them to create trouble. Secret meetings were held and a plan formulated whereby arrangements were made to call a mass meeting of miners, and at this meeting declare all the acts of the original miners' meeting that organized the district invalid, and to throw open all claims for relocation. This nefarious scheme leaked out, and word was sent to Captain Walker at St. Michael, who promptly dispatched Lieutenant Spaulding with a detachment of troops to Nome. A few days after their arrival the projected mass meeting was called. Here the agreed-on resolutions were offered, which, if passed, would have created bloody riot. Lieutenant Spaulding dispersed the meeting, receiving the thanks of the entire mass of law-abiding citizens of Nome and vicinity for this act, * * * and had it not been for the military, who proved themselves to be the true men to the American Government, much riot and bloodshed would have resulted from the conduct of the aforementioned parties.

    *   *   *   *   *   *   *

    "Yours, very truly,

                          "(Signed)   JAFET LINDEBERG.

Though it was now too late to do any mining, or even prospecting, there was abundant opportunity to stake claims, and of this all took advantage. Each man filed locations on as many claims for himself and his friends as he thought by any chance he could retain in his possession. To such an extent was the right of staking by power of attorney abused that no less than 7,000 acres were filed on by not more than 40 men. Here, as elsewhere in Alaska, the preemption under the mineral laws of many claims in one district by individuals under power of attorney or otherwise, which under a strict interpretation of the statutes would appear to be illegal, has discouraged legitimate prospecting and retarded the development of mining interests.

THE FIRST YEAR OF MINING.

When the Nome mining district was organized, Seward Peninsula was practically cut off from communication with the outside world, for the nearest post-office was St. Michael, 100 miles away, and that received mail only during the open season. However, rumors of the discovery of these rich placers gradually reached other parts of Alaska, and during the fall and winter of 1898-99 many men came from St. Michael and even from the Yukon. By the middle of May, 1899, Nome, then called Anvil City, probably had a population of 250. Meanwhile, before the ice broke, the news had been carried up the Yukon to Dawson, and thence to the outer world. This caused a general movement toward Nome among Yukon miners, but the world at large was far from being convinced of the authenticity of the published statements, for it had come to regard with considerable skepticism the wild tales emanating from the northland during the Klondike excitement. Late in June several vessels reached Nome from Puget Sound and found a population of about 400 living in tents and in a score of driftwood cabins. Mining began June 20, and the steamers returning to Seattle brought confirmatory news of placer gold in this new camp. It was not until then that people began to take the rumors seriously and that any real excitement arose about the new district. As a result, several steamers sailing for St. Michael touched at Nome to discharge passengers and freight. These newcomers, together with those that found their way down the Yukon from Dawson and other points, swelled the population of Nome to nearly 3,000. By the late fall of 1899 a veritable frenzy seized the people of the Yukon, and not a few settlements were almost depopulated as a result of the rush to Nome.

Meanwhile, in the early summer, there was anything but a contented community at Nome. The newcomers had found the whole region covered with location notices and very little mining being done. The professional claim stakers had followed their usual practice of blanketing the creeks with location notices, under powers of attorney, and then holding many claims without doing any prospecting, in the hope of being able to take advantage of any discoveries made by the labors of others.   In the early part of July probably'less than seven hundred men were actually engaged in mining, while upward of a thousand were idle, with neither prospect of employment as miners nor opportunity to prospect in the district.   It should be remembered that at that time gold had been found in only a very small area adjacent to Anvil Creek.   These idle men believed that many of the locations were illegal, as they unquestionably were under a strict interpretation of the statutes, for as the law requires an actual discovery of gold on each claim it is obvious that a man who staked twenty to thirty claims in a few days could not have determined the presence of gold in them.   It was also charged that many claims had been located by aliens and were therefore not legal preemptions. Under these conditions it is not to be wondered that an era of " claim jumping " began, during which practically every property of any prospective value was restaked.   It was then not uncommon to find a claim corner marked by half a dozen stakes, each of which represented a different claimant.

The nearest United States commissioner was at St. Michael, and there was therefore practically no means of enforcing civil law.   In fact, there were no representatives of the Government at Nome except an officer and a small detachment of soldiers which had been sent over from the army post at St. Michael in the spring.   On the commandant of this handful of soldiers rested the responsibility of maintaining law and order among a thousand discouraged and angry men, a task made all the more difficult because he was without any actual legal authority.   He deserves credit for meeting the situation as far as it lay in his power by patrolling property to which there were rival claimants and by attempting to settle the constantly rising disputes. Discontent was rife, and matters went from bad to worse.   July 10 a so-called " miners' meeting " was called for the purpose of discussing the situation, and a resolution was there presented [a] setting forth the grievances of those who believed that the claim locations had not been made in accordance with the United States statutes.   While it must be admitted that the unlimited staking was undoubtedly illegal, yet this meeting was mainly attended by those who, for one reason or another, had not succeeded in getting hold of placer claims.   Had these discontented men spent less time in protesting and airing their grievances and more in prospecting they would have been better off, as subsequent operations have shown that there remained much valuable placer ground which had not then been preempted.   This meeting, though

---

[a] Dunham, S. C., The Yukon and Nome gold region : Ann. Rept. Commissioner of Labor, 1900, pp. 849–850.

no doubt tending to increase the dissatisfaction, was entirely within the legal rights of the individuals who believed that they had been wronged. Therefore the peremptory dispersing of the crowd attendant at the meeting [a] by the commandant of the troops was a high-handed proceeding, entirely unwarranted either in law or equity. The tension grew day by day, and conflicts between rival claim owners became not infrequent. To rectify matters, the military promulgated the following order on July 13:

To put an end to apparent misunderstandings, the following statement is published:

All disputed titles, whether to mining claims or town lots, shall at once be brought before the civil authorities for settlement. So long as the civil authorities can handle such matters the military authorities will take no action. In case it becomes necessary for the military authorities to act, the claim or lot will be held in its condition at the time, neither party being allowed to do any work to change the condition of the same.

While there exists no objection to the holding of orderly meetings for the discussion of ordinary business affairs, in any meeting held for the purpose of acting in district affairs no person is entitled to participate excepting claim holders. Any attempt so to participate by other persons is illegal, and the proper steps will be taken to prevent it.

Decisions and orders of the civil courts will be supported by the entire power and authority of the United States troops.

No persons will be allowed to carry firearms, revolvers, or pistols. Anyone violating this order will have said firearms confiscated.

---

[a] Another version of this affair is contained in the following note furnished by Mr. Hess:

"The statement of Dr. A. N. Kittleson about this miners' meeting as he made it to me this last summer (1905), as nearly as I can give it from memory, was as follows:

"In the summer of 1899 he and the other original stakers on Anvil Creek who were then attempting to work their claims practically had to stand over them with guns all the time to keep from being overrun by parties of gamblers, professional jumpers, and other riffraff. Circumstances became so bad that Doctor Kittleson went over to St. Michael and asked that a detachment of soldiers be sent to Nome for the protection of property. This was done, and a great deal of hard feeling followed, especially toward Doctor Kittleson, but it seemed to be the only way in which the original stakers could hold their property. The 'outsider' had at first held that the original claims, which had been laid out 1,320 by 660 feet, according to the statute, should be reduced to 500 feet in length, and this it was proposed to do by force. Shortly after this the other parties who had been attempting to force the original stakers off their claims called a 'miners' meeting,' before doing which they conspired to offer a resolution declaring all existing locations void. Meanwhile men had been stationed upon Anvil Mountain with instructions that on the passing of the resolution a bonfire should be started in Nome, at which signal the men were to hurry down from the mountain and restake the claims on Anvil Creek, thus getting ahead of the rush which would follow from Nome. A rumor of this intended action had come to Doctor Kittleson, who communicated with the military authorities, and a lieutenant with two or three men went to the meeting and took places on the platform. When the resolution was introduced, declaring all the locations void and the land open for relocation, the lieutenant ordered that the resolution be withdrawn within two minutes, stating that he considered it not for the good of the community, and that if it was not withdrawn he would clear the hall. The men tried to argue with the lieutenant, but he was firm and at the end of the two minutes ordered the sergeant accompanying him to clear the hall, which was done. Later the claims were jumped and long litigation followed. One man told me that the company with which he was connected had alone spent over $200,000 in lawyers' and court fees in holding its property."

This order, though undoubtedly intended to relieve the situation, was far from having this effect. The ownership of practically every placer claim which had been exploited was rightfully or wrongfully disputed, and had this order been strictly enforced mining at Nome would have ceased. Such an eventuality would have worked great hardship on many bona fide property owners who depended on the summer yield of their mines for capital to make more extensive development in the following season. To meet this condition, a modification of the order was made, as follows:

The instructions contained in the order of July 13, 1899, posted at Anvil City [Nome], will be amended so as to permit original locators at work on their claims to continue their work in the event that anyone jumps the claim. The matter can afterwards be settled by the civil authorities.

The situation was suddenly relieved in an unexpected manner. It was accidentally discovered that the beach sands were rich in gold. It appears that the beach placers were found almost simultaneously by a soldier of the barracks and John Hummel,[a] an old Idaho prospector who was too sick to leave the coast. Within a few days the mutterings of discontent were almost silenced because it was found that good wages could be made with rockers on the beach. All the idle men went to work as fast as they could obtain implements. As it gradually became known that the beach sands for several miles were gold bearing and could be made to yield from $20 to $100 a day to the man, a veritable frenzy seized the people of Nome. A large part of the population went to work with shovels and rockers. During the height of the excitement it is estimated that there were 2,000 men engaged in beach mining. The yield of the beach placers is estimated at more than $1,000,000, and this was practically all taken out with hand rockers in less than two months.

There was one legal complication relative to beach mining which threatened to be serious, but ended rather ludicrously. Previous to the discovery of the beach gold many so-called "tundra claims" had been staked, which stretched inland from the ocean. A group of these, including the richest beach deposits, has been segregated and passed into the control of one company. When beach mining began this company claimed that it owned the beach and warned off all trespassers unless they paid a royalty of 50 cents a day for the privilege of mining along the water front. Most of the miners, however, contended that a 60-foot strip from high water was public property and paid no heed to the warning against trespassing. The company thereupon appealed to the commandant of the troops, and he warned off all beach miners. The order was not obeyed, and he finally arrested about three hundred men. At this time the situation

[a] Schrader, F. C., and Brooks, A. H., Preliminary report on the Cape Nome gold region, a special publication of the U. S. Geol. Survey, 1900, p. 33.

reached the point of absurdity. There being no civil magistrate at hand before whom these men could be tried, no building in which they could be confined, nor any funds from which they could be supported while awaiting trial, the perplexed officer was forced to discharge all his prisoners, who all promptly returned to their rockers on the beach. Later decisions of the Land Office have not upheld the claims of this company to the gold in the beach, for a 60-foot strip of the beach has remained open for mining to all comers.

During this same summer a few enterprising men struck out to seek new fields. Some went eastward and found gold on a number of southward-flowing streams, including Bonanza and Solomon rivers, but were unable to prove its presence in commercial quantities. Others went westward and named and staked Cripple and Penny rivers. Gold was also found during this season in the gravels of some streams near Cape York, and the York mining district was organized.

In the fall of 1899 Nome's [a] population of more than 3,000 was sheltered in a few score of frame and galvanized-iron buildings and in several hundred tents and low driftwood cabins. These were irregularly distributed along both sides of a muddy thoroughfare, a mile in length, which extended close to the edge of the tundra. Late in September, when the writer first visited this region, the aspect of the town and its environment was anything but attractive. Back of the settlement stretched the bleak tundra, and the front was bounded by the surf-swept beach.

The yield of the beach placers gave the settlement great prosperity. Prices were high, as there was a scarcity of nearly everything except gold. Before the last outgoing steamer, lumber cost from $100 to $150 a thousand and coal from $50 to $100 a ton. Fresh meat, mostly reindeer brought from Siberia, commanded $1 a pound and eggs $3 to $4 a dozen. The cost of meals varied from $1 to $2.50. Small cabins, on lots staked on the tundra back of the town with nothing but squatter's title, found ready market at $600 to $700, while corner lots, also with uncertain titles, on the main street sold as high as $10,000. The Nome News, the first newspaper, began publication early in October. By fall a mayor and town council had been elected, and, though without any legal standing, made and enforced city ordinances by common consent. Of more importance was the organization of a fire department and police force. Meanwhile the Federal Government had taken cognizance of the new settlement by establishing a post-office and appointing a United States commissioner. Wages were $10 a day, but at times rose to $2 an hour. The abundance of gold was reflected in the thriving business carried on day and night by a score of saloons and gambling rooms. Many a man spent his days in work-

----

[a] The name Anvil City had been changed to Nome during the summer of 1899.

ing beach sands and his nights in losing his gains at the roulette table. On the whole, the town was very orderly, for the first comers embraced but few of the criminal class. The greatest need was a hospital, as the surface waters of the tundra (universally used in the town) had become infected with typhoid germs and there was an epidemic of that disease. It is estimated that there were as many as 200 cases at one time, and as there was only one small army hospital it was impossible to give the patients proper care. and hence there was a large mortality.

The results of this first season of mining were as follows: Up to January 10, 1900, about 4,500 claims were recorded in the Cape Nome district, but probably not more than 50 claims were developed and not more than 100 even prospected. These 50 claims probably yielded over $1,500,000 in gold. The beach-mining operations described above were more of a dramatic incident in the history of the region than of permanent commercial significance, for the richest part of the Nome beach was worked out the first year. The only other district of Seward Peninsula in which any mining was done in 1899 was on Ophir Creek, where gold variously estimated at $50,000 to $100,000 in value was taken out. Surface prospects were also found in the Solomon River region and near Cape York. A result of far greater importance than the actual mining was that some knowledge of the character of the deposits and the condition of operations had been gained. Although this information availed little to the more inexperienced men who were to invade the peninsula during the following year, yet it was of great practical benefit to those who did the actual mining in 1900.

THE RUSH OF 1900.

With the return of the steamers in the fall of 1899, bringing probably 1,000 people from Nome, began the excitement which was to culminate in the following spring. It was not until the summer's output of gold of about $3,000,000 was brought from Nome that the outside world realized the importance of this new placer field. During the following winter interest in Nome grew rapidly, and it was evident that a rush comparable to that of the Klondike two years before would begin with the breaking of the ice.

Professional promoters and stock jobbers were not backward in taking advantage of this excitement, and there was the usual crop of flamboyant prospectuses. Scores of companies were incorporated to mine gold at Nome and much stock was sold. Though not a few of these ventures were intended to be legitimate enterprises, practically all of them were doomed to failure because of the complete ignorance on the part of many of the promoters of the character of the deposits, suitable methods of mining, and general commercial conditions. Beach-mining enterprises were the favorite because of the supposed richness of the placers, and especially because no capital was required to purchase claims. The almost incredible record of the first year's beach mining appealed to the popular mind, and its interest was maintained through the newspapers and through transportation and mining companies' circulars, which published the most preposterous statements. Not a few so-called mining experts asserted that the gold in the beach was inexhaustible because the supply was constantly renewed by the waves from the ocean bottom. It was easy to maintain that if a man with a rocker could make $20 a day on the beach that a plant which could handle twenty times as much material would yield untold wealth. There was a flood of gold-saving devices, varying from a patent gold pan hung on a pivot and turned by a crank to complex aggregates of wheels, pumps, sieves, and belts, which required a 100-horsepower engine for their operation. "The golden sands of Nome" was the slogan which inspired thousands to engage passage for the El Dorado months in advance of the sailings. Reaching Nome was far easier than going to the Klondike, for the gold seeker could be landed at his destination from an ocean steamer. Here there was no winnowing of the persevering and enterprising from the shiftless and indolent as at the Chilkoot Pass (the gateway of the Klondike). In consequence, the crowd of men that reached Nome were less well fitted for frontier life than those who went to Dawson.

In 1900 the ice on Bering Sea broke early, and some small vessels skirting the shoreward side of the ice floes dropped anchor at Nome the latter part of May, but the large steamers did not arrive until the middle of June. By July 1 upward of 50 vessels had discharged passengers and freight on the beach. It is estimated that the first and second sailings brought over 20,000 people to the peninsula. There was then a solid row of tents stretching along 5 miles of the beach, and the water front was piled high with freight of all kinds. The newcomers found little to encourage them. Those that had wintered in the peninsula had industriously extended their stakes so that a man could travel for days and hardly be out of sight of a location notice. To add to the discouragement and confusion, smallpox was introduced from one of the vessels, and had it not been for the prompt action of Capt. D. H. Jarvis, of the Revenue-Cutter Service, it would have become a serious epidemic. The inexperienced men who landed at Nome, not finding the El Dorado their fancies had painted, were loud in their denunciation of the region. Many in the course of a few days' tramping of the beach became self-styled experts on placer mining and strenuously announced that the auriferous gravels of the peninsula had practically been exhausted.

During the month of July every conceivable kind of gold-saving appliance was installed on the shore, but few except those of simplest design paid even running expenses. Nevertheless there can be no question that a strong company controlling a considerable strip of the beach could by the use of steam shovels have profitably extracted what gold had been left in the sands. But under the conditions of public ownership of the beach, if values were found in any given locality, men swarmed in with rockers and quickly worked it out. This made it impossible to extract the beach gold at a profit by other than light equipments readily movable from one rich spot to another.

Probably the most ill-conceived enterprises were those planned to dredge gold under the sea. Though the upper layer of these sands is more or less auriferous, the difficulties of excavation are such as to make it improbable that it can be profitably mined. The severe storms and lack of shelter prevent the use of dredges, except possibly during one month in the year. Many of these dredging schemes were based on a theory (held by some who were entirely ignorant of the origin of the beach gold and who refused to be instructed) that the auriferous sands are swept in from the sea. The fallacy of this conclusion had already been pointed out in a report [a] which had a wide circulation. On August 9 a severe southwesterly storm practically demolished the more elaborate appliances for gold saving and strewed the beach for miles with débris. This ended beach mining for that year except where the simplest apparatus was in use.

In 1899 the complaint had been made at Nome that, there being no civil court, important enterprises were retarded because of the impossibility of having disputes about ownership of mining property settled by law. During the winter this had been brought to the attention of Congress, and a statute was passed forming a new judicial district which included Seward Peninsula. A newly appointed judge and other court officials reached Nome in the early summer. It appears that from that time on, the district was abundantly provided with court decisions, even though the legality and justice of many of them could be brought into question. [b] Among many charges brought against the court were those asserting that receivers had been placed on valuable mining property from which they extracted gold, in spite of the fact that they were without bond and that the rightful owners had no check on the amount of gold being taken out. Excitement over these cases ran high, and at one time bloodshed was narrowly averted by military intervention. One of these receivers was ejected from a rich

---

[a] Schrader, F. C., and Brooks, A. H., Preliminary report on the Cape Nome gold region, Alaska, a special publication of the U. S. Geol. Survey, 1900, pp. 16–24.

[b] An interesting account of the legal complications at Nome in 1900, by a lawyer who played an important part in them, is contained in a book entitled The Land of Nome, by Lainier McKee, New York, 1902.

mine which he was working by the decision of a higher Federal court. It is impossible to recount here the legal strife in which practically all the rich claims were involved, but it will suffice to state that there was hardly one which had not to run the gamut of lawsuits.

During the winter of 1899–1900 some mining had been carried on in the tundra and also some through the ice on the beach. The phenominally rich beach at the mouth of Daniels Creek, 50 miles east of Nome, had been discovered and its yield of $600,000 had practically been taken out before the newcomers had landed. A very significant discovery of 1900 was that of the rich placers on a high bench between Anvil and Dexter Creeks. It had been pointed out by the Geological Survey [a] the previous year that these high gravels were likely to be gold bearing, and such proved to be the case. The presence of an older drainage system, of which these high gravels bear witness, has an important bearing on the future of placer mining in this field.

During this season Anvil, Dexter, and Glacier creeks were very large producers. Their exploitation was hastened by a narrow-gage railway constructed by Charles D. Lane from Nome to Anvil. On Ophir Creek the installation of larger plants had been begun, but the output was about the same as in the previous season. Considerable gold was taken out from the lower course of Daniels Creek, but all this mining was very crude.

Though but a small percentage of the newcomers left the Nome beach, there was in the aggregate considerably more prospecting than in the previous year. Productive placers were developed on Casadepaga River, as well as on a number of smaller tributaries of the Niukluk. The same holds true of the Solomon and Cripple river valleys. The discovery of gold in 1900 in the Bluestone and Kougarok valleys showed that the auriferous area was much larger than had been previously supposed.

An excellent example of the ineffective character of the search for gold is to be found in the discovery of gold in the Kougarok placers. In March, 1900, there was a wild rush to the Kougarok River region, which was duplicated in August, but in neither were placer values found. While the last straggler was returning to Nome after the second fruitless chase, a few more experienced men opened up some good placer ground. The York gold placers, discovered the previous year, proved of small extent, but the discovery of stream tin in this district by the Geological Survey [b] gave new hopes to the prospectors.

During the height of the excitement there was a good deal of lawlessness at Nome and life and property were none too secure. For a time it appeared as if the authorities would not be able to cope with

---

[a] Preliminary report on the Cape Nome gold region, 1900, p. 20.
[b] Brooks, A. H., Reconnaissances in the Cape Nome and Norton Bay regions, Alaska, in 1900, a special publication of the U. S. Geol. Survey, 1901, p. 136.

the great multitude which was crowded together at Nome, especially
as it included a large number of professional criminals. Conditions
gradually improved as the crowd thinned out toward fall, when the
last steamer was packed with disappointed men. Not a few of the
indigents were shipped out either through public or private charity.
On the whole, considering the inexperience of the mass of the people
who came to Nome in 1900, progress was better than could be anticipated. The following table, taken from a previous report,[a] gives an
estimate of the gold production of the individual creeks in 1900:

Esimated output of placer gold from Seward Peninsula, 1900.

Anvil Creek	$1,750,000
Glacier Creek, including Snow Gulch	750,000
Dexter Creek	300,000
Extra Dry Creek	15,000
Dry Creek	25,000
Newton Gulch	10,000
Bourbon Creek	5,000
Saturday Creek	10,000
High bench placers near Nome	145,000
Nome beach	350,000
Oregon, Hungry, and Mountain creeks	50,000
Solomon River	10,000
Topkok beach	600,000
Daniels Creek	200,000
Ophir and Sweetcake creeks	100,000
Crooked Creek	25,000
Elkhorn Creek	30,000
Goldbottom Creek	10,000
Casadepaga River	15,000
Kougarok district	50,000
Bluestone River	75,000
York River	1,500
Miscellaneous and undetermined, about	200,000
	4,726,500

Later information led to the raising of this estimate to $4,750,000.

PROGRESS IN 1901.

The fact that thousands returned from Nome in the fall of 1900
discouraged and disgusted with this mining field gave assurance that
the rush of 1900 would not be again repeated, and this was a decided
advantage to the district. Unfortunately, however, the judiciary
was generally regarded as corrupt and this no doubt deterred capitalists who would gladly have taken a part in establishing large mining enterprises had they been assured of a just administration of the

---

[a] Brooks, A. H., op. cit., p. 69.

laws. Besides these conditions, mining was also retarded because a late spring and dry weather interfered with placer operations. The arrival of a new judge in September restored confidence, but it was then too late to undertake any large enterprises.

During the previous winter some of the high-bench and tundra placers had been exploited successfully by drifting. Though their output was not large, yet their operation was important in proving the feasibility of winter mining. The completion of a railway from the head of navigation on the Niukluk to Ophir Creek during the summer of 1901 did much to promote commercial interests in that district. At Daniels Creek work was suspended during this season because of the lack of water. Considerable prospecting was also carried on in the Solomon River region, but there was little actual mining. In the Bluestone and Kougarok districts primitive methods of extraction still held sway and the yield of gold was small.

Probably the most noteworthy event of this season was the discovery of gold placers in the northeastern part of Seward Peninsula. Prospecting had in 1900 established the fact that the gravels of the several rivers tributary to Kotzebue Sound were auriferous, but it was not until 1901[a] that any actual mining was done. This proved that the gold-bearing area of Seward Peninsula was much larger than had been previously supposed. The construction of the Miocene ditch was also of great importance, as this was the first long water conduit in the whole field. Some experienced placer miners, after attaining great familiarity with the local conditions, came to recognize that the primitive "pick and shovel" methods then almost universally employed were applicable only to very rich placers. An improvement of methods demanded a larger water supply. Though there are here no great water resources like those of the Sierra Nevada, which made California placer mining the cheapest in the world, yet some water is to be had in the Kigluaik Mountains and in other highlands of the peninsula. The credit for both the conception and construction of this first ditch belongs to W. L. Leland and J. M. Davidson, who were among the earliest to obtain a broad grasp of the mining problem.

The construction of pumping plants, begun in 1901, was another, but less successful, attempt at reducing the cost of mining. By means of two extensive steam pumping plants water was delivered at high levels in the Anvil Creek region, and this made it possible to work some rich placers which formerly could only be sluiced for a few weeks in the early spring by impounded snow water. As the use of water pumped by steam power is economically possible only in a

[a] Collier, A. J., Reconnaissance of the northwestern portion of Seward Peninsula, Alaska : Prof. Paper U. S. Geol. Survey No. 2, 1902, p. 43. Moffit, F. H., The Fairhaven gold placers, Seward Peninsula, Alaska : Bull. U. S. Geol. Survey No. 247, 1904, p. 49.

region of very cheap fuel, it would appear to find no permanent place in placer mining in Alaska.

This season witnessed considerable activity in ditch construction on Ophir Creek, where large plants were being installed. Improvements in methods were also introduced in nearly all the smaller districts. A dredge was operated for a part of the season on Solomon River.

The discovery of rich placers on Candle Creek [a] in July, 1901, led to the usual rush from the near-by creeks, and hence had a retarding influence on mining in the northeastern part of the peninsula. This was particularly true of the Kougarok, whose production was far less than was anticipated. In the Bluestone region, where a few rich claims had been almost worked out in 1900, little was accomplished toward exploiting lower-grade deposits in a more systematic way. Lack of water prevented any considerable amount of mining on Daniels Creek. The fact was that during the excitement in 1900, many inexperienced men were working placers at an actual loss. By 1901 business equilibrium had been in a measure restored and such operations ceased.

In 1901 there was a decrease in the gold output of Nome. The reason for this is not far to seek. In 1900 the Nome beach placers yielded $350,000, and in 1901 not over $50,000, while the gold from the Topkok beach, which in 1900 amounted to $600,000, in 1901 was practically nothing. This reduction of nearly a million dollars in the yield of the beaches could hardly be expected to be made up by the creek placers, where there had yet been no considerable progress in methods of extraction. As the actual decrease in the total production was only $600,000, the output of the creek placers gained nearly $300,000 over that of the previous year.

In the subjoined table the gold output of the entire peninsula has been distributed among the various districts in accordance with the best data available. As there are no authentic statistics, the figures presented are only approximate.

Gold production of Seward Peninsula, 1901.

Winter drift mining (1900–1901)	$300,000
Nome district, creek placers	3,000,000
Nome district, beach placers	20,000
Council district	600,000
Kougarok district	40,000
Fairhaven district (Candle Creek, $20,000)	50,000
Bluestone, Solomon, Cripple, and other districts	120,000
	4,130,000

[a] Moffit, F. H., The Fairhaven gold placers, Seward Peninsula, Alaska : Bull. U. S. Geol. Survey No. 247, 1904, pp. 49–50.

GROWTH OF MINING FROM 1902 TO 1904.

In the winter of 1901–2 drift mining was much increased, especially in the high benches near Anvil Creek, in the tundra, and in the Solomon River region, and also to a more limited extent in many of the smaller camps. These operations promoted prosperity by assuring employment to a certain number of men throughout the year, and to that extent cutting down the migratory population.

Before navigation opened in 1902 confidence had been restored by the just administration of the laws, and considerable capital was brought for investment in mining property and plants. This season was characterized by great activity in ditch construction. The Miocene ditch was extended, and before the end of the season about 34 miles of it were in use. A number of other ditch-building projects were inaugurated. Among the most important of these was a ditch to supply water for exploiting the rich placers of Daniels Creek, which had been practically untouched. As a result of this extensive conduit construction many new legal controversies arose regarding water rights, for it came to be recognized that the water in many localities was as valuable an asset as the gold-bearing gravel.

Though the gold in the bed of Anvil Creek was practically exhausted by 1902, the discovery of a parallel channel led to a continuation of mining in the valley. Several ditches and two pumping plants now furnished water for exploiting the auriferous gravels of the Anvil, Glacier, and Dexter creek basins. Outside of this district Ophir Creek was the only large producer. Mining operations on this stream had progressed steadily, but it was not until 1902 that it came to be recognized as one of the richest creeks in the whole peninsula. It is estimated that in 1902 its production reached $1,000,000.[a]

No considerable advance was made in the Kougarok and Bluestone regions, both of which produced only a small amount of gold. In the Fairhaven district the Candle Creek placers were the only considerable producers. Placers were found on Iron Creek and on a number of other smaller tributaries of Kruzgamepa River, but they were too inaccessible to permit the exploitation of any but the richest deposits.

In spite of the improvements which had been made in mining methods, the increase in gold production for this season was only about $400,000, and most of this came from Ophir Creek. The production of Anvil Creek had become less, and in none of the other districts had any improved methods been introduced. With the increase in the number of gold-producing creeks, the distribution of the output among the different districts becomes increasingly difficult and multi-

---

[a] Brooks, A. H., Placer-gold mining in Alaska in 1902: Bull. U. S. Geol. Survey No. 213, 1903, p. 40.

plies the chance for error. The following figures can therefore be considered only an approximation:

Gold production of Seward Peninsula, 1902.

Winter drift mining, 1901–2	$350, 000
Nome district	2, 800, 000
Council district	1, 000, 000
Kougarok district	50, 000
Fairhaven district	150, 000
All other districts	210, 000
	4, 560, 000

In the winter of 1902–3 there was a decided increase in the number of claims worked by drifting. It is estimated that more than $500,000 was taken out of the winter accumulation of gravel. The amount of ocean freight bore witness to the commercial prosperity in Seward Peninsula. Twenty-seven steam and seven sailing vessels landed about 75,000 tons of cargo on the peninsula, most of it consigned to Nome. The Council City and Solomon River Railroad was among the important enterprises inaugurated in the summer of 1903, and before the close of the season about 9 miles of standard-gage track had been laid. The Nome Arctic Railway (now called the Seward Peninsula Railway), running to Anvil Creek, was extended to the head of Dexter Creek. A few roads were built connecting with these railways, but most of them are little more than wagon tracks. During dry weather, however, it is possible to drive over the tundra, which in a wet season becomes almost impassable.[a]

As in the previous year, ditch construction was actively pushed during the summer of 1903. The Miocene ditch was extended to the head of Nome River and a branch was carried to Snake River. In the Cripple, Solomon, and Ophir regions extensive ditches were laid out and in part completed, and a number of shorter conduits were constructed in other districts.

It appears that the matter of ditch building is overdone in Seward Peninsula. The striking success of several long ditches has led less conservative and less experienced operators to lose sight of the fact that certain classes of placers can be mined at lower cost by other methods. When thousands of dollars are invested in water conduits to exploit shallow placers, as has been the case in many localities which might have been much more cheaply mined, it is time to call a halt to the injudicious construction of ditches. No one who has watched the maturing of the mining industry in this field will deny the important part which the ditches have played and will

---

[a] For a discussion of the roads and transportation of Alaska, see Purington, C. W., Methods and cost of gravel and placer mining in Alaska: Bull. U. S. Geol. Survey No. 263, 1905, pp. 217–228.

play, but it is equally patent that there have been many misapplications of this method of exploitation. This is because the less experienced operators have come to regard the ditch as a panacea for all difficulties in placer mining.

In 1903 hydraulic lifts were installed on Glacier, Anvil, and Ophir creeks. One steam shovel and one dredger were operated for a short time on Niukluk River, and two dredges were used in the bed of Solomon River for most of the season. On many other creeks steam hoists, scrapers, and other mechanical appliances for cheap mining were installed. Extensive prospecting with churn drills was carried on in the tundra belt lying between Nome and the mountains. (Pl. II, B, p. 178.) As expected, the returns indicated extensive deposits of auriferous gravels.[a] Considerable work was done on an old line of beach deposits near Nome, whose discovery had been anticipated by the geologic investigations.[b] In the Topkok region 12 miles of ditch were completed, and before the close of the season some hydraulic mining was done. Prospecting in the Kougarok region showed much more extensive placers than had been previously supposed, but little actual mining was done. Nearly all the developments except in the older districts were in the nature of dead work.

The quartz mine, in operation since 1903 in the Solomon River region, is a significant feature of the gold-mining industry. Auriferous quartz veins are not uncommon in the peninsula, but in only this one instance has a lode mine actually been developed. It should be said, however, that prospectors have thus far paid small heed to lodes except in the way of location notices. In hardly half a dozen localities has any attempt been made to search systematically and intelligently for auriferous veins. Another important feature of the season's work was the finding of lode tin in the York region[c] by Mr. Collier and Mr. Hess, under the guidance of Charles Randt, Leslie Crim, and W. J. O'Brien, who had found float ore from the ledge. In this field the tin-bearing gravel had received considerable attention since 1900, and some lode-tin deposits had been reported by prospectors, but these reports were not verified until 1903.

The net results of all these operations (though possibly disappointing to those who had not a personal familiarity with the conditions) were as good as could be expected. The production was practically the same as in the previous year, though the estimates presented show

[a] Brooks, A. H., Placer mining in Alaska in 1903: Bull. U. S. Geol. Survey No. 225, 1904, pp. 48–55.

[b] Preliminary report on the Cape Nome gold region, U. S. Geol. Survey, 1900, p. 23; A reconnaissance of the Cape Nome and adjacent gold fields, U. S. Geol. Survey, 1901, pp. 150–151.

[c] Collier, A. J., Tin deposits of the York region, Alaska: Bull. U. S. Geol. Survey No. 229, 1904.

a falling off of $100,000, an amount which is probably less than the errors in the figures. In view of the great amount of dead work accomplished and the fact that some of the richest shallow placers had been exhausted, while the newer districts had not reached a productive stage, no increase in the total production was to be expected. Though there are no accurate statistics available, it seems probable that there was a considerable falling off in the Nome district proper, and that the output of practically all the other districts was higher than in previous years. It is on this basis that the following distribution of the totals has been made:

Gold production of Seward Peninsula, 1903.

Winter drift mining, 1902–3	$500, 000
Nome district	2, 400, 000
Council district	1, 000, 000
Kougarok district	75, 000
Fairhaven district	200, 000
All other districts	290, 000
	4, 465, 000

Drift mining[a] during the winter of 1903–4 was more extensive than ever before, and the yield probably was nearly $1,000,000. Of this amount about half was taken from the high bench deposits near Nome. Some low-lying gravels near the base of the hills were also exploited, as were the ancient beach placers in the tundra. In the Solomon River region probably $200,000 was taken out during the winter, and Ophir Creek yielded about $100,000. Several of the other districts yielded smaller amounts as the result of the winter's work.

The summer season opened very inauspiciously. Up to July 10 more than half the mines were idle because of lack of water due to dry weather. Fortunately heavy rains came early in July, and mining began. The summer's sluicing was, however, practically confined to two months, and even during this short period there was a considerable shortage of water in many districts. Ditch building continued, notably in the Nome, Cripple, Flambeau, and Solomon river valleys. Much of the work in these localities was directed toward the installation of plants and conduit construction, and the summer's output of gold was comparatively small. On Anvil Creek a steam shovel was successfully operated. On Glacier and adjacent creeks hydraulic elevators continued to be used. A sensational discovery of rich placers on Little Creek near the base of the highland back of Nome was significant because it proved that the limit of dis-

[a] Brooks, A. H., Placer mining in Alaska in 1904; Bull. U. S. Geol. Survey No. 259, 1905, pp. 19–24.

coveries had not yet been reached, even in the best-known parts of the peninsula.

The Council City and Solomon River Railway extended its road about 10 miles. The Topkok Ditch Company operated its hydraulic plant on Daniels Creek throughout the season. This well-managed enterprise is among the most successful operations of this class and has clearly demonstrated that with a sufficient water supply and enough grade for the disposal of tailings it is possible to hydraulic even frozen gravels. Ophir Creek continued to be the heaviest producer. One ditch was completed in the Kougarok region, and several more were planned or in construction.

Little was done in the northeastern part of the peninsula except some rather primitive mining. Though the presence of workable placers in this region has been demonstrated, yet with the present difficulties and expense of transportation capital has been chary of entering the field. There are, however, some enterprises on foot for ditch building in this district.

The general and healthy growth in the commercial interests continued. There were fewer legal complications and less promotion of "wild-cat" mining schemes. This community has, however, still to learn the necessity of intrusting large enterprises to the hands of well-trained men rather than to promoters. Probably two-thirds of the incorporated companies have made failures or achieved only partial success because of poor management. As has been stated, much of the activity was directed toward dead work, and the output of the placer mines was probably no greater than in the season before. This is what might have been expected, and it will be several years before any very great increase in the gold output is likely.

In distributing the totals shown in the following table it has been considered that there has probably been a considerable decrease in the production of the Nome district proper outside of the high-bench placers. This progressive decrease will probably continue until some systematic exploitation of the high-bench and tundra gravels has been introduced, when Nome should again lead all the other camps of the peninsula.

Gold production of Seward Peninsula, 1904.

Winter drift mining, 1903–4	$900,000
Nome district	2,000,000
Council district	1,000,000
Kougarok district	75,000
Fairhaven district	125,000
All other districts	400,000
	4,500,000

MINING DEVELOPMENTS IN 1905-6.

During the winter [a] of 1904–5 much mining was done near Nome along the so-called " second beach line," which was traced from Hastings to Bourbon Creek, stretching parallel to the present coast. What appears to be an extension of the same deposit was discovered at Jess Creek, 10 miles west of Nome. These elevated beach placers are comparatively shallow, and, though not as rich as some described in succeeding paragraphs, yielded handsome profits.

In the fall of 1904 considerable excitement had been caused by the discovery of some rich placers in the tundra gravels near Little Creek. A group of claims in this locality proved to carry enormous values, yielding, it is said, more than $1,500,000 in gold during the first twelve months that they were operated. Although these proved to be only the forerunners of still more important discoveries, they were of the utmost importance, first, because they showed that the limits of new discoveries in the best-known parts of Seward Peninsula had not been reached, and second, because they stimulated the prospecting of the tundra gravels. In addition to the mining operations on the Nome tundra and in the high gravels of the vicinity, winter mining was also carried on in other localities, notably at Candle Creek. Operations requiring fuel at the latter locality, as well as at other points in the Fairhaven district, utilized to a considerable extent the lignite coal mined at Chicago Creek.

It is estimated that the gold production of the winter of 1904–5 exceeded $1,000,000 in value, but these figures can be regarded only as an approximation.

The open season of 1905 was much shortened by a late spring and an early fall, the freeze-up coming about September 18. The most important event of the summer was the installation of a large dredge on Solomon River. Its successful operation furnished the final proof that dredges were to play an important part in the mining industry of Seward Peninsula. Other dredges had been tried with more or less success, but this was the first to be operated in a large way.

A notable advance was made in ditch construction during 1905, when thirteen ditches were either completed or building. Among the most important was the Seward ditch (37 miles long), which carries water from the head of Nome River to the Nome tundra claims. About 8 miles of the Pioneer ditch was completed, to carry water from Nome River to the south slope of Anvil Mountain. The Cedric ditch (24 miles long), which proved to be a less well-planned enterprise, was built to carry water to Arctic Creek, west of Nome. Smaller ditches were also constructed in the Osborn, Flambeau, Solomon, and Ophir basins, as well as in other parts of the peninsula.

---

[a] Moffit, F. H., Gold mining on Seward Peninsula: Bull. U. S. Geol. Survey No. 284, 1906, pp. 133–144.

On the whole, a very large amount of dead work was accomplished during the summer, and this, together with the fact that it was an adverse climatic season, made the production of $4,800,000 of gold very considerable. This amount is approximately distributed as follows:

Gold production of Seward Peninsula, 1905.

Nome district	$3, 400, 000
Council district	500, 000
Kougarok district	400, 000
Fairhaven district	300, 000
All other districts	200, 000
	4, 800, 000

The discovery of rich gravels at Little Creek led to great activity in prospecting the tundra, and as a result the so-called third-beach line was discovered during the winter of 1905–6. This discovery bore out the prediction made by the Geological Survey as early as 1899, and established the value of its investigations.

These third-beach placers yielded so much greater profit than the creek placers, partly due to cheapness of operation and partly to higher values, that for the time being mining activity was centered on their development. During the winter so much prospecting was done with steam thawers and hoists, requiring a large amount of fuel, that Nome was threatened with a coal famine. This was avoided only by drawing on the Government stores of coal at Fort Davis. The spring clean-up of the gravels mined during the winter yielded probably $2,500,000 in gold. Considerable winter work was done in other parts of the peninsula, but the total output from these sources probably did not exceed a few hundred thousand dollars in value.

The opening of navigation saw no abatement in tundra mining, for work was continued on the third beach line. It was fortunate that this was so, for the extraordinarily dry summer of 1906 worked havoc with nearly all creek mining. The operations supplied by water from Nome River were continued for the most part, but the miners on the smaller streams were forced to close down throughout much of the summer.

Dredging operations continued as in the previous year, notably at Solomon. In the aggregate there was considerable ditch building, especially in the Kougarok and Fairhaven districts. In spite of the shortage of water, there was considerable mining in the Topkok, Ophir, and Iron creek regions.

A new management took over the narrow-gage railway running to Dexter and extended it over the pass at the head of Nome River to Salmon Lake and thence to Lanes Landing, on the Kuzitrin. This line, called the Seward Peninsula Railway, now running into the heart of the peninsula, will accelerate the development of some of the inland districts. The Council City and Solomon River Railway was extended into the Casadepaga basin, and promoted mining activities in this field. Some wagon roads were built under the direction of the Alaska road commission.

During the summer of 1906 Nome was exceedingly prosperous. Many substantial buildings were erected and the whole town began to assume an appearance of greater permanency than it had previously shown. Business was facilitated by the improvement in steamboat connection with Puget Sound.

The estimated value of the gold production of the peninsula in 1906 is $7,500,000, of which probably nearly half was taken from the old beach line. Practically all the other districts showed a falling off in yield, due both to the insufficient water supply and to the fact that the energies of many men had been diverted from creek to tundra mining. It is impossible at the present time to distribute this total among the different districts.

Lode mining in 1905 and 1906 was still confined to the one property on Big Hurrah Creek, but considerable prospecting was done for auriferous quartz. However, until placer mining proves a less attractive field it is not to be expected that much attention will be paid to the finding of gold in bed rock. In addition to the auriferous lodes, some antimony and bismuth ores and graphite deposits have received attention. Some work has also been done on the Fish River galena deposits, which have already been referred to.

During these two years there has been great activity in the York tin district. One lode mine near Cape Prince of Wales has made some small shipments of tin ore, as have also some placer mines on Buck Creek. Most of the work in the tin district has been directed toward prospecting.

SUMMARY.

The facts presented in the foregoing pages make it evident that real progress in the industrial advancement within the peninsula has come chiefly during the last five years. The first years after the gold discovery were given over to skimming the cream from some of the richest creeks, only such placers being developed as assured immediate returns and required but a small investment of capital. Such operations, while they made for quick returns and attracted a large population, have been of little permanent benefit to the district. No placer camp whose output is derived solely from bonanzas ever has had or can have a long successful history. It has been shown that the absence of the judiciary, followed by what was believed to be a corrupt court, was a serious handicap to progress in the two years following the discovery. Since then faults of government have been rectified, and confidence has thereby been restored to the community.

The statistics presented on pages 30–37 show that from 1901 to 1905 the variation in the total output of gold from the peninsula was less than 10 per cent. This does not, however, indicate a uniform yield from the different districts that make up the peninsula, but rather that, by chance, when one bonanza was exhausted, another would be found. For example, by the time the beach placers had been exhausted, the creek placers at Nome had become large producers, and when in turn the richest of these had been mined out, new ones were discovered in other parts of the peninsula. As mining enterprises become better established and a larger amount of gravel is handled annually, less fluctuation is to be expected.

It was not until 1902, when at least one of the richest creeks was supposed to be nearly exhausted, that systematic effort was made to lower mining costs, for up to that time a large majority of the operators believed that only gravels of a high gold tenor could be profitably exploited. A few far-sighted and experienced men, however, realized that, by the installation of proper equipment, a reduction in costs was possible, which would make available for mining the large deposits of auriferous gravels carrying lower values, and that by such means only the life of the district could be prolonged.

The first successful enterprises looking toward this end were the construction of high-level ditches in 1901 and the installation at about the same time of various mechanical devices. The introduction of the churn drill for prospecting about this time is also worthy of note. A second great advance was made in 1904, when the first large dredge to be successfully operated was installed. The evolution of mining methods is still in progress, for many problems remain unsolved. One of the most important is the economic handling of the frozen gravels which underlie much of the tundra. It seems probable that utilization of some of the water powers not available for hydraulicking, by transforming them into electricity, will be one of the early lines of development.

As in most other mining camps, the necessity of intrusting the management of large enterprises only to men of technical training, experience, and proved ability has not by any means been fully realized. It is common knowledge that the failures of mining companies are in a large measure chargeable solely to the ignorance of the men responsible for their management. This is because the public has not been educated to the point of regarding mining as anything but a gamble, and holds in small esteem those who would put it on a sound business basis.

GEOGRAPHY AND GEOLOGY.

By Arthur J. Collier.

INTRODUCTION.

In the foregoing pages Mr. Brooks has outlined the progress of geographic and geologic surveys in Seward Peninsula. Except along the coast line the surveys began in 1899 and have been continued during every season but 1902 up to the present time. Nearly all of this work has been of a reconnaissance character [a] and much of it has been directed toward an investigation of the placer deposits. This was necessarily so, for detailed surveys must be preceded by a general knowledge of the whole province.

It is the purpose in this section to present only the salient features of the geography and geology, and especially those which bear more directly on the occurrence of placer gold. The new matter presented, as has already been stated, is based on studies carried on during three months of the summer of 1903, supplemented by some observations made during 1904, when a few weeks were devoted to a reexamination of some of the more important mining districts. The itinerary which follows shows that the writer was not by any means able to cover the entire field indicated by the title of this volume. This work could not have been completed were it not that free use has been made not only of the reports of Brooks, Mendenhall, Richardson, and Moffit, but also of their unpublished notes and manuscripts. The writer is also fortunate in having access to the notes made by C. W. Purington [b] and Sidney Paige, who investigated some of the gold placers of the peninsula in 1903. Finally, he is under deep personal obligation to the mine operators of the peninsula, who almost without exception have shown a cordial appreciation of the work of the Survey and have aided it by every means in their power.

---

[a] In 1904 T. G. Gerdine made detailed surveys in the vicinity of Nome and in the following year F. H. Moffit and F. L. Hess studied in detail the geology of the same region.

[b] Methods and cost of placer and gravel mining in Alaska: Bull. U. S. Geol. Survey No. 263, 1905.

The party organized for the purposes of the investigation in 1903 included Frank L. Hess, assistant, and a packer and a cook. It was provided with a light camp equipment, which was carried by a pack train of six horses. The disembarkment at Nome was made on June 16. After spending a few days in the Anvil Creek region, a route was laid up the Nome River valley to the Kigluaik Mountains, thence westward and around their seaward end to the Bluestone district. From Teller an excursion was made into the York tin-bearing region by boat, the results of which have already been published.[a]

From Teller the party took a northerly course and visited the gold-bearing region of the Agiapuk and the upper Kougarok basin. Turning southward through a gap in the Bendeleben Mountains the party reached the Council district, where a study of most of the producing placers was made. Thence the party proceeded to Nome by way of the Solomon and Eldorado river valleys, whose placers were examined en route. At the close of the season about ten days were devoted to some detailed studies of the auriferous gravels near Nome, special attention being given to the high-bench deposits. The party embarked for Seattle on September 13. In the course of a journey of nearly 500 miles, occupying about one hundred days, the party had examined the placers on upward of 150 creeks, besides carrying on geologic reconnaissance mapping. It will be evident, therefore, that while there were better opportunities than had fallen to any previous investigators in this field, yet much of the work was necessarily hasty and not made in the detail desirable for an exhaustive study of the auriferous gravels.

GEOGRAPHY.

GENERAL OUTLINE.

Seward Peninsula, cut out from the mainland by Norton Sound on the south and Kotzebue Sound on the north, has an outline which suggests a crudely shaped arrowhead pointing to the west toward Cape Nuniamo, on the Siberian coast. (See fig. 2.) The total length from east to west is about 200 miles. From Norton Sound to Kotzebue Sound the distance is only 80 miles, but the width of the peninsula along a line from Cape Nome to Cape Espenberg is 150 miles. The total area is approximately 20,000 square miles. The main mass of the peninsula is included between meridians 161° and 168° west longitude and parallel 64° north latitude and the Arctic Circle.

The dominant topographic forms of the province are flat-topped uplands from 800 to 2,500 feet high, broken by broad valleys and

---

[a] Collier, A. J., Tin deposits of the York region : Bull. U. S. Geol. Survey No. 229, 1904.

basin lowlands (Pl. I, *B*).   In the southern half of the peninsula the
Kigluaik, Bendeleben, and Darby mountains form a broken range
along a crescentic axis.   The York Mountains and several other highland masses form isolated groups in the northern half of the peninsula.   The watercourses, as a rule, follow broad valleys with gentle
slopes.   About one-fourth of the drainage finds its way northward to
the Arctic Ocean; the rest flows southward into Bering Sea and its
connecting bodies of water.   The shore line is characterized by long,
straight beaches with gentle slopes, broken by rocky bluffs and by

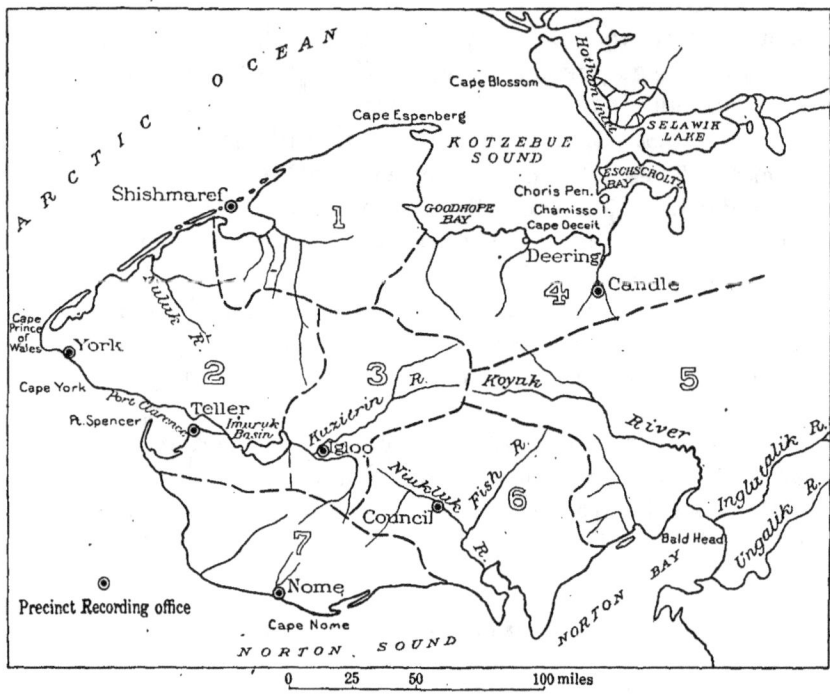

Fig. 2.—Precinct map of Seward Peninsula. 1, Goodhope; 2, Port Clarence; 3,
Kougarok; 4, Fairhaven; 5, Koyuk; 6, Council; 7, Nome.

deep embayments and inlets. (See topographic maps, Pls. VIII and
IX, in pocket.)

SHORE LINE.

The coast line of Seward Peninsula is remarkably even and regular
as compared with that of southern Alaska.   Nearly the whole peninsula is bordered by slightly curving beaches, which mark the seaward margins of coastal plains, broken by numerous headlands, but
with some deep inlets.   The sea floor slopes off at a very gentle angle,
and at only a few places are depths of 10 fathoms found within 5
miles of the shore.   Bering Sea and the Arctic Ocean in this vicinity

are so shallow that no soundings above 35 fathoms have been reported within 100 miles of the coast.

Golofnin Bay, which breaks the coast line 100 miles east of Nome, has a depth of more than $3\frac{1}{2}$ fathoms at the entrance, but large vessels can not approach nearer than within 3 miles of the landing at the head of the bay.

Port Clarence, about 80 miles northwest of Nome, is the only harbor of the peninsula. It is nearly circular in outline, 15 miles in diameter, and cut off from the open sea by a long, low sand spit. The Coast Survey charts show a depth of 9 fathoms near the entrance, and good anchorages for large vessels can be found over much of its area. From the head of the bay Grantley Harbor, navigable for light-draft vessels, extends inland for 15 miles, and this in turn connects at its head, by way of a narrow tidal inlet called the Tuksuk Channel, with the large body of brackish water called Imuruk Basin or Salt Lake. By using these waters light-draft steamers can penetrate well into the heart of the peninsula.

Along the north coast the shore line is almost continuous from Cape Prince of Wales to Kotzebue Sound, but there are a number of shallow lagoons lying between barrier beaches and the mainland and connected with the Arctic Ocean by narrow channels. The largest of these lagoons, known as Shishmaref Inlet, can be used as a harbor by vessels of light draft.

Goodhope Bay, which breaks the northern coast line and cuts Seward Peninsula off from the mainland on the north, is about 30 miles wide. A depth of water of more than 5 fathoms can be found over much of its area, but owing to its great size it affords little more protection to seagoing vessels than does the open coast of the peninsula. There is, however, a good anchorage protected from nearly all winds south of Chamisso Island. This was used as winter quarters by some of the vessels of the Franklin relief expedition.[a]

Headlands that break the continuity of the beaches are common features of both the north and south coasts east of a line running from Rocky Point to Cape Espenberg. Rocky Point, Cape Darby, and Bald Head are notable promontories on the north shore of Norton Bay, and Cape Deceit is the most prominent of the rocky points on the south shore of Goodhope Bay. The limestone cliffs west of Cape Deceit are famous sea-bird rookeries and during the nesting season supply the town of Deering, 2 miles east of the cape, with abundance of fresh eggs. West of Rocky Point on the south coast Topkok Head and Cape Nome, the only promontories which reach the coast in a distance of 100 miles, scarcely break the continuity of the

---

[a] See Reconnaissances in Cape Nome and Norton Bay regions, Alaska, in 1900, a special publication of the U. S. Geol. Survey, 1901, pp. 19–26.

beach line.   For about 5 miles on either side of Cape York limestone cliffs rise from the sea to an elevation of 600 feet.   West of these cliffs lie beaches backed by a narrow coastal plain for 15 miles to Cape Prince of Wales, where an isolated peak called Cape Mountain rises to an elevation of 2,300 feet above the sea.   On its south and west sides the slopes fall off to the sea in cliffs several hundred feet high. On the north coast of the peninsula there are no headlands from Cape Prince of Wales to the south shore of Goodhope Bay.   The coastal plain slopes to sea level, and through a considerable portion of the distance probably descends slightly below sea level, since the beaches are built on the seaward side of sand spits, the product of wave and current action.

RELIEF.

Although the detailed topography of the peninsula presents great irregularity, induced chiefly by the lack of uniformity of the drainage system, yet practically all the land forms belong to one of three types. These present little variation throughout the province.   In order of importance they are as follows: The uplands, the lowlands, and the mountains.   A fourth type, of subordinate importance in extent, yet containing the key to the evolution of the topography, consists of the rock and gravel floored terraces which occur throughout the peninsula.

In much the greater part of the peninsula the broad, flat-topped and rounded hills, here collectively called the uplands, are the dominating topographic features.   The summit levels, which locally have a striking uniformity of altitude, range from a few hundred feet near tide water to 3,000 feet near the heart of the peninsula.   The evolution of this topographic type will not be here discussed, but, in general terms, this upland can be considered a dissected plateau, whose summit level marks a peneplain.   Attention will be drawn to a number of base-levels of lower altitude, indicating later epochs of erosion, remnants of which are preserved as minor plateaus and benches.

The lowlands embrace three types.   The most extensive are the coastal plains, which nearly everywhere fringe the shore line and in places reach a width of 20 miles or more.   These plains merge inland with the lower slopes of the uplands and form the typical tundra of the circumpolar province.   The extensive basin lowlands are striking topographic features of the peninsula and form a second type of this group.   They are in many places nearly surrounded by uplands and are drained through comparatively narrow valleys.   Like the coastal plains, their floors merge with the slopes of the uplands.   The third type of this group includes the valleys, which are characteristically flat floored with gently sloping walls.   A less common valley type consists of the steep-walled rocky canyons, usually to be accounted for by recent changes of drainage, brought about by local warpings. Valleys of still another type are those which occur in a few of the higher mountain masses and which have been formed by the erosive action of glaciers. These have the typical U form and find their sources in glacial cirques.

The third group of topographic forms embraces the mountains, which, as has been stated, include only one considerable range within the peninsula. There are, however, several other isolated mountain masses, which rise above the general summit level of the upland. The mountains are characterized by a rugged topography and sharply cut drainage channels. As will be shown, a number of them have been the scene of recent glaciation.

To speak broadly, the peninsula is divided into two topographic provinces by the range whose various parts have been called the Kigluaik, Bendeleben, and Darby mountains. South and west of these mountains is an upland region with a relief ranging from 800 to 3,000 feet in altitude. North of the mountains is a second upland region with about the same relief.

The Kigluaik, Bendeleben, and Darby mountains lie along an axial uplift, which is of a crescentic shape. This axis stretches northeastward from a point near Cape Woolley to the one hundred and sixty-third meridian, then bends toward the south, embracing within its sweep the drainage basin of Fish River, and extends southward to Cape Darby, east of Golofnin Bay. The highest peaks of the Kigluaik Mountains reach an altitude of nearly 5,000 feet; those of the adjacent Bendeleben Mountains do not exceed 3,700 feet; and the Darby Mountains are of still lesser altitude, ranging from 2,500 to 3,000 feet. Both the Kigluaik and Bendeleben ranges are rugged, with sharply cut valleys, and both include many glacial cirques. These mountains have been deeply dissected, and the valley walls rise precipitously from the floors. Within the mountains proper the stream gradients are steep and the watercourses torrential. A striking feature of the drainage system of the Kigluaik Mountains is the remarkable straightness of many of the valleys, the larger of which divide the mountains into a number of irregular masses, some of them forming subordinate ranges.

The Kigluaik Mountains are nearly everywhere separated from the uplands by extensive lowlands. On the north lie the extensive flats which surround Imuruk Basin; these flats stretch eastward and sweep around the end of the range and are continued by the valley of Kruzgamepa River. On the south the headwaters of Kruzgamepa and Stewart rivers flow in broad depressions. The west end of the Bendeleben Mountains is similarly isolated, but at the east end of the range there is a more gradual transition between the upland and the

mountains, and this is also true of the Darby Mountains, where there
is in many places no sharp line of demarcation between mountains
and upland.

To the south of the mountains is, as already stated, a highland mass
whose summits range from 800 to 3,000 feet in elevation. The slopes
of this upland are in many places broken by well-marked benches,
which up to an altitude of 800 feet are plainly due to stream erosion.
There are also some still higher benches whose origin has not been
definitely determined. This highland area is essentially one of irregular topography, with no well-defined system of ridges. The watercourses flow in broad, deeply cut valleys, whose slopes ascend gradually to the divides. The summits are rounded, but are broken by
numerous rocky knobs, many of which are carved into fantastic
shapes. These castellated peaks are very characteristic features of
the topography and their preservation plainly indicates the absence
of regional glaciation. Their outline is determined both by the lithologic and the structural character of the rock from which they have
been carved. The general trend of the larger valleys is north and
south, and these block out broad ridges whose margins are
scalloped by the minor tributaries. Studied in detail this feature
would appear to indicate a north-south trend of the topography, but
study of a larger area shows such an interpretation to be at fault.
The highlands, as a rule, fall off to the coastal plain by a series of
well-defined terraces.

As has been stated, the highlands do not in general fall off directly
to the sea, but are separated by a coastal plain of varying width.
Many of these lowlands have a general crescentic form, terminating
at either end in bluffs whose bases are washed by the sea.

The Kigluaik and Bendeleben mountains are separated by lowlands
on the north from the upland region which constitutes the northern
two-thirds of the peninsula. This upland, like the similar one to
the south, is characterized by flat-topped ridges and hills rising to
altitudes ranging from 600 to 2,500 feet. Here, too, the monotony
of the summit level is broken by numerous minor peaks of irregular
form. Besides these there are a number of more extensive mountain
masses that are comparable to the mountains to the south, though of
lesser altitude and not so rugged a character. Of these subordinate
ranges the York Mountains are the most extensive. These form an
irregular mass near the western limit of the peninsula and reach their
maximum altitude in Brooks Mountain, about 2,900 feet above the sea.
The general trend of these mountains is northeastward, but their
extension appears to merge with the general summit level of the upland. On the south and east they fall off to a well-marked plateau,
as they do also on the west. This physiographic feature has been de-

scribed in detail elsewhere.[a] On the north the York Mountains slope off gradually to the upland, which in turn falls off to an extensive coastal plain. The York Mountains proper are rugged and their stream valleys are sharply cut. Among the high summits evidence of some glaciation is to be found, but this is of much less extent than in the mountains to the southeast.

In the northeastern part of the peninsula there are a number of isolated mountain masses, but these appear to be local elevations of the plateau.[b]

The northern upland, as has been stated, reaches an altitude of 2,900 feet. Between the summit level and the lowland there is in many

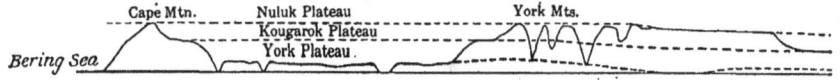

FIG. 3.—Profile of the western part of Seward Peninsula, showing old erosion levels.

places a series of broad benches, which represent distinct epochs of erosion. These have been described in detail elsewhere;[c] briefly stated, four epochs of erosion have been recognized, each of which marks a stage of stability during a general period of uplift. The highest and, therefore, oldest of these plateaus is preserved in benches lying at altitudes of 2,000 to 2,700 feet. This erosion level has been termed the Nuluk Plateau (figs. 3 and 4). A second erosion period, the result of which has been called the Kugruk Plateau, is marked by a well-preserved bench, ranging in altitude from 400 to 1,200 feet.

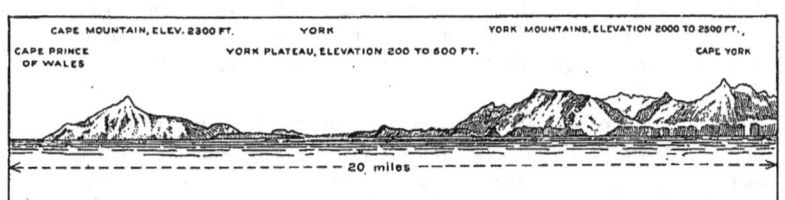

FIG. 4.—Sketch of the coast from Cape York to Cape Prince of Wales.

The so-called York Plateau gives evidence of a third local base-level and stands 300 to 700 feet above the sea. The extensive coastal plain is the result of the fourth and most recent epoch of erosion. While some if not all of these base-levels are probably represented throughout the peninsula they have been correlated only in the northwestern part.

[a] Reconnaissance in the Cape Nome and Norton Bay regions, Alaska, in 1900, a special publication of the U. S. Geol. Survey, 1901, p. 52.

[b] Moffit, F. H., The Fairhaven gold placers, Seward Peninsula, Alaska: Bull. U. S. Geol. Survey No. 247, 1905, pp. 42–44.

[c] Collier, A. J., A reconnaissance of the northwestern portion of Seward Peninsula, Alaska: Prof. Paper U. S. Geol. Survey No. 2, 1902, pp. 35–40.

To speak broadly, the northern upland is of greater irregularity than the southern. There is, however, here the same general trend of the drainage system, which has produced ridges stretching irregularly north and south. This upland on the north falls off by a gradual slope to the broad lowland that skirts the Arctic Ocean, a plain much more extensive than any of those along the southern coast of the peninsula. The transition between the two seems to be more gradual than in the south, where numerous benches break the seaward slope of the upland. As in the southern part of the peninsula, extensive basin lowlands form important topographic types in this region.

DRAINAGE.

The following account of the drainage is quoted from a recent report by Brooks:[a]

The Bering-Arctic watershed follows a sinuous line along the longer axis of the peninsula to Cape Prince of Wales and sends the waters of over two-thirds of its area southward to the Bering Sea. The highlands consist of rolling upland 1,000 to 2,500 feet above the sea, and there are some isolated mountain masses which stand higher, but the latter in no case determine the east and west divide. The headwaters of the streams flowing into the Arctic and Bering seas interlock irregularly within the upland, throwing the watershed first north, then south, and giving it the irregularity which has been described.

The Koyuk River, one of the largest of the peninsula, enters a small tidal estuary at the head of Norton Bay. Its source is in a gravel-floored basin, which lies well toward the center of the peninsula and is bounded on the south by the Bendeleben Mountains, which rise 2,000 feet, and by highlands on the north rising 1,500 feet above its floor. The Koyuk flows in a tortuous course and with sluggish current across this basin, which it leaves through a narrow valley with a steep gradient. This canyon-valley type continues eastward for about 20 miles, to where the river enters another broad, level-floored valley which extends east and south, gradually opening out and merging into the coastal plain of Norton Bay.

The Kwik River enters Norton Bay 20 miles west of the Koyuk. It flows southward through a broad and almost featureless depression which northward is connected with the Koyuk basin by a low pass.

Eight miles west of the Kwik is a lagoon, into which flows the Tubutulik River. This stream rises in a group of low mountains 30 miles from the sea, traverses a small basin, then takes a tortuous course through the hills and enters the coastal plain 10 miles from the sea. The Kwiniuk River, which empties into the same lagoon, rises in low hills 10 miles from the coast and flows in a broad valley from which it emerges on the coastal plain.

At Cape Darby the shore line suddenly retreats inland, and to the northwest are Golofnin Bay and Golofnin Sound. At the head of the sound is the broad delta of Fish River, and the valley of the latter stretches northward for 10 miles as a broad depression, then contracts for a few miles only to open again to a second extensive lowland, which also includes the lower reaches of the Niukluk River, the principal tributary of the Fish. The Fish rises in a basin which is typical of the basins at the headwaters of many rivers of northern Alaska. These basins are level, gravel-filled depressions encircled by uplands

---

[a] Brooks, A. H., Geography and geology of Alaska: Prof. Paper U. S. Geol. Survey No. 45, 1906, pp. 92–95.

whose slopes often rise abruptly. Within them the streams have low grades and flow with tortuous courses, but below them flow through narrow canyon-like—often rockbound—valleys, with straight courses, steep gradients, and frequent rapids. The basin of the Fish is of rectangular outline, 30 miles long and 20 miles wide, and below it is a constricted valley which is about 10 miles long, less than a mile in width, and has abrupt walls. Within the basin the river is tortuous and has a sluggish current, while in the canyon it is comparatively straight and descends through a series of rapids.

Niukluk River, the largest confluent stream, joins Fish River about 20 miles from the sea. Near the mouth its valley has a width of about 6 miles, which gradually decreases to less than a mile near Richter Creek, while 8 miles above the stream flows for 2 miles through a steep-walled rockbound canyon 50 feet deep. Above the mouth of the Casadepaga the valley broadens out to a basin separated by only a very low divide from the Kruzgamepa, which flows into Port Clarence. The two basins connected by a sharply incised valley, are striking features. The valley of Casadepaga River, the chief tributary of the Niukluk, is broad, with gentle slopes, broken by gravel terraces up to an altitude of 600 feet. The headwaters of the Casadepaga are connected by a low, gravel-filled divide with Solomon River, which flows southward.

West of Golofnin Bay the Solomon and Eldorado rivers, as well as many smaller streams, carry the drainage southward through broad, open valleys, whose slopes are often terraced. The Eldorado and Nome rivers rise in gravel-filled basins of the type already described, which to the north are connected by broad passes with Kruzgamepa waters and to the south are succeeded in turn by a constricted valley and a broad valley whose floor merges into the coastal plain. The Snake, Penny, and Cripple river valleys merit no special description. Near the coast they are broad, and their floors are extensions of the coastal plain, above which they become constricted.

Sinuk River,[a] which empties into the sea about 30 miles west of Nome, is one of the largest rivers of the southern watershed of the Seward Peninsula, and receives the drainage from the southern slope of the Kigluaik Mountains. It emerges from the mountains in a narrow gorge and flows in a broad depression parallel to the range for about 15 miles, turns southward and reaches the sea through a broad, flat valley. Its waters, as well as the headwaters of its chief tributary, Stewart River, are connected by a broad, gravel-filled pass with the Kruzgamepa Valley.

Fairview and Feather rivers are streams of minor importance which flow westward from the southern flanks of the Kigluaik Mountains. Tisuk Creek is somewhat larger and drains the northern slopes of the Kigluaik Mountains, from which it emerges in a narrow valley which broadens out. In its lower course the current meanders sluggishly over a flat valley floor and finally empties into a lagoon of Bering Sea. A gravel-filled divide, about 200 feet high, separates the Tisuk from Canyon Creek, which flows northeastward into Imuruk Basin.

A number of small streams flow into Port Clarence and Grantley Harbor from both the north and the south, but the Bluestone River drains the larger part of the area lying between the Kigluaik Mountains and Port Clarence. Its headwaters are in a basin-shaped valley, but at the mouth of the Alder it enters a rock-cut canyon, below which it flows through a broad valley tributary to the Tuksuk Channel.

The head of Imuruk Basin is bounded by a flat, swampy area, through which meanders a broad river, sometimes called the Kaviruk, which forks about 20

---

[a] Locally often called the Sinrock River.

miles from the bay. The southern fork, called the Kruzgamepa, rises well within the Kigluaik Mountains, flowing southward till it leaves them; then takes a northeasterly course and encircles the east end of the range. This part of the valley lies in the same lowland that includes the upper basin of the Niukluk. The northern fork of the Kaviruk, called the Kuzitrin, has its source in a broad, flat basin in which it is joined by its chief tributary, the Kougarok. Below the basin it flows through a rather narrow valley for about 20 miles, and debouches on the plain at the head of Imuruk Basin. A third large area is drained into Imuruk Basin from the north by the Agiapuk River.

Between Port Clarence and York are several streams which have in general a southerly course, but which are insignificant in size. West of Cape York the first important river is the Kanauguk, which lies well within the York Mountains.

The Arctic drainage of the Seward Peninsula, embracing probably not over a third of its area, is all of practically the same general type. The headwater valleys are broad and open and the watersheds separating them from the valleys of the streams flowing south are often very low. The passes are broad and gravel filled and suggest recent changes in drainage. The valleys of the northward-flowing streams open out as they approach the Arctic, the gradients become less, and finally the floors merge into the coastal areas, through which the streams meander with sluggish currents. Many of the Arctic streams empty into large lagoons, which are cut off from the ocean by long barrier beaches. The streams on the Arctic coast are straighter than those flowing into Bering Sea. The easterly streams flow northeast, the westerly ones northwest. West of Cape Espenberg the Serpentine, Arctic, Kougarok, Pinguk, and Mint are the chief rivers, while the Goodhope, Cripple, Inmachuk, Kugruk, Kiwalik, and Buckland drain the northeastern part of the peninsula.

The Buckland River, which flows into Eschscholtz Bay, a part of Kotzebue Sound, through a long tidal estuary, rises about 75 miles southeast of the bay, and opposes the drainage flowing on one hand into Norton Bay and on the other into the lower Koyukuk. The lower 60 miles of this river were explored in 1849 [a] by expeditions sent out from H. M. S. *Herald* and *Plover*. The reports of these parties show that dead water extends for about 30 miles from the mouth, and that there are no serious rapids for 30 miles farther upstream. There is a native settlement near the head of the river which is connected by portage trail with the Kateel River, a tributary of the Koyukuk.

CLIMATE.[b]

GEOGRAPHIC LOCATION.

The climate of Seward Peninsula is clearly under the control of its outline and location. Located between Kotzebue Sound of the Arctic Ocean and Norton Sound of Bering Sea, its summer and winter temperatures are much less extreme than those characterizing the other portions of Alaska under the same parallels. The proximity of these two water bodies also supplies more moisture to the atmosphere over the peninsula than is found farther east. The high

[a] Mendenhall, W. C., Prof. Paper U. S. Geol. Survey No. 10, 1902, p. 12.
[b] Prepared in collaboration with Cleveland Abbe, jr. Compare section on climate by Mr. Abbe in Geography and geology of Alaska: Prof. Paper U. S. Geol. Survey No. 45, 1906, pp. 133–200.

latitude of the region makes both seas almost icebound for half the year, and this reduces their equalizing influence to some extent.

OBSERVATIONS.

Very few reliable instrumental observations of the climatic elements have been made on Seward Peninsula, and none have been carried on consecutively during a considerable number of years. Most of the records are for fragments of years, or even of months, and the only ones from the interior of the region have been made by geologic exploring parties, moving camp almost daily.[a]

TEMPERATURE.

As already stated, the temperature conditions of Seward Peninsula are less severe than those experienced in the interior of Alaska, but are subject to greater extremes than those of southern Alaska. During quiet days in winter the temperature has fallen to about —50° F., but residents do not consider the usual extreme winter temperature of about —40° F. dangerous except when accompanied by high winds from the northeast. The lowest temperatures occur during January and February, and so far have been reported from points near the coast. It is probable that the interior of the peninsula is subject to lower winter temperatures, but no records are known to have been kept there during that season. It is remarkable that the coldest months are subject to sudden thaws and warm rains which open all the streams and flood them. Such winter thaws have also been accompanied by open water in Norton Sound. The summer temperatures rarely rise above 80° F. On one occasion only has there been recorded a temperature as high as 84° F., and that was in August, 1900, at a point on Fish River. Probably the maximum temperature is below 80° over most of the peninsula. In 1903 the highest temperatures experienced over the northeastern portion of the peninsula were 69° in July, 66° in August, and 62° in September; the most common temperatures lay between 45° and 60°, with extremes of 10° in September and 69° in July.

The daily minimum temperatures are 32° or lower from about the first week in September until about the middle of May. The first fall frosts occur about the middle of September, and the last frost in spring toward the middle of June.

The sea temperature of Port Clarence has been found to average 38.9° F. in September, and the waters of Kotzebue Sound averaged from 50.9° in August to 38.3° in October. Both bodies of water thus

---

[a] See Collier, A. J., Reconnaissances in Cape Nome and Norton Bay regions, Alaska, in 1900; a special publication of the U. S. Geol. Survey, 1901; Prof. Paper U. S. Geol. Survey No. 2, 1902. Mendenhall, W. C., Prof. Paper U. S. Geol. Survey No. 10, 1902. Moffit, F. H., Bull. U. S. Geol. Survey No. 247, 1905.

range lower than Norton Sound at St. Michael, where an average temperature of 56° in July and August and 49.6° in September has been observed.

PRECIPITATION.

The actual rain and snow fall of the peninsula is probably small. The nearest station with a good record is St. Michael, where the annual fall amounts to about 18 inches. Comparison of the records there with fragmentary records from the peninsula indicates an annual fall of about 20 inches of water over its southern portion and of perhaps 25 inches of water over the northern portion. So far as experience and observation go, the greater portion of this water falls as rain during the months from June to October. During the winter months an actually small depth of snow, perhaps 3 feet, falls, but it is so badly drifted by the frequent severe northeastern gales of this season that it has proved impossible to obtain accurate measurements of the amount. There are no known measurements of the water equivalent of the winter snowfall of the peninsula; it may be estimated at 3 inches, allowing something less than 0.1 inch water per inch of snow. The snowfall is reported sufficient to permit sledding throughout the winter, and the heavy drifts which form annually in valleys and on hill slopes remain far into the summer or even until the succeeding winter.

With so small an annual supply of rain water it is fortunate for the miner that the snowdrifts thus preserve the scanty winter fall until the season when he can use it to the best advantage—the summer. The spring thaw, however, with its attendant May freshet, due to melting drifts, is often followed by a six weeks' drought, since the rainy season does not begin until July. This rainy season is usually characterized by frequent showers rather than continued rains. Another factor which saves the rainfall for the miner is the perennially frozen condition of the gravels and bed rock throughout most of the peninsula. This circumstance gives a maximum surface run-off and a minimum loss of ground water, so that while the frozen gravels are difficult to work, yet the resources for their exploitation are for the same reason greater than they would be if the temperature were uniformly higher. The frozen ground extends to an unknown depth and embraces both gravels and fissured country rock. Large bodies of gravel deposits, however, have been found always free from ice. It seems probable that these unfrozen gravels rest upon a more porous bed rock that permits the ground water to drain away more rapidly than is possible in the surrounding frozen regions.

The rain which falls during the summer is rapidly soaked up by the surface layer of moss where other vegetal cover is lacking. This covering protects the frozen ground beneath from the sun's rays, summer and winter, so that the soil remains frozen to the very surface as long as covered. The absorbed rain water flows into the streams at a fairly uniform and constant rate, because of the cover, and thus the streams are less variable than they would be if vegetation were absent from the frozen soil.

STORMS.

The peninsula is subject to frequent severe north-northeast gales during the winter, in which the dry snow is blown along in blinding clouds and heaped into drifts. Such storms often continue for three or four days and are usually accompanied by only small quantities of fresh snow and temperatures between — 30° and — 40°. They seem to occur with some regularity at intervals of about a week from December to April, and occasionally in the other months. Records from the Teller reindeer station indicate that high winds are sometimes accompanied by very moderate temperatures. Although most of the storm winds are from the northeast to north, in September and March, there are occasionally severe storms from the south and southeast, bringing higher temperatures, and, during September, heavy rains. Such storms are found to lower the level of Norton Sound, and their winds are heavy with moisture and produce raw, uncomfortable weather.

RIVER AND HARBOR ICE.

All the streams of the peninsula freeze across in winter, and the smaller streams frequently freeze to the bottom. They close about the end of September in the vicinity of Fish River, and open about the middle of May. When a stream is not frozen to the bottom, the running water often breaks through its ice arch and flows out over the surface, where it soon freezes. Such winter overflows sometimes completely floor the valley with successive ice sheets. These ice sheets are locally known as "glaciers" and are quite distinct in character and origin from the lenticular masses of ground ice, or crystospheres, due to the freezing of ground water, which are also colloquially known as "glaciers."

Bering Strait is rarely closed by an ice bridge, but Bering Sea is often frozen to a distance of 5 or 6 miles from land, and Arctic floe ice may reach as far south as the Pribilof Islands, which lie in the latitude of Newfoundland. The ice of Bering Sea begins to retreat northward early in the spring. St. Lawrence Island has become accessible as early as April 22. St. Michael Harbor, on Norton Sound, which closes early in November, usually is not open until June 10 to 15, or a little later, and Nome seems to be open about the same date. The shallower waters of Port Clarence keep Teller imprisoned for some time longer, as is also the case in Kotzebue Sound. Point

Hope, on the Arctic coast, is not clear of ice before July 1, or even August 1 in exceptional years, and is again icebound by November 15.

VEGETABLE AND ANIMAL LIFE.

VEGETATION.

A description of the vegetation of the peninsula has already been given by the writer in a previous report,[a] hence only a brief statement will be inserted here. In general, the vegetation is stunted and dwarfed, a condition to be expected in a subarctic climate. As shown by the sketch map (fig. 5), the greater part of the area is occupied by tundra and timberless uplands; but in proceeding from east to west the country becomes progressively more barren and desolate. In the eastern part, more especially in the region north of Norton Sound, there is a scattering growth of small spruce timber confined to the river valleys and valley slopes, the higher hills being above timber line. The largest trees are not over 16 inches in diameter and 50 feet high. Lumber for the first sluice boxes used in the peninsula was whipsawed from this timber, and a small steam saw-mill at Council is still able to compete in the local market with the product of outside mills. The majority of the buildings at Council are log houses, to which use the spruce trees are well adapted. Many of the trees are of suitable size for tie timber and will doubtless be used in railroad building. The western limit of the spruce is approximately represented by a line extending from Golofnin Bay, northwestward of the headwaters of Niukluk River, and thence northeastward to the east end of Kotzebue Sound, as shown in fig. 5. Along the streams west and north of this line cottonwoods, willows, and dwarf alders grow in sufficient quantity to afford a scanty amount of fuel, which, though it is of rather poor quality, is used for camping purposes and for thawing frozen ground, and has made possible the exploration of the region and the development of the placers. Of these trees the willows are most widely distributed, as they occur over nearly the whole peninsula. The cottonwoods are of larger size, but are confined to a few localities, reaching as far west as Kruzgamepa River. The alders form numerous tangled thickets on the valley slopes and hillsides as far west as Sinuk River. In favorable localities the larger willows attain a thickness of 6 inches and a height of 20 feet, but most of them are smaller, a large proportion being less than 2 inches in diameter and 5 feet in height.

Along creek and river bottoms and valley slopes grass is abundant and is suitable for pack animals, but most horses used for heavy teaming are fed on imported hay and grain. During the field season of

---

[a] Reconnaissances in Cape Nome and Norton Bay regions, Alaska, in 1900, a special publication of the U. S. Geol. Survey, 1901, pp. 164–174.

1900 the writer collected for identification specimens of six species of grass suitable for forage, but the collection was incomplete, and it is probable that as many as a dozen different grasses are represented in this region. Here and there are natural meadows where hay could be harvested in considerable quantity, but owing to the short season it does not mature properly and has not the strength of hay grown in the States, and as it is almost impossible to cure it, on account of rainy weather in haying time, it has not been used to any considerable extent.

The most common plants are mosses and lichens, which are present everywhere. On the lowland plains and portions of the upland where

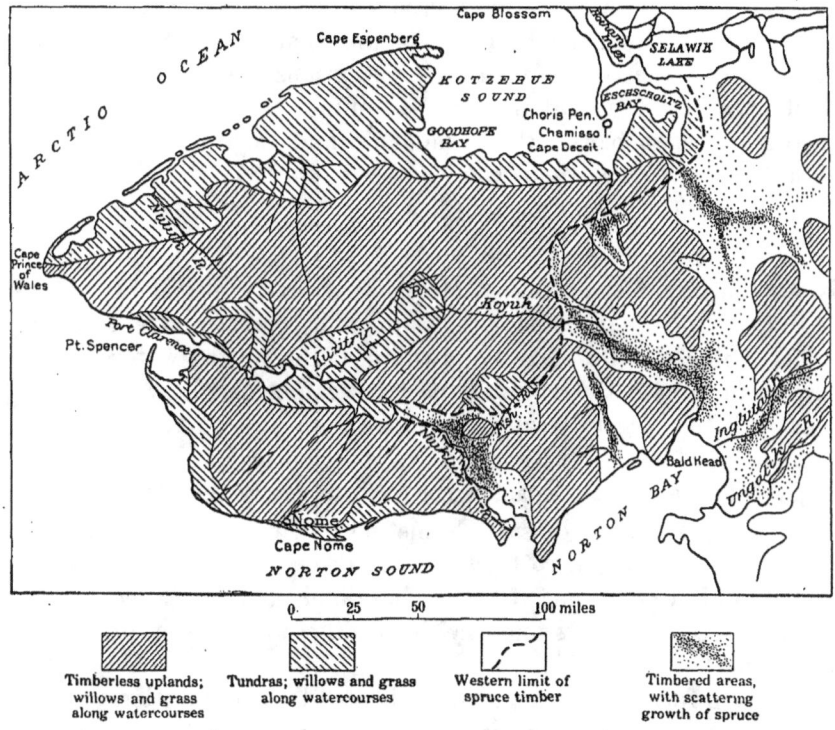

FIG. 5.—Map showing distribution of vegetation in Seward Peninsula.

Timberless uplands; willows and grass along watercourses

Tundras; willows and grass along watercourses

Western limit of spruce timber

Timbered areas, with scattering growth of spruce

drainage is imperfect a thick mat of vegetation, composed of mosses, lichens, sedges, dwarf birches, and some grass, overlying peat beds, covers the surface and forms the tundra. The underlying soil is perpetually frozen, as the mat of vegetation and peat protects it from changes of temperature, but during the open season the tundra is difficult to traverse on account of its soft, swampy surface. Good road-beds for wagons can not be built across the tundra for less than $2,000 per mile. The larger tundra areas are shown on the sketch map (fig. 5). Over the greater part of the upland, especially along the ridges, the vegetable mat is comparatively thin, travel is easier, and fair road-beds can be constructed at slight expense. The peat found on the tundra, although it is usually not very pure, might, by some briquetting process, be made into valuable fuel, but up to the present time it has received little attention. Reindeer moss is distributed over the whole region as an important constituent of the vegetable mat, making possible the maintenance of large deer herds. The number of domestic deer in the three herds kept on the peninsula in 1904 was estimated at 3,485.[a]

Two species of edible berries, the marsh blueberry and the " salmon berry," are fairly abundant over the whole peninsula, and often form a welcome addition to the prospectors' fare. Mushrooms, some of which are edible, are seen occasionally, but, so far as the writer knows, they are never plentiful and have not been used as food by prospectors and miners. There are also a few herbs which are sometimes used as food or medicine by the natives. With these exceptions, there are no native food plants in the region, but near the mining camps some of the hardier vegetables, such as lettuce, radishes, and turnips, have been successfully raised in small gardens.

FISH AND GAME.

In nearly all the larger streams fish are comparatively abundant and can be caught during the summer. Salmon in great numbers run up the larger rivers and die after spawning, polluting the water with their decaying carcasses. Grayling and trout can be caught with a line in nearly all the larger streams through the open season, and pike and whitefish are caught with gill nets on some of the larger rivers. In the waters along the coast, also, fish are fairly abundant. During certain seasons schools of smelt sometimes swim in the surf and wash up on the beach, where they can easily be caught in the hands. The natives have depended for food in winter on a fish called tomcod, which they catch in nets and dry during the open season, or later catch through holes in the ice.

That caribou were formerly abundant is indicated by many antlers scattered over the country and by large piles of bones found near some of the deserted villages of the interior. The wild caribou now, however, is practically extinct, though a few of them are to be seen in the northern part. As noted above, three large herds of domestic caribou, or reindeer, under Government supervision, are maintained on the peninsula.

Around the coasts polar bear are occasionally killed in winter, and brown and black bears are occasionally reported.

Of game birds, ptarmigan, snipe, plover, and curlew—plentiful near the settlements in 1903—are now scarce. Geese, ducks, sand-

---

[a] Rept. Commissioner of Education, 1903, vol. 1, 1905, p. lxiv.

hill cranes, and swans abound along the rivers and lakes, more especially in the northern part of the peninsula, but they are not usually killed by prospectors unless a special hunt is made for them, as the regions frequented by them are apt to be remote from the placer-mining districts.

COMMERCIAL FEATURES.

The lines of industrial advancement and settlement in the peninsula have been less influenced by its physical advantages or disadvantages for such movements than by the distribution of the gold placers. Settlements and precincts [a] have been established with little regard to the geographic conditions which in other fields would determine their location. Indeed, except for its mineral resources, Seward Peninsula would certainly not for many decades and probably never have supported its present population. From an industrial standpoint, therefore, the distribution of mineral resources is more important than the geographic features.

For administrative purposes the peninsula is divided by arbitrary boundaries into a number of precincts or districts which have been created and modified to conform with the progress of settlement and development of mining. Their boundaries are subject to change at the discretion of the United States court, located at Nome, and although some of them have remained unchanged for several years they can not be regarded as permanent. In June, 1903, there were eight of these precincts, and since that time a number of minor changes have been made, though the general location of the precincts remains the same. These frequent changes in local boundaries are indices of the industrial progress, as they are made to meet the convenience and immediate needs of miners and prospectors. The precincts, as they existed in June, 1903, are shown in the sketch map (fig. 2, p. 42), and the descriptions of placers given on later pages are grouped according to these boundaries.

The local government of each precinct is administered by a commissioner, who is appointed by the United States district court at Nome and who is also the recorder of mining claims. These seats of local government are in general located at the more important settlements and distributing points of the mining districts.

The principal settlement and largest town in the peninsula is Nome, situated on the southern coast several miles west of Cape Nome. It has a population of about 3,000 in winter and 5,000 to 6,000 in summer. Although Nome has been in existence only about seven years,

---

[a] The limits of the precinct, the administrative and judicial unit in Alaska, are determined by the district judges. A precinct may be compared with a county, though the absence of all self-government in the Territory makes the similarity only a very general one.

it has many of the conveniences of a city, such as planked streets, a water-supply system, telephone lines, and banks. Though there is no harbor, nearly all supplies and passengers for the peninsula are landed at Nome. During the heavy weather in the late summer and fall all communication between the shore and the ships in the roadstead is often cut off for several days at a time. No permanent wharves or other conveniences for landing can be maintained on account of the heavy shore ice in winter and the moving ice floes in the spring, and all supplies and passengers are landed on the beach through the surf. (See Pl. I, A.) The Seward Peninsula Railway leads inland from Nome to the important mining district that centers at Anvil Creek, where the first discoveries of gold were made, and to Lanes Landing, in the Kougarok district. Nome is also the starting point for many overland trails and roads to other mining camps farther inland.

To the east along the shore there are several smaller towns similar to Nome in their situation on the open coast which have aspired to be its rivals. Dickson, 30 miles east of Nome, at the mouth of Solomon River, is the starting point of the Council City and Solomon River Railway, which extends into the Casadepaga Valley, as well as the center of a mining district of some importance. Bluff, a village and post-office several miles farther east, is the distributing point for a district that has been producing gold in large quantities since 1900. Council, the second town in importance of the peninsula, is the central point of the Council precinct and lies about 25 miles from the coast, on Niukluk River. In production of placer gold this precinct almost equals the Nome precinct. Supplies for the Council precinct are landed at Chenik (Golofnin post-office), a small village on Golofnin Bay. They are then taken in small steamers to White Mountain, on Fish River near the head of Golofnin Sound, where they are transferred to flat-bottomed river boats that are towed by horses up Fish and Niukluk rivers to Council. The completion of the railway from Dickson as originally planned will give more direct communication. Another means of communication would be by building a branch line to Council from the Seward Peninsula Railway near Iron Creek.

Teller, located on Port Clarence at the entrance to Grantley Harbor, is the supply point for the Bluestone and Kougarok mining regions and several others of less importance. It occupies the best town site in the peninsula, as the ground on which it is built is dry, and the harbor and landing facilities are the best. The deep water of Port Clarence, however, does not extend to the town, and the larger vessels are obliged to anchor about 2 miles away and discharge freight and passengers by means of lighters. Moreover, the ice in the bay does not break until several weeks after the roadstead at Nome is open.

A. LIGHTERING FREIGHT AT LOW TIDE, NOME.

B. CHARACTERISTIC CREEK VALLEY: THOMPSON CREEK, A TRIBUTARY OF STEWART RIVER.

58A

The supplies for the mines of the Kougarok region up to 1906 were usually sent by small steamers from Teller to Igloo, a village and post-office 50 miles inland at the head of steamboat navigation on Kuzitrin River, where they were transferred to flat-bottomed river boats that were towed up Kuzitrin and Kougarok rivers. Since the completion of the Seward Peninsula Railway to Lanes Landing more direct communication with Nome has been established. The mining centers are beyond the reach of navigable waters and the railway, and the last stages of the journey are made by pack train or wagon. Sullivan, the center for the Bluestone region, is 16 miles southeast of Teller. It has a summer population of about 50, a post-office, and several roadhouses, and is connected with Teller and a landing at the mouth of Tisuk Creek by a regular line of stages.

The York region, embracing about 400 square miles in the western extremity of the peninsula, has not been productive of placer gold, but there are several small settlements maintained by prospectors engaged in a search for tin ores. These are supplied by small coasters from Nome. York, a small collection of cabins and tents at the mouth of Anikovik River, 12 miles east of Cape Prince of Wales, has been in existence since 1899. Tin City, established in 1903, is a camp at the southeast base of Cape Mountain, 10 miles west of York and about 4 miles west of the old native village Palazruk. At Cape Prince of Wales there is a large native village called Kingegan, which has been in existence since prehistoric times. The location has been a favorable one for sealing and whaling operations; and as it is the nearest point to the Siberian coast, the Kingegan natives have controlled the trade between Siberia and Alaska. The village supports a population of about 500. A Congregational mission has been maintained here for ten years or more, and a Government school was established in 1904. There is also a store and post-office. These natives have learned something of the ways of civilization. Many of them own herds of reindeer, and one of them keeps a general-merchandise store which does not differ greatly from similar stores kept by white men at Nome. The post-office here is called Wales. The town is contiguous to the tin region and is the seat of the commissioner's court for the Wales precinct and headquarters for a number of prospectors.

On Kotzebue Sound there are two thriving towns, Deering and Kiwalik, the distributing points for the Inmachuk and Candle Creek districts, both in the Fairhaven precinct. The commissioner for the precinct has headquarters at Kiwalik. Both towns are supplied from Nome by small coasting vessels.

Seward Peninsula is accessible by sea only during the months from June to November. For the remainder of the year navigation is closed, and the only communication with the outside world is by the

Yukon River route, involving a journey over the ice by dog team
to Dawson for upward of 1,000 miles. Mails and a few passengers
follow this route. The cost of the winter journey to Nome from
Seattle by way of Skagway and Dawson is not far from $1,000, and
it ordinarily consumes about three months. Telegraphic communication over the military lines, partly by a wireless system, has been
maintained since 1904. The most-frequented routes of travel
between points on the peninsula are along the coasts and navigable
rivers, and, as a matter of course, development has been most rapid
in districts thus made accessible. All the yet notable mining camps
are within easy reach of such navigable waters. Wagon roads and
trails lead from the railways and from the coast to the mining districts of the interior, but these are difficult for summer travel on
account of the soft road beds, and freight rates are almost prohibitive.
Travel over the snow is easier, however, and supplies may be sledded
to the mines located inland more cheaply in winter than in summer,
but for points remote from the coast the freight rate is still often
too high to permit successful mining operations. Mr. Brooks [a] estimated that in 1903 the cost of summer overland transportation
ranged from $10 to $16 per ton per mile, and the cost of water
transportation between coastal points from 70 cents to $1.50 per ton
per mile, including the embarking and disembarking. Though these
costs have been somewhat lowered, yet they are still prohibitive to
mining anything but high-grade placers, except in regions close to
tide water. There are but few wagon roads in Seward Peninsula,
but during dry weather it is possible to haul loads over the tundra
and along the watercourses. Purington [b] has estimated that $3,000 to
$4,000 a mile is the cost of wagon-road construction. He believes that
narrow-gage railways can be built at a cost of $6,000 a mile. It is
probable that the railways will be preferred for long hauls. Several
short lines are already in operation, and experience has shown that
narrow-gage roads with light equipment can be maintained at no
very great cost. The topographic maps indicate that easy grades
can be found to all parts of the peninsula.

GENERAL GEOLOGY.

INTRODUCTION.

The geologic surveys in this field have all been of a reconnaissance
character, more especially directed toward a study of the gold placers.
Such reconnaissances are a necessary preliminary to detailed surveys
of any part of the field, because the results obtained from critical

[a] Brooks, A. H., Placer mining in Alaska in 1903: Bull. U. S. Geol. Survey No. 225,
1904, p. 51.
[b] Purington, C. W., Methods and costs of gravel and placer mining in Alaska: Bull.
U. S. Geol. Survey No. 263, 1905, p. 221.

examinations of small areas are unsatisfactory without a general knowledge of the problems presented by the whole province.[a]

When this work began practically nothing was known of the geology of northwestern Alaska, and hence the conservative policy was adopted of publishing only such general conclusions as were demanded to elucidate the economic problems. In the present treatment the same policy will be adhered to, though the accumulation of many additional data will make it possible to amplify the previous reports. The discovery of organic remains in some of the terranes has given a key to the stratigraphy that was not available when the earlier reports on the geology were prepared.

The surveys of 1900 resulted in the publication of two preliminary geologic maps covering the southern half of the peninsula—one of the southwestern part, or Nome region, by Brooks,[b] and the other of the southeastern part, or Norton Bay region, by Mendenhall.[c] Mendenhall, whose journey was rather of the nature of an exploration, did not attempt to differentiate the various groups of metamorphic sediments, but Brooks recognized three terranes, named in ascending order the "Kigluaik series," "Kuzitrin series," and "Nome series."[d]

The two lower terranes are described by Brooks as rather well defined stratigraphic units, while a great complex of metamorphic rocks which overlie them was thrown together as a cartographic unit under the name "Nome series," since the field work had not gone far enough to indicate any logical basis of subdivision.

During the following season the writer extended these subdivisions to the rocks of the northwestern portion of the peninsula, where he was able to differentiate within the Nome a massive limestone member containing fossils which were at that time considered to be of Ordovician age. This member was called the Port Clarence limestone. During the same season fossils referred to the Silurian were found in limestones and schists that more nearly resemble the rocks of the "Nome series" as exposed at Nome, and the rocks containing them were called the "Kugruk group."[e] The "Nome series" in the northern portion of the peninsula, therefore, was regarded as comprising a lower member, called the Port Clarence limestone, and an upper one, called the "Kugruk group."

In the course of the investigations in 1903 larger collections of fossils were made at a number of new localities, both in the northern

---

[a] In the summer of 1905 Fred H. Moffit and Frank L. Hess carried on detailed geologic mapping in the Nome region; this work was continued by Mr. Moffit, aided by Philip S. Smith, in 1906.

[b] Reconnaissances in the Cape Nome and Norton Bay regions, Alaska, in 1900, a special publication of the U. S. Geol. Survey, 1901, Pl. III.

[c] Idem, Pl. XXI.

[d] Idem, p. 27.

[e] The spelling of the name of the river from which this group derives its designation has since the previous report been changed by the United States Geographic Board to "Kougarok," as there is another Kugruk River in Seward Peninsula.

and southern parts of the peninsula, which have made it possible to reconsider the determinations made in 1901. Near the western extremity of the peninsula, Mississippian ("Lower Carboniferous") fossils were found in some metamorphic limestones and schists which on lithologic grounds had previously been correlated with the "Kigluaik series." All the fossils from the "Nome series," however, are now regarded as Silurian or Ordovician, but the differentiation of the Silurian from the Ordovician horizons can not be demonstrated, as some forms heretofore regarded as Ordovician are associated with forms known to be Silurian, and for this reason the limestones of the "Kugruk group" are correlated with the Port Clarence.

The more critical examinations which have been made of a section of the highly altered rocks between Nome and the Kigluaik Mountains have demonstrated that there is here a great series of schists, including some calcareous beds, overlain by an almost equally great series of beds, for the most part limestone, which may be tentatively correlated with the Port Clarence. Somewhat similar conditions exist northwest of the Kigluaik Mountains, where a broad belt of schists containing some calcareous beds intervenes between the mountains and the known area of Port Clarence limestone.

In the northeastern part of the peninsula schists of slightly different character from those near Nome overlie the Silurian limestones of the "Kugruk group," now correlated with the Port Clarence.

The "Nome series," as mapped by Brooks in 1900 and by the writer in 1901, may be provisionally regarded as comprising (1) a basal formation, composed for the most part of schists, but including some limestones; (2) a middle formation, composed mainly of massive limestones to which the name Port Clarence limestone has been applied; and (3) an upper formation, composed mainly of schists. In the following discussions and accompanying maps this so-called series will therefore be designated the Nome group.

In general it can be said of the metamorphic rocks of Seward Peninsula (1) that they comprise a complex mass of more or less altered sediments, intrusives, and eruptives, the greater part of which is probably of Paleozoic age, and (2) that with some modification the divisions made by Brooks in the southern part of the peninsula are well grounded and can be recognized over much of the adjoining region.

It must be remembered, however, that this subdivision is based largely on the lithologic character of metamorphic rocks. The degree of metamorphism changes greatly in different parts of the region, producing various kinds of rocks from the same original material, and it is also probably true that some lithologically similar schists may be due as much to peculiar conditions of alteration as to original texture.

Here and there in the northern part of the peninsula there is evidence of a certain amount of folding before the schistosity was developed, and at such places the foliation and bedding are very divergent. As a general rule throughout the peninsula, however, the foliation and bedding seem to coincide; but inasmuch as this condition would result if, before metamorphism, the beds were closely folded and overturned, the evidence afforded by the foliation planes must be used with caution in interpreting the structure. For this reason the stratigraphic succession here outlined, where not supported by paleontologic determinations, must for the present be regarded as provisional.

Since all of the geologic investigation has been directed primarily to a study of the placer deposits, the unconsolidated Quaternary terranes containing them have naturally been examined in the most detail.

The rocks of the Nome group also have been examined more thoroughly than those of the other metamorphic terranes, first, because they are more widely distributed, and, second, because most of the gold placers have been derived from them. These formations are therefore described at some length in this section, and many specific details regarding them are given in the descriptions of localities producing placer gold. The other metamorphic rocks, on the contrary, have been examined only cursorily at a few localities, and their descriptions are necessarily brief. For this reason the general geologic discussion here presented may seem ill balanced, but this result is unavoidable under the conditions governing the investigation.

DESCRIPTION OF GEOLOGIC MAPS.

The distribution of the sedimentary and metamorphic formations and of the igneous rocks of various types is shown on the two geologic reconnaissance maps (Pls. X and XI, in pocket). Nearly all the colored areas have been examined by United States geologists, but as this examination was in the nature of a reconnaissance, the interpretation of the facts obtained as well as the boundaries mapped must be regarded as subject to revision after more thorough study. A more detailed account of the local geology of certain areas will be found in the descriptions of the placer deposits.

In addition to the metamorphic rocks, several groups of igneous rocks are differentiated: (1) Unaltered basalts and basaltic lavas in dikes, necks, and surface flows are of abundant occurrence in the northern portion of the peninsula. Most of these rocks were erupted during Pleistocene time, and the whole group is believed to be of comparatively recent origin. (2) Altered basic rocks, here termed greenstones, in stocks, dikes, and sills, are widely distributed within the areas of metamorphic rocks in both the northern and southern parts of the peninsula. As in many places they have been affected by the same influences as the metamorphic rocks, they are believed to be nearly as old. (3) Acidic igneous rocks, including granites, pegmatites, and granitic porphyries, have their greatest development in the Kigluaik, Bendeleben, and Darby mountains, but in smaller masses appear in all of the Paleozoic formations represented. Some of these rocks were intruded previous to the final metamorphism of the Kigluaik group and have a gneissoid texture, but most of them are unaltered. It is the writer's opinion that most of these granites were intruded during Mesozoic or early Tertiary time, and reasons for this belief will be advanced. If this is true the granites are intermediate in age between the greenstones and the basalts.

The boundaries of the granite areas in the Bendeleben and Kigluaik mountains are not known in detail, as the structure of these ranges is complex and few sections have been carefully examined, but the general type of the intrusions has been determined and their representation on the map is diagrammatic. The correlation of the massive limestones and larger schist areas of the Nome group is based on lithologic resemblance and stratigraphic succession, corroborated in a few localities by paleontologic evidence. The mapping of many of the Quaternary areas has been extended beyond the limits of actual observation, on evidence offered by the topographic maps.

The stratigraphic position of the various sedimentary formations mapped is shown in the accompanying tabular statement:

Stratigraphy of Seward Peninsula.

System.	Series.	Formation.	Lithology.
Quaternary, including some Pliocene?			Tundra deposits; peat, capped by mat of living moss and other vegetation. Alluvium in beds and flood plains of existing streams; often frozen. Glacial moraines and extramorainic bowlder beds. Frozen silts, sands, and gravels of coastal and interior basin plains. Silts, sands, gravels, and residual deposits resting on high benches and plateaus of ancient erosive cycles.
			Unconformity.
Tertiary?	Eocene?	Kenai formation?	Conglomerates, sandstones, clays, and coal.
			Unconformity.
Carboniferous?	Mississippian?		Massive crystalline limestones interbedded with black phyllites and schists.
			Conformity?
Carboniferous, Devonian, or Silurian.			Dark-colored, usually graphitic slates and phyllites, in many places cleaving into pencil-shaped fragments. Including some beds of earthy limestone.
			(?)
Devonian or Silurian.		Undifferentiated schists.	Schists of the Fairhaven precinct. Differentiated from schists of the Nome precinct by stratigraphic position overlying Port Clarence limestone and by smaller percentage of chlorite.
			(?)
Silurian and Ordovician?		Port Clarence limestone.	Massive limestones varying from unaltered more or less earthy limestones in the type locality to crystalline marbles and calcareous schists in more metamorphosed portions.
	Nome group.		Conformity?
Ordovician or pre-Silurian.		Undifferentiated schists.	Various schists and phyllites, including quartz-chlorite schist, quartz-chlorite-albite schist, quartz-graphite schist, calcite-muscovite schist, thin limestones, etc., differentiated from Port Clarence limestone by stratigraphic position and lithologic character.
			Unconformity?
Pre-Silurian?		Kuzitrin formation.	Graphitic quartzites and schists, maximum thickness probably less than 2,000 feet.
			Conformity.
Pre-Silurian?		Kigluaik group.	1. Biotite schists and thin limestone bands. 2. Heavily bedded limestone. 3. Biotite gneiss. These are cut by granite in dikes, stocks, and sills. Total thickness over 4,000 feet. Differentiated from Nome group by stratigraphic position and by universal presence of biotite in the schists.

METAMORPHIC ROCKS.

KIGLUAIK GROUP.

At the base of the geologic column there is a group of highly crystalline rocks consisting of limestones, schists, and gneisses, which were described by Brooks under the name " Kigluaik series." [a] There is reason to believe that, with the exception of some of the gneisses whose origin is in doubt, most of these rocks are altered sediments. They differ from the overlying schists and limestones mainly in the degree and character of the metamorphism to which they have been subjected and in the nature of the resultant minerals, though it is probably also true that the beds were distinct in their original composition. The rocks of this group have been recognized only in the Kigluaik and Bendeleben mountain areas, and in both ranges they are so intimately associated with intrusions of granite, in the form of stocks, sills, and dikes, that it has been impossible to differentiate on the geologic map the igneous from the sedimentary rocks.

The limestones are in general comparatively pure and white or bluish in color. They are always highly crystalline and are in many places specked with graphite in disseminated tabular grains. The beds range in thickness from 1 to 20 feet or more. These limestones, although very prominent in some localities, seem to comprise only a minor part of the group. They are interbedded with schists, and in the few places where sections have been examined in some detail seem to occur near the base of the Kigluaik group.

The schists which comprise the greater part of this group are generally dark gray in color and consist essentially of quartz and biotite, with various accessory minerals in the different beds, such as graphite, pyrite, magnetite, garnet, staurolite, hornblende, augite, orthoclase, plagioclase, etc. In some of the beds quartz predominates to such an extent as to make the rock essentially quartzite, while in others plagioclase and orthoclase feldspar are so abundant as to give the rock a gneissoid character. Between the quartzose and gneissoid phases there seem to be all gradations. Graphite and pyrite are important accessories in many of the more siliceous beds, giving them a black color. In some beds graphite occurs to the exclusion of all other minerals. Though graphitic schists occur in some of the later formations, as will be shown, no beds of pure graphite comparable to these have been found in other formations. The presence of these graphitic members and of the limestones already described is regarded as conclusive evidence of the sedimentary origin of the greater part of the group. It is probably true, however, that some at least

---

[a] Reconnaissances in the Cape Nome and Norton Bay regions, Alaska, in 1900, a special publication of the U. S. Geol. Survey, 1901, p. 27.

of the gniesses were originally igneous and represent intrusive sills or possibly extrusive beds of acidic rocks which have been recrystallized under the same influences that produced the metamorphism of the schists and limestones. It will be shown that many of the intrusive granite masses which occur in the Kigluaik group also have gneissoid phases that are difficult to distinguish from the older metamorphosed rock, though it is believed that after a more refined geologic and petrographic examination all these later intrusives will be differentiated.

In general structure the Kigluaik and Bendeleben mountains are elliptical dome-shaped uplifts, and although this is in numerous places complicated and obscured by intrusions of granite, all the sections across these ranges thus far examined indicate anticlinal conditions. In a section across the Kigluaik Mountains along Grand Central River and a pass leading northward from its head the anticlinal structure of the range is most evident. Crystalline limestones are exposed in the heart of the range and are apparently overlain by several thousand feet of nearly black quartz-biotite schist containing sills and dikes of massive granite. The upper valley of Windy Creek, which lies in the heart of the mountains and west of this pass, is cut for several miles in a mass of granite that probably represents a great intrusive stock. The greater part of this granite mass is slightly gneissoid, but it is cut by younger dikes which have a granitoid texture. A section across the range by way of the lower part of Windy Creek and Cobblestone River, although is it broken by a great intrusion of granite, also shows the anticlinal structure with a similar stratigraphic arrangement of massive limestones and schists. In this exposure the cliffs, which are not less than 1,000 feet high, are composed of limestone interbedded with biotite schist and gneiss. At the west end of the range the biotite schists dip toward the west, indicating a dome-like termination of the mountain uplift. In this region the crystalline limestones are exposed at only a few points near the heads of some of the streams.

Along the mountain crests the many jagged outcrops of granite have given rise to the local name Sawtooth Mountains, which is descriptive not only of the Kigluaik but also of the Bendeleben Range. The prominence of these granite outcrops is due to their ability to resist erosion, and though the granite intrusives probably form considerably less than half the mass of the mountains, on hasty superficial examination they often seem to make up a much larger part.

The only section across the Bendeleben Range which has been examined by the writer is along Parantulik River and Ella Creek, between the heads of which there is a low pass. The structure here appears to be anticlinal and the prevailing rocks are dark-colored

quartz-biotite schists and gneisses similar to those of the Kigluaik
-Range. Sills and dikes of coarse-grained granite or pegmatite are
also present. White crystalline limestones containing scattered grains
of graphite occur in beds 20 feet or more in thickness interbedded
with the schists, along Parantulik River from a point near its head
to the edge of the Fish River lowland. Mendenhall,[a] who examined
a section across this range at the head of Fish River in 1900, described the geologic condition as follows:

About the head of Fish River the mountains are chiefly granitic, but along
their flanks and sometimes extending through them in belts of varying breadth,
which mark the passes, are areas of schistose sediments.

About 1 mile above the camp of July 20, along the creek bed, is an outcrop of
rusty and very graphitic schist associated with more calcareous phases. Five
miles farther along the right bank of this same branch of Fish River is an outcrop of buff slates with very little calcareous matter, while 2 miles farther
northwest, in a gap between two branches of Fish River, the series is represented
by a white, coarsely crystalline marble. This narrow belt of the crystalline
series, between two great intrusive granitic masses, expands to the northwest
and forms a series of hills somewhat lower than the more resistant masses on
either side, separating the Fish River drainage basin from a low area to the
north, which is probably a part of the Kuzitrin Valley. East and west from
the camp of July 20 these sediments appear as remnants on the flanks of
many of the spurs running from the high mountains out toward the Fish
River basin. Whether or not these outcrops form a continuous belt westward from the camp of July 20 could not be determined during the hasty reconnaissance, but eastward they seem to extend uninterruptedly to the broad
belt which runs northeastward toward the basin of Tubutulik River.

From this description it is inferred that the geologic conditions here
are not different from those in other parts of the Bendeleben Range
except that there are few, if any, intrusions of granite exposed in
this valley.

Where this range has been crossed near its west end there are few
good exposures of bed rock in place. The surface is covered with tundra vegetation and residuary soil through which some masses of
coarse-grained granite or pegmatite crop out in the form of pinnacles. Fragments of biotite schist and gneiss are found in the surface débris and there are a few exposures of the rock in place, but
not enough to determine the structure of the whole range.

There is no direct paleontologic evidence regarding the age of
these rocks. The limestone and graphite beds mentioned, which are
too highly metamorphosed to retain any recognizable fossils, are the
only materials found in them of possibly organic origin. Their
stratigraphic relation to the rocks of the Nome group, which, as
will be shown, are in part Silurian, indicates that they are either
lower in the Silurian or are pre-Silurian. The apparently great thick-

---

[a] Mendenhall, W. C., Reconnaissances in the Cape Nome and Norton Bay regions,
Alaska, in 1900, a special publication of the U. S. Geol. Survey, 1901, pp. 200–201.

ness of schists which intervenes between the known Silurian beds and the rocks of this series seem to point to the latter conclusion.

KUZITRIN FORMATION.

Overlying the biotite schists and gneisses of the Kigluaik group in some localities is what seems to be a well-defined formation, whose type rocks are graphitic quartzites or quartz schists and which was described by Brooks in 1900 under the name "Kuzitrin series." These rocks have not been reexamined in their type locality since 1901 and the following description is quoted from the report of reconnaissances in 1900:[a]

The type rock of this series is a well-jointed graphitic quartz schist, but there are many variations from this type. Intercalated strata of graphitic slates are not uncommon, and the schists are sometimes very calcareous, approaching an impure limestone in composition. The calcareous phases are more particularly characteristic of the top of the series. Locally the Kuzitrin rocks contain beds of mica and chloritic schist whose sedimentary origin it is not always possible to prove. In many localities amphibole schists and chloritic schists are abundant in the Kuzitrin series. These are usually intruded parallel to the bedding plane and are often difficult to differentiate from the sediment; they will be discussed below in more detail. On Kuzitrin River some pegmatite dikes were found cutting the graphitic schists.

Though these rocks are usually well bedded, in some localities the bedding planes are nearly obliterated by a secondary development of schistosity and jointing. In attempting to obtain a measure of their thickness several difficulties were encountered. The intercalated beds of graphitic argillites are often intensely crumpled and folded, and the inclusion of such crumpled beds in a section must necessarily introduce a large element of error. In other localities, again, the large intrusion of greenstones gives rise to another source of error. Moreover, the base of this series is not well defined, as it was not found in contact with the underlying limestone except in the York region. In view of the above facts it is clear that an estimate of thickness can be regarded only as a very rough approximation. The several sections studied would go to show that the Kuzitrin beds are between 2,000 and 3,000 feet thick.

The Kuzitrin series is typically exposed in the Kigluaik Mountains, near their northern limit, where it occurs in a belt which runs parallel to the axis of the range and is cut by the northerly tributaries of Kruzgamepa River. The rocks strike about east and west, and dip with great regularity to the south. These rocks are again found on the upper portion of Ophir Creek, where they strike a little north of east and dip to the south, and their topographic relation to the Bendeleben Mountains is similar to that which the western belt bears to the Kigluaik Mountains. The series takes its name from Kuzitrin River, along whose lower course it is well exposed. Here the rocks in the fresh cuttings of the river channel show well-marked bedding planes. The series as mapped in this part of the area includes some calcareous schists which may possibly belong to the overlying Nome series.

From this description it will be seen that these rocks were not clearly differentiated from the Kigluaik group on the one hand or

---

[a] Reconnaissances in the Cape Nome and Norton Bay regions, Alaska, in 1900, a special publication of the U. S. Geol. Survey, 1901, pp. 28–29.

from the Nome group on the other. In the type locality along Kuzitrin River the graphitic quartzites are in contact with the biotite
schists and gneisses which they overlie. They form a belt here several miles wide in which the rocks seem to be distinct lithologically
from those of the Nome group. South of the Kigluaik Mountains at
their east end, also, there is a somewhat similar belt of graphitic
quartzites overlying the Kigluaik schists, though here the boundary
between them and the schists has never been clearly defined. The
continuation of this belt west of Grand Central River has not yet
been recognized, though it may exist there as an undifferentiated
member of the Nome group, which seems to rest directly on the Kigluaik. Outside of these two localities this formation has not been
positively identified, though it is possible that it occurs in the Nome
group at a number of localities where it can not now be differentiated.
In 1900 Brooks correlated the slates of the York region with this formation on lithologic grounds, and in 1901 the writer extended this
correlation to include some dark siliceous slates or schists which lie
northwest of the Kigluaik Mountains and some quartzites in the
vicinity of Ear Mountain. There are also some dark siliceous schists
near the head of Kougarok River which resemble these and may ultimately be correlated with them. In the present report the Kuzitrin
will be regarded as a formation composed for the most part of graphitic quartzites and siliceous schists, which overlies the biotite schists
and gneisses of the Kigluaik group and underlies the schists of the
Nome group. It is not differentiated on the geologic map from the
Nome group. The absence of this formation in the contact of the
Kigluaik and Nome groups at some points is regarded as an indication of unconformity in the metamorphic series at the base of the
Nome. The slates of the York region, which were formerly correlated
with the Kuzitrin, will be described separately, as they comprise a formation which is well defined locally and of some economic importance,
and whose correlation with the Kuzitrin can not be established beyond
question.

There is no direct evidence regarding the age of the Kuzitrin formation. Like the Kigluaik group, it underlies a great thickness of
schists which are in part Silurian. The formation may belong near
the base of the Ordovician, though the probabilities would seem to
place it either in the Cambrian or Algonkian.

NOME GROUP.

GENERAL DESCRIPTION.

Overlying the Kigluaik and Kuzitrin is a complex assemblage
of altered sediments and intrusives, which was first described by

Brooks under the name "Nome series."[a] These are the most widely distributed metamorphic rocks in the peninsula and are also of the most economic importance, as they contain nearly all the veins from which placer gold has been derived. For this reason they have been examined more closely than the other formations; but owing to their complicated structure in many places and the inadequate field work on them they are still only imperfectly understood.

These rocks are here described as the Nome group. The group is not regarded as a stratigraphic unit, as the lithologic constitution of its beds is heterogeneous, and on the map some portions of overlying and underlying formations that are elsewhere differentiated may be and probably are included. The general conclusion from the observations made during the last several years is that this group of metamorphic rocks may be roughly divided into three parts. The lower portion, which is characteristically exposed near Nome and again in a wide belt south of the Kigluaik Mountains, consists of several thousand feet of schists of various types, including some limestones in beds less than 100 feet thick. These schists are overlain by massive limestones several thousand feet thick, which over a large area in the western part of the peninsula are comparatively unaltered and contain fossils of Silurian age. This formation has been called the Port Clarence limestone. The limestones are themselves overlain in the northeastern part of the peninsula by schists of somewhat different type from those forming the base of the group.

The prevailing rocks of the basal portion of this group are schists consisting essentially of quartz and chlorite, with albite, muscovite, and epidote in varying amounts. Locally the formation contains beds of graphitic mica schist, graphitic quartzite, limestone, and calcareous schist, as well as intrusive masses of greenstone that are also more or less schistose. Among the many thin sections examined, specimens of graphitic chlorite schist and feldspathic calcite schists are not uncommon. Magnetite, pyrite, ilmenite, leucoxene, and garnet occur in many of the specimens. The elements from which such metamorphic rocks are formed are afforded by sedimentary as well as igneous rocks, and probably the group includes rocks derived from both types.

In the section from Nome to the Kigluaik Mountains these schists, which here contain some massive limestone beds up to 100 feet or more in thickness, extend northward from the coast for 16 miles, to a point where they dip under massive limestones that are interbedded with thin beds of schist and lie in a synclinal trough extending east and west. These limestones are tentatively correlated with the Port

---

[a] Brooks, A. H., Reconnaissances in the Cape Nome and Norton Bay regions, Alaska, in 1900, a special publication of the U. S. Geol. Survey, 1901, pp. 29–31.

Clarence. On the south side of Stewart River, 22 miles from the coast, the schists reappear and are continuously exposed to the base of the mountains. In the southern part of this section they are faulted at so many places that no estimate of their thickness is possible. North of Stewart River, however, the faults are not so numerous, and the prevailing dips seem to indicate a thickness of about 7,000 feet. Whether the beds in this area are repeated by close folding or by unobserved faults could not be determined in the rápid reconnaissance. A large allowance for any such folding or faulting would seem to indicate that the minimum thickness possible can scarcely be less than 2,500 feet; the probabilities are that the thickness is much greater.

In the region northwest of the Kigluaik Mountains these rocks are less altered than in the Nome region. They here contain many sills of greenstone, and some of the sedimentary beds are but little altered clay slates. They overlie the biotite schists of the Kigluaik group and appear to be themselves overlain by the Port Clarence limestone. In their comparatively unaltered condition they apparently represent a series of rocks such as would by a little greater metamorphism produce the schists of the Nome region.

North of the Kigluaik and Bendeleben mountains schists similar to those of the Nome region occur in the basins of American and Kougarok rivers, where they were mapped in 1901 by the writer, who, on the evidence of obscure fossils collected from associated limestones, regarded them as younger than the Port Clarence formation and applied the name " Kugruk group " to them.[a] The larger collections made in 1903, however, justify the correlation of these limestones with the Port Clarence and of the schists which underlie them, with those of the Nome precinct.

The clay slates and phyllites of the York region apparently underlie the Port Clarence limestone and may represent a less-altered portion of the same body of sediments that produced the schists of the Nome group. The rocks were provisionally correlated with the Kuzitrin[b] in 1900 and 1901, but inasmuch as it is certainly true that they are lithologically distinct from any extensive mass of rocks represented elsewhere in this peninsula, they will be described separately (pp. 79–80). Although these phyllites underlie the Port Clarence limestone, the stratigraphic relation is not definitely known, as there is evidence of a thrust fault along the contact. This fault, which was not detected in 1901, led the writer to infer an unconformity between

---

[a] Collier, A. J., Reconnaissance of the northwestern portion of Seward Peninsula, Alaska : Prof. Paper U. S. Geol. Survey No. 2, 1902, pp. 21–24.

[b] Brooks, A. H., Richardson, G. B., and Collier, A. J., Reconniassances in the Cape Nome and Norton Bay regions, Alaska, in 1900, a special publication of the U. S. Geol. Survey, 1901, pp. 28–29.

the two formations.[a] Two small areas of phyllites resembling these occur east of this contact; they are surrounded by Port Clarence limestone, which they seem to underlie conformably.

In the east end of the Nome precinct and in the Council precinct the schists resemble those of the Nome region in their degree of meta-morphism. They are extensively developed and have been studied by the writer along Fox River and the eastern tributaries of Solomon River. A wide belt of such rocks, including the lower drainage of Eldorado and Bonanza rivers, extends northeastward across the head-waters of the Casadepaga. These rocks were examined near the coast by Richardson in 1900 [b] and along the upper Casadepaga and Eldorado by the writer in 1903. Although they are apparently identical with the schist of the Nome region, they overlie the massive limestone correlated with the Port Clarence, near Solomon River, and may be in part equivalent to the schists which overlie the Port Clarence limestone in the northeastern portion of the peninsula, de-scribed on page 79. It seems more probable, however, that they are either overthrust on the Port Clarence limestone or that the correla-tion of the massive limestone with the Port Clarence is at fault.

The age of the schists at the base of the Nome group is probably Ordovician, though they may be in part older, for there is no direct paleontologic evidence regarding it. They underlie the Port Clar-ence limestone, which is known to be Silurian or Ordovician. No definite evidence of unconformity between the formations has been discovered and they are therefore provisionally regarded as conformable.

PORT CLARENCE LIMESTONE.

The massive limestone member of the Nome group is comparatively unaltered over a large area north of Port Clarence and was described and named the Port Clarence limestone by the writer after his visit to that region in 1901.[c] In some of the sections examined, unless there are faults which were not detected, a thickness of at least 5,000 feet is exposed. In this vicinity the unaltered limestones form a continuous outcrop over an area of 1,400 square miles. In the west end of this area they are cut by bosses of granite and dikes of rhyolite and other acidic rocks. The limestones surrounding these intrusions are locally marmorized and contain many contact minerals, among which are vesuvianite, wollastonite, garnet, and tourmaline. They also contain veins of cassiterite, zinnwaldite, fluorite, tourmaline,

[a] Collier, A. J., Prof. Paper U. S. Geol. Survey No. 2, 1902, p. 18.

[b] Brooks, A. H., Richardson, G. B., and Collier, A. J., Reconnaissances in the Cape Nome and Norton Bay regions, Alaska, in 1900, a special publication of the U. S. Geol. Survey, 1901, p. 31.

[c] Collier, A. J., Prof. Paper U. S. Geol. Survey No. 2, 1902, pp. 18–19.

wolframite, and quartz. The limestones become progressively more metamorphosed and locally contain thin beds of graphitic and other schists toward the east. In the divide between American and Kougarok rivers and also east of the Kougarok they exist as massive crystalline marbles, but retain traces of fossils by which their correlation with the Port Clarence is rendered certain. In the southern part of the peninsula they form the large area of white marble hills at White Mountain, north of Golofnin Bay. Obscure fossils were collected here in 1900 which were regarded as of possible Mesozoic [a] age, but the larger collections made by Hess in 1903 permit their correlation with the Port Clarence limestone. No estimate of the thickness of the formation at White Mountain has been attempted, as its upper and lower limits could not be defined. It comprises a great thickness of crystalline limestones which contain some minor beds of calcareous schist and overlie schists similar to those of the Nome region.

West and north of White Mountain many extensive areas of massive crystalline limestone have been differentiated, as shown on the geologic map (Pl. X, in pocket). In general, these rocks are more highly altered than those already described, and are destitute of fossil evidence. They are correlated with the Port Clarence on the grounds of stratigraphic and lithologic similarity. Near Casadepaga River Richardson measured a thickness of 2,500 feet.[b] South of American Creek and west of the Kruzgamepa, near Iron Creek, the thickness is probably as great, though no section has been measured.

In this region the limestones contain numerous sills of basic igneous rock, some of which are so much altered that they can not be distinguished from the schists.

In the region between Nome and the Kigluaik Mountains the thickness probably does not exceed 2,000 feet, but the upper limit is not defined. The limestones contain many small beds of schist, also some definite sills and dikes of greenstone, and locally they are capped by schists which may also be altered greenstones.

Near Cape Woolley the limestone covers a large area which has not been examined in detail. The thickness, however, doubtless exceeds 2,000 feet. The rock is not so much metamorphosed as in the Nome region, though tremolite is locally developed, and it should yield fossil evidence if carefully searched.

The Port Clarence limestone has been determined from paleontologic evidence to be probably of Silurian age, though forms heretofore regarded as Ordovician are present in some of the collections.

---

[a] Mendenhall, W. C., Reconnaissances in the Cape Nome and Norton Bay regions, Alaska, in 1900, a special publication of the U. S. Geol. Survey, 1901, pp. 203–204.

[b] Brooks, A. H., Richardson, G. B., and Collier, A. J., Reconnaissances in the Cape Nome and Norton Bay regions, Alaska, in 1900, a special publication of the U. S. Geol. Survey, 1901, p. 109.

In 1901 the writer obtained small collections of fossils from many localities north of Port Clarence; on these Dr. Charles Schuchert, of the United States National Museum, reported as follows: [a]

Loc. 26. Two miles southwest of forks of the Don River. *Porambonites*, probably *P. intercedens* Pander; *Columnaria* with large corallites, and *Bythotrypa?* sp. undet.

Loc. 28. Mountain 4 miles north of Rapid [Lost] River. *Illænus* near *I. tauricornis*, but much smaller; *Illænus* sp. undet.; and a lithistid sponge on the order of *Calathium*.

Loc. 45. Don River 4 miles north of Tozier Creek. *Maclurina*, probably *M. manitobensis* Whiteaves; *Columnaria* with small corallites; *Halysites catenularia* Linné; *Syringopora* sp. undet.; *Streptelasma?;* and an undetermined *Lophospira*.

Loc. 77. Bluff above Nuluk River, latitude 65° 41', longitude 166° 20'. *Orthis*, probably *O. parva* of European Russia.

Loc. 78. Foot of talus slope, Nuluk River. A large section of an undetermined gasteropod.

Loc. 76. Bowlder from gravel of Nuluk River. Fragments of a large trilobite. Undeterminable.

Loc. 185. Pebble from sand spit 1 mile north from Teller. *Girvanella* and *Raphistoma?*.

The above localities represent the middle of the Lower Silurian system. This is the first proof of the occurrence of rocks of this age in Alaska. Localities 26, 28, and 77 are particularly interesting, as species found there were first described from European Russia about St. Petersburg. This fauna has not before been ascertained to occur in America, nor has *Porambonites* been noted in any American Arctic region.

Locality 45 is also very interesting and perplexing—perplexing in that the association of *Halysites* with *Syringopora* would under ordinary circumstances be accepted as proof for Upper Silurian age. However, since both are here associated with such unmistakable Lower Silurian forms as *Maclurina manitobensis* and *Columnaria*, the latter outweighs the evidence of the corals. On the other hand, *Halysites* is often found in the Lower Silurian and while I know of no *Syringopora* in this system, yet I learn from Mr. Ulrich that this genus has been noticed by him twice in association with Lower Silurian species.

* * * * * * *

Locality 163 [summit of Baldy Mountain] indicates a second formation younger than any of the above-described lots. The rock has undergone considerable alteration, so that the fossils are mostly undeterminable. However, the presence of a coral like *Cladopora* or *Striatopora* indicates either Upper Silurian or Devonian. No coral of this nature is known to me in the Lower Silurian.

In 1903 larger collections were made at a few localities, on which Doctor Schuchert reports as follows:

Loc. 3AC146 (near the forks of Rock Creek, a northern tributary of Agiapuk River) has—

Stromatopora sp.

Favosites favosus Goldfuss.

Favosites cfr. *niagarensis* Hall or *F. gothlandicus* Fought.

---

[a] Collier, A. J., Prof. Paper U. S. Geol. Survey No. 2, 1902, pp. 20–21, 22.

Columnaris? apparently closely related to *Favistella reticulata* Salter of the Arctic Silurian.

Halysites catenularia feildeni Etheridge.

Halysites catenularia harti Etheridge.

Diphyphyllum cfr. *multicaulis* (Hall).

Alveolites sp.

Gasteropods, one a *Lophospira*, another a *Hormotoma*.

Orthoceras sp. undeterminable.

Loc. 3AC170 (Harris Creek, about 3 miles north of Baldy Mountain, where No. 163 was collected in 1901). *Cladopora* sp., resembles *C. seriata* Hall of the New York Silurian; *Striatopora* sp.; *Alveolites* sp., a branching form.

The species of these two localities indicate clearly the age of the limestones to be Niagaran or the middle third of the Silurian. This same horizon with the same species is known to occur in the Arctic region at Dobbin Bay, Cape Hilgard, Cape Louis Napoleon, Offley Island, etc.

These two collections of Mr. Collier also shed light on his earlier collections, numbered 1AC45 and 1AC26. The former has an assemblage of species suggesting both the Ordovician and Silurian formations, and a similar occurrence was noted by Etheridge [a] for the Arctic localities mentioned above. Either there are two dolomitic superposed horizons in the far north, one of Ordovician age, the other of Silurian, causing the fossils of different ages to be mixed, or we have here indications that there was no break in the sedimentation between the Ordovician and Silurian permitting species usually accepted as restricted to the Ordovician to pass upward into the Silurian. Reference is here made to my former identification of *Maclurina manitobensis*. Etheridge also finds a similar shell which he labels *Maclurea magna* associated with Silurian fossils at Cape Hilgard and Cape Louis Napoleon. Further, Salter and Etheridge report *Favistella-Columnaria* in the same association. This is a genus of corals elsewhere known only in the Ordovician. Collier also appears to have specimens of this genus at his loc. 3AC146 associated with unmistakable Silurian forms. My identifications of lot 1AC26 also should be changed to *Favosites* sp. undet., *Conchidium? coppingeri* Etheridge?, and *Bythopora?*. Loc. IAC77 is either Ordovician or Silurian.

There may be Ordovician strata on Seward Peninsula, as lot 1AC28 seems to have fragments of *Illænus*. However, the material is so fragmentary that no positive identification can be made, and the age of the strata in question is either Ordovician or Silurian. Lot 3AH51 (South Fork of Budd Creek) has a *Syringopora* different from other Seward Peninsula species and yet it may be of Silurian age.

In 1900 some obscure fossils were collected by Mendenhall [b] from the limestones at White Mountain, about 10 miles north of Golofnin Sound, which, though they could not be identified by paleontologists, seemed to indicate either a Mesozoic or Tertiary age.

In 1903 Hess made a thorough search for fossils on several low, rounded limestone buttes which stand as islands in the Quaternary deposits of the Fish River delta near White Mountain. His search resulted not only in a collection of fossils from Mendenhall's original locality, but in additional collections from a hill called Black Moun-

---

[a] Quart. Jour. Geol. Soc. London, vol. 34, 1878.

[b] Mendenhall, W. C., Reconnaissances in the Cape Nome and Norton Bay regions, Alaska, in 1900, a special publication of the U. S. Geol. Survey, pp. 203–204.

tain, which lies several miles north of White Mountain. The collections were submitted to Doctor Schuchert, who reports on them as follows:

Lot 3AH103 (Black Mountain, near Fish River, 6 miles above White Mountain) has a *Diphyphyllum* quite unlike the other Seward Peninsula species. So far as the fossils go the horizon may be either Silurian or Devonian. However, I am told that the specimens from loc. 3AH102 (on the southeast slope of Black Mountain, which are undoubtedly Silurian) are from a horizon above 103, and as the species from the former seems to be the abundant *Cladopora* of loc. 3AC170 it follows that if no overturn of the beds occurs here loc. 103 also is Silurian.

Loc. 3AH106 (Mendenhall's original locality at White Mountain). The limestone of this locality is much metamorphosed, shattered, and squeezed, so that one hesitates to state what genus the white calcite sections in the rocks represent. However, they appear to be of a large bivalve, and as the strata not far away from this locality seems to be certainly of Silurian age the suggestion is arrived at that *Megalomus* may be represented. This suggestion would not have offered itself if we did not know of a similar occurrence on Drake Island, Glacier Bay, Alaska. At this locality the natural sections indicate that they belong to a thick-shelled bivalve like *Megalomus*, and as there are associated with them unmistakable Silurian fossils the fact is established that a horizon of about the age of the Silurian Guelph is present on Drake Island. For the present it may be best to accept the age of the rocks at loc. 3AH106 as Silurian. Accepting this as true, then 106 should hold a somewhat higher horizon in the section than AH102 and 103.

The above remarks lead to the conclusion that there may be Ordovician strata on Seward Peninsula and that there is there an extensive development of Silurian. The places where Silurian fossils occur are 1AC26 (2 miles southwest of the forks of Don River), 1AC45 (Don River, 4 miles above Tozier Creek), 3AC146 (Rock Creek, tributary of Agiapuk River), 3AC154 (forks of Budd Creek), 3AC170 (Harris Creek, 3 miles north of Baldy Mountain), 3AH52 (South Fork of Budd Creek, tributary of American River), and 3AH102 (southeast slope of Black Mountain, near Fish River, 15 miles above Golofnin Bay).

The following localities may also be Silurian, as they seem to have single species of apparently the same forms as occur at the localities just cited: 3AC156 (head of South Fork of Budd Creek, near Kougarok Mountain), 3AC155 (mountain south of forks of Budd Creek), 3AH104 (river face of Black Mountain, 6 to 8 miles above White Mountain).

The geologists of the Harriman expedition, who touched at a point not definitely located on the east shore of Port Clarence in 1899, describe the rocks here as slates intercalated with brown sandstones. The slates are described as " dark-gray fissile argillites, which are often fossiliferous and some of which are calcareous and run into thin beds of limestone." [a]

These rocks were correlated with the Jurassic-Triassic " Vancouver series." Although this coast has been traversed by geologists of the Geological Survey, no rocks were found which could be distinguished from the Port Clarence limestone and the undifferentiated schists of the Nome group. Moreover, some of the fossils collected by mem-

---

[a] Emerson, B. K., Harriman Alaska Expedition, vol. 4, 1904, pp. 44, 53–54.

bers of the Harriman expedition seem to have been mixed or mislabeled, so that no fossils which are known to have been collected at Port Clarence appear to have been brought back. Fucoid forms which may have been confused with some of the fossils collected in the Yakutat formation[a] are common in the less-metamorphosed earthy Port Clarence limestone, and a small collection of these was made in 1904 near Mount Merrill for comparison with the Yakutat fucoids. These were submitted to Mr. Ulrich, whose report is as follows:

The fossils collected by Mr. Washburne at Merrill Mountain, 3 miles north-northeast of the mouth of Lost River, Seward Peninsula, Alaska, comprise, all told, 17 specimens in an extremely unsatisfactory state of preservation. On consultation with Mr. Collier I learned that the main purpose of the collection was to determine whether any of the fucoid-like markings, which are said to be abundant in the Port Clarence limestone, are comparable to the Yakutat formation fucoids recently described by me. This probability may be set aside at once, since the "fucoids" of the two formations are in no wise similar. Indeed, the organic nature of the Port Clarence markings is doubtful, while I can entertain no such doubt concerning the Yakutat fucoids. Nine of the specimens from Merrill Mountain were collected on account of the markings in question.

Of the remaining eight specimens one is a light-colored banded magnesian limestone; the others are in a dark, compact limestone very different lithologically from the first. The banded or laminated specimen may not be organic, but it agrees in appearance and general structure very well with those doubtful fossils called Cryptozoan which are so widely distributed in the early Ordovician or Cambrian magnesian limestones in the United States. The fact that the Alaskan specimen is dolomitic is probably significant in this connection. However, if on investigation the bed from which this specimen is derived is not distinctly beneath the dark, purer limestones containing the other fossils, I would be inclined to compare it with the similarly laminated *Stromatocerium rugosum*, a fossil characterizing the Black River limestone of New York.

The fossils in the dark limestone are:

Sections of two specimens of an undetermined gasteropod having about the shape of *Hormotoma bellicincta*, only much smaller—one-half to three-fourths of an inch in length.

Two specimens of a low-spired, round-whorled gasteropod of uncertain affinities. *Trochonema* is suggested by the broad umbilicus and general form of the shell, but it is so imperfectly preserved that identification is impossible.

A cephalopod with a curved shell like these commonly referred to *Cyrtoceras*—perhaps a species of *Cyrtorizoceras* according to Hyatt's classification. However, the material is so poor that it is not safe to say more than that it represents a simple *Cyrtoceras*-like cephalopod with close septa and apparently a nearly smooth surface.

The other (3) specimens consist (1) of a part of a radiately marked impression, the relations of which are very uncertain, and (2) of fragments of a thick-walled hollow cylinder, possibly indicating siphuncles of some small *Endoceras*. They are about one-fourth inch in diameter.

[a] Ulrich, E. O., Fossils and age of the Yakutat formation: Harriman Alaska Expedition, vol. 4, 1904, pp. 125–146.

Despite the imperfection of the material there is little doubt in my mind as to its Ordovician age. It must come from a lower horizon than that in the same vicinity from which the probably Silurian corals reported on by Mr. Charles Schuchert were found.

The collections regarded by Messrs. Schuchert and Ulrich as possibly Ordovician or "Lower Silurian" are 1AC28 (near mouth of Lost River) and 1AC77 (Nuluk River, latitude 65° 41′, longitude 166° 20′). From a review of the stratigraphic evidence the writer is of the opinion that these horizons are not lower than the other fossil-bearing horizons of the Port Clarence formation and that, therefore, they also are Ordovician or Silurian. Larger collections from these localities will, it is believed, verify this conclusion.[a]

UNDIFFERENTIATED SCHISTS OF FAIRHAVEN PRECINCT.

The schists which overlie the massive Port Clarence limestone find their greatest development in the northeastern part of the Peninsula. In this region Moffit found extensive areas of mica schists overlying the massive limestones believed to be Port Clarence. These rocks differ from the schists which underlie the Port Clarence limestone in the Nome region in that they are calcareous or siliceous and in many places graphitic mica schists, but are rarely feldspathic.[b] Their stratigraphic relation to the Port Clarence limestone places them either in the Silurian or higher. In lithologic character they are distinct from the slates in the York region, which will be next described and which are also regarded as of possible Devonian age.

SLATES IN THE YORK REGION.

Northwest of the York Mountains there is a large area of slates or phyllites which seem to be distinct in lithologic character from any of the other large rock masses in the peninsula, though rocks somewhat similar occur in the Nome group south of Grantley Harbor. These rocks have been differentiated on the geologic map (Pl. XI, in pocket). Similar slates which are provisionally correlated with them occur in two small isolated areas surrounded by Port

---

[a] In regard to fossil collections the writer begs leave to make the following suggestions to miners and prospectors in Alaska:

First. Fossils furnish the most definite evidence of the relative age of rock formations.

Second. Though the relative age of rock formations does not seem to have a direct economic bearing, as rocks of all ages carry gold, and, according to the old adage, "gold is where you find it," it is equally true that in many districts the gold-bearing rocks are confined to definite horizons, which are often indicated by the fossils contained in them or in contiguous rocks.

Third. The men engaged in United States Geological Survey work in Alaska invite the cooperation of miners and prospectors in collecting such evidence, either by furnishing information regarding fossil-bearing localities or by sending specimens. Such correspondence should be addressed to the Division of Alaskan Mineral Resources, U. S. Geological Survey, Washington, D. C.

[b] Moffit, F. H., The Fairhaven gold placers, Seward Peninsula, Alaska: Bull. U. S. Geol. Survey No. 247, 1905, pp. 20–24.

Clarence limestone north of Brooks Mountain and east of Don River, and in a large area at Ear Mountain.

The type rocks of the formation are fine-grained graphitic are- naceous slates which are in many places broken into rhomboidal blocks and pencil-shaped fragments, but there is considerable varia- tion from this type. Some beds of coarse-grained dark sandstone have been observed and here and there calcareous phases are mani- fested in beds of impure flaggy limestone.

On the west these slates are in contact with some crystalline lime- stones and interbedded phyllites which are of Mississippian (" Lower Carboniferous ") age. The relation of the slates to the limestones is difficult to determine, mainly on account of the obscure bedding of the slates and the few exposures within the limestone area near the contact. In 1900 Brooks was of the opinion that the limestones underlie the slates and are therefore the older.[a] Although the writer has twice crossed this contact since that time, he is unable to advance any positive evidence either confirmatory or otherwise of this opinion.

In regard to the stratigraphic position of these slates, either one of the three following hypotheses is possible in view of the facts obtained: (1) They may be of the same geologic series as the lime- stones exposed near Palazruk, as they are not perceptibly more altered. Moreover, fossils similar to those obtained from the lime- stone have been found in beach pebbles several miles east of the line of contact, and though these may have been transported by ice floes it is equally possible that they came from some minor bed of impure limestone included in the slate formation. (2) They may be older than the limestone and underlie it conformably. In a general way they resemble some highly jointed slates and sandstones which con- formably underlie rocks of the Mississippian (" Lower Carbonifer- ous ") series in the Cape Lisburne region.[b] The close association of these slates with the Mississippian limestones therefore suggests their correlation with the supposed Devonian rocks of the Cape Lisburne region. (3) These slates may be equivalent to the lower undifferen- tiated schists of the Nome group. If this be true, there must be a fault of several thousand feet which has not yet been detected along their contact with the limestone exposed near Palazruk. The rela- tion of the slates, and more especially of one of the smaller outlying masses correlated with them, to the Port Clarence limestone argues strongly for this hypothesis.

---

[a] Brooks, A. H., Reconnaissances in the Cape Nome and Norton Bay regions, Alaska, in 1900, a special publication of the U. S. Geol. Survey, 1901, p. 28.
[b] Collier, A. J., Geology and coal resources of the Cape Lisburne region, Alaska: Bull. U. S. Geol. Survey No. 278, 1906, pp. 17-18.

LIMESTONE NEAR PALAZRUK.

West of the area of the slates there is a belt of crystalline limestone and interbedded dark mica schists or phyllites. These rocks are exposed along the Bering Sea coast for about 5 miles, between the native village of Palazruk and the granite outcrop of Cape Mountain. Near Palazruk they are intensely crushed and probably faulted. These limestones were correlated in 1900 with the limestones of the Kigluaik group.[a] As has been noted, they lie immediately west of the typical area of the slates, their relation to which has not yet been clearly determined, though they form a distinct lithologic unit.

In 1903 a small collection of rather poorly preserved fossils was obtained by the writer along the beach between Palazruk and Cape Mountain. Similar fossils were found in 1904 in the limestones south of Lopp Lagoon and also in beach pebbles several miles east of Palazruk, but were lost in shipment. The collections made from beach pebbles in 1903 were referred to Dr. George H. Girty, who reported on them as follows:

Only three species seem to be represented in the collection, and these belong to the corals, a group which serves ill as a means for correlation or for age determination. The material is also so metamorphosed that almost all structure has been effaced. There is a single small zaphrentoid coral whose generic position has not been determined. Several specimens of *Syringopora*, apparently all one species, have been identified by their characteristic appearance and mode of growth. All internal structure in these fossils has been obliterated. A third form, represented by several specimens, appears to be related to *Lithostrotion* or *Lonsdalea*. The corallites are moderately large and form bushy fasciculate colonies. Septa are present, and also a central axis is simple or compound. The presence of this form seems to make it necessary to refer this fauna to the Carboniferous, and it is more probably Lower than Upper Carboniferous. Were it not for the species last mentioned (*Lithostrotion?*) I would have regarded this as one of the Devonian coral faunas. Its age can hardly be older than Devonian.

In regard to the same collection, Doctor Schuchert, who also examined it, says:

Regarding locality 3AC136, it is certain that these corals are wholly different from any others of Seward Peninsula and that they do not suggest any horizon below the Carboniferous. In other words, these corals are of Carboniferous age. A very similar coral, *Lithostrotion*, was collected by Schrader in 1901, on East Fork of John River, at a locality on Contact Creek.

A large collection of Mississippian fossils in a good state of preservation was obtained in 1904 at Cape Lisburne, which is about 200 miles northeast of Cape Prince of Wales. After these had been deter-

---

[a] Brooks, A. H., Richardson, G. B., and Collier, A. J., Reconnaissances in the Cape Nome and Norton Bay regions, Alaska, in 1900, a special publication of the U. S. Geol. Survey, 1901, p. 28. Collier, A. J., Reconnaissance of the northwestern portion of Seward Peninsula, Alaska: Prof. Paper U. S. Geol. Survey No. 2, 1902, p. 16.

mined the fragmentary fossils from the York region were compared with them by Doctor Girty, from whose report the following note is added to those already given:

Another collection was that made by Mr. Collier in 1903 near Cape Mountain. These fossils have been so altered that absolute identification of the species is in most cases impossible, but I feel little doubt that the horizon will prove to be that of the Lisburne series.

From this evidence the correlation of the limestone exposed near Palazruk with the Mississippian limestone of Cape Lisburne seems well established. Fossils similar to those from the Cape Lisburne region were also found on St. Lawrence Island 160 miles southwest of Cape Prince of Wales, though their source in the bed rock has not been located. Mississippian rocks are extensively developed in a broad belt extending eastward from the Cape Lisburne region [a] across northern Alaska. The same horizon is also represented on Chichagof Island,[b] southeastern Alaska.

SUMMARY OF THE METAMORPHIC FORMATIONS.

The metamorphic rocks of Seward Peninsula consist of inter-bedded schists and limestones and are among the oldest known rocks in Alaska. In general the schists are of sedimentary origin, though they doubtless include many undifferentiated igneous masses. Litho-logically they present a variety of types, the most common being quartz-biotite schists grading into gneisses; quartz-chlorite-muscovite schists, many of which contain secondary feldspar; graphite schists, hornblende schists, epidote schists, calcite schists, quartzites, and clay slates. The degree of metamorphism varies, but as a general rule is greater in the eastern than in the western part of the peninsula. In many places the schistosity appears to be parallel with the bedding, but locally there is evidence of close folding developed previous to the schistosity, and it is probable that this condition prevails in many of the more highly altered rock masses where it has not been detected in the reconnaissance examination. The most extensive limestone formation, which is called the Port Clarence limestone, is believed to belong to the Silurian, though several of the collections of fossils contain forms heretofore regarded as Ordovician associated with Silurian forms. Concerning the schists of the Nome group, which underlie the Port Clarence limestone, it can only-be said that they and the Kuzitrin and Kigluaik are pre-Silurian. Some undifferen-tiated schists which overlie the Port Clarence limestone may be either Silurian or Devonian. So far as the evidence at hand shows, the

---

[a] Collier, A. J., Geology and coal resources of the Cape Lisburne region, Alaska: Bull. U. S. Geol. Survey No. 278, 1906, pp. 18-26.
[b] Communicated to the writer by Dr. G. H. Girty.

slates in the York region may belong in either of the Paleozoic systems. The youngest rocks known in the metamorphic series are the limestones near Cape Mountain, which are of Mississippian age.

Briefly stated, the metamorphic rocks of Seward Peninsula include sediments which range in age from pre-Silurian to Carboniferous. The formations that have been differentiated are lithologic and probably also stratigraphic units, though as a general rule the stratigraphic features are obscured by metamorphic structures.

Where paleontologic evidence is wanting any correlations are apt to be at fault, as the undifferentiated schists, although mainly of pre-Port Clarence age, may include infolded masses of younger Paleozoic rocks, and some of the limestones which have been mapped as Port Clarence may belong to an undifferentiated horizon above or below the Port Clarence.

UNALTERED SEDIMENTARY ROCKS.

Unaltered sediments composed of conglomerates, sandstones, and shales, which are in places coal-bearing, overlie the metamorphic rocks unconformably in a few small areas, most of which are east of the region under discussion in this report. Several such areas are reported along Tubutulik and Koyuk rivers, in the Koyuk precinct,[a] and a similar area which has produced some coal occurs on Kiwalik River, in the Fairhaven precinct.[b]

In the Nome precinct a small area of unaltered sediments, including some coal beds, occurs on Coal Creek, a western tributary of the Sinuk, about 14 miles from the coast. The surface here is deeply covered with erratic bowlders, peat, and moss, and the bed rock, even in stream channels, is exposed in but few places, so that it was not possible to outline definitely the area occupied by this formation. From the meager exposures which occur, however, it seems probable that the greater part of the plateau surface in which Coal Creek is trenched is underlain by crystalline limestones and schists resembling those near Nome, and that the younger sediments do not extend for more than half a mile along the creek. The most prominent outcrops consist of a conglomerate containing pebbles of schist and vein quartz and some large, well-rounded bowlders of greenstone that has been slightly sheared. A diligent search of this conglomerate failed to reveal any pebbles of granite, although granite pebbles are common in the surficial deposit. In addition to the conglomerates, the coal-bearing formation contains finer sediments made up largely of

[a] Mendenhall, W. C., Reconnaissances in the Cape Nome and Norton Bay regions, Alaska, in 1900, a special publication of the U. S. Geol. Survey, 1901, pp. 205–206.

[b] Moffit, F. H., The Fairhaven gold placers, Seward Peninsula, Alaska: Bull. U. S. Geol. Survey No. 247, p. 18.

schist pebbles that are very much decomposed. These beds have been slightly crushed and sheared subsequent to their deposition, making it difficult to distinguish the clastic material from the schists which are in place. The formation also contains a number of seams of finer material in the condition of fire clay. The beds strike nearly northwest and dip to the southwest at an angle of about 30°.

The attention of prospectors was first called to the coal deposit by Eskimos from the village at the mouth of Sinuk River, and systematic development was attempted in 1902. A tunnel driven into the west bank of Coal Creek across the strike of the coal-bearing strata exposes a number of thin seams of coal with beds of white fire clay between them. It is reported that while the tunnel was in construction seventeen stringers or thin seams of coal were exposed, the thickest of which measured about 16 inches, and the thinnest about 3 inches. Below the lowest bed the floor consists of white fire clay similar to that between the beds. No analysis has been made, but the appearance of the coal indicates that it is bituminous and of fair quality. A blacksmith at Nome reported that he found it very satisfactory for welding purposes, but it is evident that the deposit has little if any value on account of the small size of the beds.

Although it is not believed by the writer that the whole area of coal-bearing land is more than one-half of a square mile, it is possible that its extent is somewhat greater, for it is natural to suppose that the coal-bearing rocks, being softer than the schists and limestones, would be less likely to crop out through the surficial deposits, and for this reason their presence would not be detected.

As was noted, the conglomerates of the coal measures contain a great many pebbles of a coarse-grained greenstone. Greenstones of this character occur about 10 miles to the northwest and there they make up a large part of the bed rock. It is entirely possible that they may also occur at many places nearer to the basin, as dikes of finer-grained greenstones are known in the vicinity. These sediments were probably deposited in a fresh-water basin that was of small extent. Subsequent to their deposition they have been folded with the older rocks and eroded to base levels with them, so that the original extent of the basin can not be determined. The character of the pebbles of this formation shows that it was deposited after the metamorphism of the Nome group and after the greenstone intrusions. The absence of granite pebbles suggests that the sediments may be older than the granite intrusions, though it may be due to lack of drainage from that direction.

No paleontologic or paleobotanic evidence regarding the age of any of the coal-bearing rocks of Seward Peninsula has been obtained, but from their lithologic resemblance to other coal-bearing formations of Alaska a late Mesozoic or Tertiary age is inferred. The coal-bearing series of the lower Yukon, which includes both marine and fresh-water deposits, is mainly of Upper Cretaceous age, whereas the isolated coal basins of the upper Yukon, which these basins most resemble, are Eocene.[a]

Coal-bearing sediments of Eocene age also occur on St. Lawrence Island, about 160 miles southwest of Seward Peninsula.

UNCONSOLIDATED DEPOSITS.

INTRODUCTION.

An accumulation of detritus, composed of gravel, sand, silt, ground ice, and residuary clay, or glacial moraines and erratic bowlders overlain by peat and moss, forms a widely spread mantle of varying thickness that effectually conceals the bed rock over much of the surface. Except for their frozen condition these deposits are unconsolidated, and, though their deposition may have begun in the late Tertiary, they are, for the most part, of Quaternary age, including both Pleistocene and Recent. They are of the greatest economic importance, inasmuch as they probably everywhere contain traces of gold and no productive gold placers are found except in them. On the geologic map the Pleistocene and Recent sediments are not differentiated from each other, since most of the Recent deposits overlap and merge with the Pleistocene. The areas in which gold has actually been found, but not in commercial quantities, and the areas which have produced placer gold are differentiated from those in which the presence of gold has not yet been proved. Glacial deposits that are probably of Pleistocene age have not been differentiated from the sediments. They are not important economically, as no productive mines occur in them, though they are believed to contain some disseminated gold.

The Quaternary sediments are most conveniently classified according to their extent and topographic relations, as follows:

1. Gravel-plain deposits, consisting of gravels, sand, and silts, covering the coastal and inland-basin plains.

2. Alluvial deposits, found in stream beds, flood plains, and benches along river and creek valleys.

3. High-bench deposits of both fluvial and residuary character, occurring in small areas and at considerable elevation above the lowland plains and the existing valleys.

---

[a] Collier, A. J., Coal resources of the Yukon: Bull. U. S. Geol. Survey No. 218, 1903, p. 17.

GRAVEL-PLAIN DEPOSITS.

CHARACTER AND EXTENT.

The deposits of the first group are most extensive in area and also in depth. They form the surficial covering of all the coastal plains and inland basins that are such marked features of the topography. In general they are all of similar character, consisting of silts, sands, and gravels, which, as a rule, become coarser near the foothills. Although they include deposits of two types—namely, marine and lacustrine—they are probably nearly identical in the method of their deposition, as shown on pages 90–91. More than half of the deposits of this type are found on the extensive lowlands that fringe both the northern and southern coasts. The Arctic coastal plain, which extends from Cape Prince of Wales to Cape Espenberg, has an average width of about 15 miles and an approximate area of 2,500 square miles. Along its southern margin the gravels are from 100 to 200 feet above sea level and seem to lap over the surface of the York Plateau, which rises toward the south by a very gentle slope. Very little is known regarding the detail of this deposit. Near its southern edge it contains some beds of gravel derived from the schists and limestones of the immediate vicinity. Along the coast the material is finer and by the action of waves and wind it has been piled up in sand dunes. The streams flowing across the plain have trenched their beds, exposing numerous sections of the upper part of the deposit, in some of which a layer of black silt, several feet thick, overlies the gravels and sands. Remains of the mammoth and other extinct mammals have been seen in this layer, but no marine fossils are known to have been discovered.

The Bering Sea coastal plain extends with some interruptions from the north shore of Port Clarence to Golofnin Bay. In some places it has a width as great as 7 miles, though it is in general much narrower. It differs from the Arctic coastal plain in that the hills rise abruptly from its inland margin. North of Port Clarence the deposit consists mainly of gravel and has considerable thickness near the shore, but thins out perceptibly near the hills, where the harder rocks are locally exposed in stream beds. Some inequalities in the elevation along the coast are best explained as the result of unequal uplifting. Between Port Clarence and Sinuk River also the deposit is very thin in some places, as is shown by bed rock outcropping along the beach escarpment. Here, too, there is some evidence of deformation, for part of the surface is depressed below sea level north of Cape Woolley.

The gravel-plain deposits have been examined more thoroughly than elsewhere on that portion of the coastal plain which lies between

Cripple River and Cape Nome.[a] This plain is roughly crescent shaped and has a maximum width of 5 miles. It is bounded on its seaward side by a low escarpment from 10 to 20 feet high, from which it rises to an elevation of about 100 feet at the foothills. The detailed topographic map [b] of this plain shows that there is considerable inequality in the surface levels, though without careful measurement this would scarcely be detected. About half a mile back of the present beach a well-defined escarpment that is evidently the inland margin of an older beach line can be recognized for about 10 miles. The western extension of the line of this escarpment falls along the seaward side of a depression, occupied for several miles by Snake River, which was probably a lagoon formed on a low portion of the plain surface and shut in by the same beach that lay along the base of the escarpment. The inequality of the surface levels may be due either to unequal deposition or to deformation. The deposit varies in thickness from 30 to more than 100 feet. It rests upon a somewhat uneven bed-rock surface that slopes gently from the foothills to the shore. Many excavations which have been made in this deposit show that it is composed mainly of stratified gravels, sands, and silts overlain by peat and moss. The stratification is more or less irregular, indicating varying water currents. The pebbles are for the most part derived from the rocks of the Nome group, though they include some granite bowlders and pebbles which could have come only from the Kigluaik region. Near the coast the pebbles are well rounded, but toward the foothills considerable semiangular material is found in places. On this account a possible glacial origin for part of the deposit has been suggested, though such material could doubtless have been contributed as talus by local slides from the higher ground north of the lowland, or could have been brought by floating ice.

No remains of mammoths or other extinct mammals have been found in this deposit, but marine fossils are common on the dumps of prospect holes within a mile of the beach. The species represented are identical with forms now living in Bering Sea. This portion of the coastal plain is described in considerable detail in connection with the economic geology (pp. 156–170), as it is known to contain gold values in many places, and the method of its exploitation is one of the difficult problems of placer mining.

Little is known to the writer of the portion of the coastal plain lying between Cape Nome and Topkok Head, though it is one of the largest areas on the south side of the peninsula. From the evidence of the topographic map it is inferred that it has been slightly

---

[a] See Brooks, A. H., Reconnaissances in the Cape Nome and Norton Bay regions, Alaska, in 1900, a special publication of the U. S. Geol. Survey, 1901, p. 81.

[b] Nome special map, U. S. Geol. Survey, 1906.

depressed in its western part, forming the inlet known as Port Safety. In parts at least the deposits are comparatively thin, so that some of the streams have cut through to bed rock.

Gravel and silt deposits resembling those of the coastal plains occur in a remarkable series of extensive valley lowland areas that characterize the topography of the interior. These deposits differ superficially from those of the coastal plains only in the fact that they are wholly surrounded by uplands. The areas are best designated by the names of the rivers which drain them or the lakes or bays which they surround, as the Imuruk, Agiapuk, Kuzitrin, Niukluk, Golofnin, and Fish River lowlands. Of these, the first three are all but connected with each other by continuous deposits, only short rock canyons cut in low plateau surfaces intervening along Agiapuk and Kuzitrin rivers. The Imuruk lowland deposit also extends without interruption up Kruzgamepa River and over a broad, flat divide to the upper waters of Niukluk River, but from that point the deposit is not continuous to the lowland basin on the lower Niukluk, as between the two the river flows for several miles through a rock-cut valley. The sediments of the lower Niukluk basin are continuous along Fish River with those which surround Golofnin Sound. The Fish River basin is more isolated than the others, and its deposits are not connected with any of them.

In some of these basins there is evidence that the vertical extent of the deposit has been much greater and that a considerable upper portion has been removed by erosion. Whether or not they have suffered deformation anywhere has not been positively proved, though the writer is of the opinion that they have been so affected and that certain mesa-like buttes and terraces described elsewhere are remnants of differentially uplifted portions.

The gravel plain of the Imuruk lowland surrounds a large body of brackish water called Imuruk Basin. This basin is connected with Bering Sea by a sinuous tidal channel called Tuksuk Channel, which lies in a precipitous canyon cut to a depth of 100 feet or more in what appears to be the elevated floor of a broader valley connecting the inland basin with the coast. The various rivers which discharge into Imuruk Basin are gradually extending their deltas, forming plains which can not superficially be distinguished from the older gravel-plain deposits. The materials deposited at the outer margins of these deltas consist of fine silt, with only here and there piles of coarse gravel floated down on cakes of ice, but near the heads of the deltas the river beds and banks are composed of gravel. The surface of the plain is dotted over with ponds, many of which are abandoned river channels. It is covered with a layer of peat and moss, below which, in a few places, there are beds of clear ice, though as a rule the peat rests upon frozen silt. There is reason to believe that some at least of these bodies of ice are frozen ponds covered with moss. Near the northern margin of this lowland there are several mesa-like gravel buttes 20 feet or more high, remnants of a higher surface removed by erosion.

Along Kruzgamepa River the gravel plain stands at considerable elevation, and sections exposed along the river's trench show coarse gravel and bowlders overlain by finer gravels and silts. No data regarding the thickness of this deposit are available except around the margins, where in some places the streams have cut through to bed rock. No fossil remains have been reported, though marine shells are known to exist in the recent sediments of Imuruk Basin.

Little is known in detail regarding the deposit covering the Agiapuk lowland, though one prospect hole is said to have been sunk to a depth of 65 feet without reaching bed rock. The section disclosed in this hole consisted for the most part of blue clay with some vegetable fragments. At its west end this deposit overlaps a plateau that is correlated with the York Plateau, and here the upper part of the deposit consists of gravel with a thin covering of peat and moss. Fossil remains of the mammoth derived from these deposits have been found in some of the river gravels.

The Kuzitrin lowland deposit is the most extensive of the basin areas. The upper layers seem to be more gravelly than those of the Imuruk deposit. Around its western and northern borders there is a terrace from 50 to 100 feet above the general level, from which it is separated by a marked escarpment. There are also many isolated gravel buttes from 25 to 50 feet high scattered over the lower portion. These features are regarded as residual parts of an older surface left behind in the erosion that has reduced the surface to the present level. Remains of the mammoth, horse, and other mammals, together with some large logs and accumulations of vegetable material, have been found in these buttes and terraces. Near the east end of this basin the gravels are overlain by lavas that flowed from vents near the head of Kuzitrin River.

The deposits of the lower Niukluk are continuous with those which surround Golofnin Sound. Excavations which have been made near Council show several feet of sandy silt overlain by several feet of peat, but in general little is known of the composition or thickness of the deposit. Fragments of mammoth bones have been found at Council and White Mountain. At the head of Golofnin Sound the gravel plain merges with the growing delta of Fish River. A portion of the gravel plain between Golofnin Bay and Golofnin Sound has been elevated from 10 to 20 feet since it was deposited.

The deposits of the Fish River basin have been described by Mendenhall as follows:[a]

Fish River basin above the gorge, already described among the topographic features as the gathering point for the upper tributaries of the river, is filled with similar deposits, coarse near the borders and finer near the center of the basin. The depth of this filling is purely conjectural, but presumably is not great. No islands of bed rock exist within it as far as known, but sand and gravel prominences, rising in some instances 30 or 40 feet above the general level, are abundant over it, and are interpreted as remnants of a slightly higher level generally destroyed by the meanderings of the stream.

At the west end Ophir Creek drains a part of this basin through a narrow canyon, cut in an ill-defined plateau which the gravel deposits seem to overlap. Mammoth remains have been found on the gravel bars of some of the streams.

ORIGIN OF THE GRAVEL PLAINS.

From the above descriptions it may be inferred that these deposits are probably all comparatively shallow and that they rest upon bed-rock surfaces produced by erosion. The gravels, sands, and silts are for the most part ancient deposits laid down by water, and the upper layer of peat and moss is the product of vegetable growth and accumulation still in progress. In the Coastal Plains the erosion of the underlying bed rock may have been either subaerial or marine, though the minor inequalities noted in some places would seem to indicate stream rather than wave action. The sediments are littoral and lagoon deposits laid down during periods of subsidence, and were contributed for the most part by streams from the upland, though along the inland margins, especially where foothills rise abruptly, local slides probably contributed some talus. While these deposits were again being elevated the level sometimes remained constant long enough for beaches like the ancient one near Nome to form. In their elevation these plains have suffered slight deformation, some portions being brought above sea level and others being still submerged. Parts of the Arctic coastal-plain deposit appear to rest on and overlap a depressed portion of the York Plateau, whereas a portion at least of the Bering Sea coastal-plain floor is certainly of more recent erosion than that plateau. It may therefore be inferred that a part at least of the Bering Sea gravel plain is of more recent origin than that along the Arctic coast.

The interior-basin plains are in all probability mainly the products of subaerial erosion and deformation. In the course of these processes broad, flat river valleys were carved from the bed rock and partially covered with alluvium, after which deformations of the surface have occurred, depressing some of the valleys below sea level to

---

[a] Mendenhall, W. C., Reconnaissances in the Cape Nome and Norton Bay regions, Alaska, in 1900, a special publication of the U. S. Geol. Survey, 1901, p. 207.

form bays of the type of Grantley Harbor and Golofnin Sound, and impounding the water of others to form lakes. Imuruk Basin is probably an example of the latter type. By the deformation of an old river valley its waters were impounded, forming a lake or system of lakes coextensive in area with the Imuruk, Agiapuk, and Kuzitrin lowlands. These lakes remained in existence from the time of this deformation until Tuksuk River could cut its gorge to the lake-bed level through the barrier at the lower end of Imuruk Basin. A later depression has permitted the sea again to invade Imuruk Basin along Tuksuk Channel. It seems probable that the greater part of the silt, sand, and gravel covering of these lowland areas was laid down in the deltas of rivers discharging into these lakes in much the same manner as such deltas form at the present time. They are therefore for the most part lacustrine deposits. Since their deposition parts of the original surface have been eroded and reduced to a lower level. From the many evidences of recent warping afforded by some of the base-leveled plains, more especially along the coasts, it is inferred that these inland-basin plains have also suffered differential move-ments and that the eroded areas are in all probability only differentially uplifted parts. Erosion of elevated parts may have begun while deposition was still in progress over depressed portions. The lower Niukluk and Golofnin gravel plains were probably laid down mainly as delta deposits over the bed of a large bay or sound formed by the subsidence of a broad river valley.

The Fish River basin is more difficult to account for than the others, and at present our information regarding it is too meager to afford a satisfactory explanation of its origin. The depression is probably the result of the deformation of an old erosion surface, though the detail of its history has not yet been ascertained.

AGE OF THE GRAVEL PLAINS.

It has been shown that though deposits of this type are still forming as deltas in some of the bays, most of them were formed previous to the latest earth movement, by which the gravel plains were brought to their present level. The presence of mammoth and other mammalian remains indicates that the deposits containing them are of Pleistocene age and the apparent conformity of the beds throughout the sections exposed suggests continuous deposition from Pleistocene to recent time.

The remains of these extinct mammals and of large trees also indicate a climatic condition vastly different from the present one. A generally warmer and moister climate must have prevailed, and as a result rock decay, erosion, and consequent deposition must have been much more rapid than at present. That the precipitation of moisture and consequently the erosive and transporting power of the rivers was greater during a previous period, presumably the Pleistocene, is also proved by the extinct glaciers of the mountain areas. The conclusion seems warranted, therefore, (1) that previous to and at the time of the deposition of most of the gravel plains rock decay was more rapid, and consequently the depth of the zone of weathering was greater; (2) that the streams and rivers were more active agents in removing, transporting, and redepositing residual materials than they are now. In brief, the sediments of the gravel plains were accumulated under conditions which were probably much more favorable to rapid deposition than those now existing.

ALLUVIAL DEPOSITS.

CHARACTER AND EXTENT.

The deposits included in the second group differ from those of the gravel plains in two particulars—namely, (1) the areas covered are of less extent and are generally confined to river and creek valleys, where they occur in the stream beds and flood plains and on terraces and benches, with local remnants of older drainage systems in the divides; (2) they are the product of ordinary stream action. In many places these deposits merge with those of the gravel plains around their margins, and in such places the differentiation of the two is necessarily arbitrary. They are confined to channels eroded in the bed rock by rivers and creeks and rest upon surfaces formed by such erosion. In general the gravels show less attrition than those of the gravel plains and become progressively coarser toward the heads of the streams. Although in an economic sense these deposits are of greater importance than any of the geologic formations yet described and are of general distribution, many of the areas covered by them are too small to appear if reduced to the scale of the geologic map. As many of them are auriferous they will be described in detail under the heading "Description of placers" (pp. 142 et seq.), but the following general statements regarding them may be appropriately inserted here:

The alluvial deposits consist for the most part of gravels, sands, and clays, and, like the plain gravels, are overlain by peat and moss except along creek and river beds. As a general rule the coarser and heavier sediments are near the bottom and the finer and lighter materials near the top. Although the gravel pebbles can usually be traced to bed-rock sources within the drainage basins, many of the pebbles and bowlders have been transported to their present positions by other agencies than the existing streams. Such transported material is as a rule confined to certain layers of the alluvium. The gold and other heavy minerals contained in the alluvium are generally found concentrated in a lower layer which rests upon bed rock and is called the pay streak, but in a few places the gravel resting directly upon the bed rock is barren and is overlain by a bed of impervious clay called a false bed rock, above which the gold and other heavy minerals are concentrated.

ORIGIN AND AGE OF THE ALLUVIUM.

The topography of Seward Peninsula is characterized by broad, flat, and comparatively straight trunk valleys in which the streams appear to be disproportionately small. Into these valleys the waste from the uplands is carried by the many tributaries, which, while the snows are melting, are overloaded with sediments partly plucked from their beds and partly moved down their valley slopes by creeping slides. Such creeping movements are more pronounced on the north and east sides of the valleys, as on such slopes the sun's heat is more intense; consequently many of the streams have been crowded to the south and west sides of their valleys and the alluvium along the north and east sides may be covered with talus or slide deposits. Wherever in mining operations abandoned channels have been found they generally lie on the north and east sides of the valleys and their gravels are overlain by slide deposits.

It has been shown that these deposits are still accumulating and also that in some places they merge with the gravel-plain deposits, whose age has been determined to be for the most part Pleistocene. Moreover, from the fact that in a few places mammoth remains have been obtained from the alluvial gravels, it may be inferred that their deposition was in process during the Pleistocene and has continued to the present time. It has been shown that during that epoch erosion was much more active than at present; the streams were larger and presumably, although heavily burdened, they were not overloaded with detritus. Such conditions would be more favorable for the concentration of heavy minerals than those at present prevailing, on account of the great activity of the streams and the more rapid disintegration of the bed rock.

HIGH-BENCH DEPOSITS.

CHARACTER AND EXTENT.

The group of high-bench deposits includes many accumulations of small extent and heterogeneous character which resemble each other mainly in their topographic location and elevation. They occur on benches, terraces, flat divides, and plateaus high above the level of the gravel plains and disconnected with the valleys of existing streams. The materials contained vary from distinctly rounded, water-laid gravels to angular, residual gravels formed in place, and between these extremes they vary in the proportion of residual and waterworn materials contained. Several of the high benches

near Nome are known to be auriferous and are described in detail
on pages 198–209, but of most of such deposits in other localities little
is known, as they are in general mantled with moss, peat, and earth,
and no natural sections are exposed. Many extensive plateau surfaces are covered with residual clay mixed with angular bowlders
and pebbles, among which some rounded pebbles can occasionally be
found. These deposits are probably not very thick and through them
the bed rock outcrops in many places. Generally they are not shown
on the geologic map. The distinctly water-laid materials are commonly of great thickness, but they are confined to small areas, many
of which are defined by channel walls and can not be shown on the
scale of the geologic map.

ORIGIN AND AGE OF THE HIGH BENCHES.

The waterworn gravels are in general confined to more or less
definite channels and can best be accounted for as the deposits formed
by the streams of older drainage systems. In many places they are
overlain by angular and other residual materials, having probably
been covered by creeping slides from higher hills. The broad plateau
surfaces are topographic features due to erosion when the land stood
at lower elevations. On such surfaces deposits of washed gravel are
to be expected in old flood plains and stream channels, but since
their elevation much of the gravel has either been washed away or
disintegrated, leaving the remnants of the old gravels mixed with the
products of bed-rock weathering.

No fossil remains that would give a clue to the age of these deposits have been found. They are certainly older than the alluvium
of the existing streams, and since much of the detritus from the high
interstream areas must have washed down to form the gravel plains,
it is probably safe to assume that the high-bench gravels are also
older than the greater part of the gravel-plain deposits; in brief,
they are Pleistocene or older. It has been shown that during the
Pleistocene the conditions were more favorable to rock decay than at
present. It is therefore probable that most of these high-bench deposits, including both the water-laid and the residual portions, were
formed before the end of the Pleistocene.

GLACIATION AND GLACIAL DEPOSITS.

Very little has been added during the last three years to the
general information on glaciation and glacial deposits published in
the report of 1900,[a] which may be briefly summarized as follows:
(1) Seward Peninsula has suffered no general glaciation at any recent
time and none of the unconsolidated deposits thus far described can

---

[a] Reconnaissances in the Cape Nome and Norton Bay regions, Alaska, in 1900, a special
publication of the U. S. Geol. Survey, 1901, pp. 43–47.

be accounted for in this way. This is proved by the mantle of residuary waste resting undisturbed upon the rock from which it was derived and by needles and crags of bed rock projecting through the residuary soil. (2) In its nonglaciated character this region does not differ from other parts of northwestern Alaska surrounding Bering Sea which have been visited and described by geologists.[a] (3) Although there has been no regional glaciation, the valleys of the higher mountain areas have been occupied by local glaciers, some of which were rather extensive. The Kigluaik and Bendeleben mountains were the two main centers of such glaciation, but there were probably also smaller glaciers which did not extend beyond the mountain gorges in several of the lower mountain areas. (4) In the glaciated region the valleys have typical U-shaped cross sections, and head in glacial cirques, and some of the higher valleys still contain remnants of these former glaciers. Around the flanks of the mountains where the valleys debouch on the lowlands, moraines and morainal lakes are common features. (5) Beyond the limits of the definite morainal deposits erratic bowlders are strewn over the surface up to elevations of 1,000 feet. These bowlders are provisionally attributed to the action of floating ice.

The explorations since 1900 have added only cumulative evidence as to the truth of the above statements, but much detailed information regarding the extent and character of the glacial phenomena has been obtained. The glaciers of the Kigluaik Mountain area were the most active and extensive. Practically every valley is glaciated and has the characteristic U shape, though in the floors of some of them V-shaped canyons have been cut by streams since the glaciers disappeared. Glacial moraines extend to the surrounding lowlands on all sides of this mountain mass, but are most extensive on the south side, where there were at least two great piedmont glaciers. During the time of greatest glaciation the ice from many mountain gorges discharged into the upper Sinuk Valley, forming there an immense glacier of the piedmont type. Apparently this valley was filled to overflowing and tongues of ice were forced through two low passes into Stewart Valley, which was also filled to a depth of several hundred feet. The southern limit of the definitely morainic deposits of this glacier is in Sinuk Valley, a short distance below the mouth of Stewart River and about $4\frac{1}{2}$ miles from the base of the mountains.

---

[a] Dall, W. H., Alaska and its resources, 1870, p. 462. Muir, John, Report of the cruise of the U. S. revenue steamer *Corwin* in the Arctic, 1881, pp. 133–145. Dawson, G. M., Geological notes on some of the coasts and islands of Bering Sea and vicinity: Bull. Geol. Soc. America, vol. 5, 1894, pp. 117–146. Stanley-Brown, J., Geology of the Pribilof Islands: Bull. Geol. Soc. America, vol. 3, 1892, pp. 496–500. Nordenskiold, A. E., The voyage of the *Vega*, New York ed., 1882, pp. 569, 583–585. Gilbert, G. K., Glaciers and glaciation: Harriman Alaska Expedition, vol. 3, 1904, pp. 186–194.

Definite moraines have not been recognized as such in Nome Valley, which heads in the Kigluaik Mountains and leads directly southward, but the valley of Salmon Lake is known to have been occupied by a great ice mass, from which a glacial tongue extended southward across the divide into Eldorado Valley, the extreme moraine being about 5 miles south of the base of the mountains.

In the Bendeleben Mountains glaciation was not nearly so active as in the Kigluaik Range. Though practically all the valleys are glaciated and many cirques and U-shaped gorges are to be seen, the fringe of moraines, which is so pronounced around the Kigluaik Mountains, seems to be wanting in this range.

In the lower mountain areas the evidences of glaciation consist only of U-shaped valleys, some of which contain small moraines. Only very local and small glaciers are indicated. In many places the snow, which accumulates to great depths in gulches on hillsides, even at comparatively low elevations, remains until late in the summer. While melting these snowdrifts often move after the manner of real glaciers, and when they have disappeared disclose paved beds which resemble small glacial cirques. In the absence of morainic deposits, therefore, it is difficult to distinguish small glacial valleys formed by perennial glaciers now extinct from those which may be due to local annual glaciers still in operation. It is evident that a climatic change producing a slight increase in the precipitation of snow might be sufficient to convert these annual glaciers into perennial ones and to restore to life the glaciers of the Kigluaik and Bendeleben mountains.

Since 1903 accurate topographic maps, on a large scale, covering a considerable area north of Nome, have been prepared.[a] This area includes several of the typical valleys, both in glaciated and nonglaciated regions, and affords a basis for their comparison. From these maps it will be seen that, except in their greater depth, the cross sections of the glaciated valleys do not differ materially in character from those of some that are nonglaciated. In the writer's opinion processes of erosion now operating in Alaska are competent to produce valleys of this type. Moreover, beyond the limits of the glacial moraines shown on the geologic map the valley floors are covered with alluvium and the slopes with residual detritus through which pinnacles of bed rock outcrop here and there. If these valleys ever contained glaciers, it was at such a remote period that all definite surficial evidences have been destroyed by subsequent erosion.

The deposits of glacial origin which are of sufficient extent to be noted on the geologic map are of two distinct types—the moraines and the extramorainic bowlder beds. They are all confined to the comparatively low areas fringing the Kigluaik Mountains. Although

---

[a] Grand Central and Nome special maps, U. S. Geol. Survey, 1906.

on the geologic map they are not differentiated from other Quaternary deposits, their extreme limits are shown approximately by broken lines. The limits of the moraine north of these mountains have not been determined, though such deposits are known to extend out for some distance on the Imuruk lowland and to lie little, if any, above sea level. On the southeast, south, and west sides of the Kigluaik Mountains the limits of the moraine have been fairly well determined. East of the range the moraines from Big and Crater creeks reach the left bank of Kruzgamepa River. The moraine from Grand Central River and Fox Creek fills the whole upper valley of the Kruzgamepa and extends southward to the head of Eldorado River. The glaciers from the gorges north of the Sinuk Valley filled the Sinuk and Stewart valleys as far down as the mouth of Stewart River. A phase of this invasion was the blocking of Sinuk River below Windy Creek, forming a lake in the Sinuk Valley that persisted long enough to permit the cutting of beach terraces. The glaciers west of Glacial Lake have left little evidence of their existence in the form of moraines, though some of them must have extended a considerable distance from the mountains.

The extramorainic bowlders and bowlder beds, so far as known, are confined to the region south of the Kigluaik Mountains. They consist of more or less angular bowlders of granite, gneiss, and biotite schist derived from the Kigluaik group and are scattered over the surface on hillsides and plateaus up to an elevation of 1,000 feet. The heaviest deposits of such material occur on the plateau between Sinuk and Feather rivers, where they extend to the headwaters of Fairview and Independence creeks, both of which streams are filled with pebbles derived from them. Similar deposits occur to an elevation of 800 feet along Cripple River, where, though they form some definite bowlder beds, they are represented mainly by scattered erratics lying on the surface. East of Cripple River no extensive bowlder beds of this kind have been noted, though erratic bowlders strewn here and there over the surface up to an elevation of 800 feet occur as far east as Osborne Creek. One bowlder of this kind was found lying on a limestone outcrop in the divide between Kruzgamepa River and Iron Creek at an elevation of 800 feet. West of the Kigluaik Mountains the extramorainic drift extends to the basin of Gold Run, but no bowlders or pebbles of this kind occur in Bluestone River above the forks.

The following hypotheses regarding the transportation and distribution of these bowlders suggest themselves: (1) They may have been washed down from the mountains by the rivers and streams of an older drainage system belonging to an earlier physiographic cycle. (2) They may represent an earlier stage of the Kigluaik glaciers than that shown by the moraines, in which they extended from the mountains to the shores of Bering Sea. (3) They may have been transported from the mountain front by ice floes or bergs in estuaries, produced by the subsidence of the land surface.

That bowlders may be carried for long distances by rivers and streams is evident. This is especially true in a region like Alaska, where at the breaking of spring the streams are filled with floating ice. A good example of granite bowlders transported in this way may be seen in Ophir Creek, which is not glaciated, but contains granite pebbles and bowlders, from outcrops near its source, distributed through its whole length. All the streams of the Nome region that head in the glaciated area contain granite pebbles that have been washed down in this way. Although it is possible, therefore, that some of the extramorainic bowlders reached their present position by river and stream action, their distribution as bowlder beds over flat hilltops and the occurrence of scattered erratics render this hypothesis in the main untenable.

In regard to the second hypothesis, it may be said that although some of the bowlder beds near Sinuk and Cripple rivers resemble glacial moraines, the region between them and the glacial center is apparently unglaciated. Some angular crags remain which could not have withstood an advancing ice sheet. Many of the bowlders also occur as solitary erratics overlying residuary soil.

The occurrence and distribution of these bowlders is best accounted for by the third hypothesis, which is that during a part of the glacial period, and presumably an early part, a subsidence of the region south of the mountains occurred and estuaries from the sea extended inland to the ice front. Over these waters the bowlders were floated by icebergs and floes. The maximum amount of subsidence required to produce this effect would be 1,100 feet in the Sinuk Valley. This hypothesis requires that the area of maximum subsidence must have been local and coextensive with the distribution of erratic bowlders. It will be seen from the topographic map (Pl. VIII, in pocket), however, that the average upland levels of the bowlder-strewn area are considerably lower than those of the remainder of the southern half of the peninsula, and hence a general subsidence of the southern part of the peninsula might be competent to account for this distribution. No elevated beach lines or other indications of such a system of estuaries, except the erratic bowlders, have been recognized. Since many of these erratics lie upon or near the surface of the mantle of residuary and waterworn waste on the uplands and near the surface of the coastal-plain gravels, it seems probable that they were carried to their present resting places after most of these deposits were formed. The subsidence must therefore have occurred in late Pleistocene time. If such an earth movement were of sufficient extent to affect materially the distribution of ocean currents, it would doubtless also have produced climatic changes of great importance, and the inauguration of the glacial epoch just described is tentatively assigned to this cause.

IGNEOUS ROCKS.

INTRODUCTION.

Igneous rocks are widely distributed in the peninsula. Though they present considerable variety both in composition and in the degree of metamorphism which they have suffered, they fall readily into three groups, two of which, the greenstones and basalts, are basic, the third being acidic. The type rocks of each group are distinct and can be readily recognized, but in some localities intermediate phases occur whose relations can be determined only by more refined petrographic and stratigraphic investigation than has yet been undertaken.

GREENSTONES.

The most widely distributed igneous rocks of the region are some granular intrusives which are characterized by a dark-green color, and which, though they exhibit many varying phases, are most conveniently grouped under the field name greenstones. Without exception these rocks are more or less altered and are composed of secondary minerals, many of which are due to dynamic as well as to metasomatic influences. Though usually massive, in many localities they present schistose phases, so that in extreme cases it is impossible to distinguish them from schists of sedimentary origin. The investigations that have been made show that the rocks from which the greenstones were derived, although generally of a basic character, were of considerable variety. It has not yet seemed feasible to attempt to subdivide the group, but the following notes regarding some of the more definite types in the various regions will give an idea as to what it includes.

A dike of greenstone in a dynamically almost unaltered condition occurs at Cape Douglas. Under the microscope the rock appears to be composed mainly of pale amber augite and feldspar, a part of which is included in the augite. The augite is changed to light-colored hornblende only around the margins, but the feldspar is too much decomposed to determine its character. There are also large grains of ilmenite bordered by leucoxene and some patches of isotropic serpentinous matter. In the region northeast of Cape Douglas there are many bodies of greenstone that show a somewhat similar general texture, but present various degrees of alteration from the original type. The pyroxene is usually changed to hornblende and chlorite, and garnet, epidote, zoisite, calcite, and quartz are common constituents. B. K. Emerson, who with the Harriman expedition touched on the east shore of Port Clarence, has described the igneous rocks found in this region as amphibolites derived from eruptive pyroxenites.[a]

West of the York Mountains there are some coarsely granular greenstones which are only slightly more altered than that from Cape Douglas, which they resemble, but in these rocks the feldspar is somewhat in excess of the pyroxene, and the rocks were probably gabbros.

Along Fox River, about 10 miles south of Council, there are several dikes of greenstone that under the microscope are found to be slightly altered diabase. They consist of labradorite and olivine phenocrysts in a groundmass of plagioclase, augite, and biotite.

Some greenstone masses which appeared to be slightly altered tuffs were found in the divide between Fox and Solomon rivers.

In nearly all places where the greenstone exists in comparatively unaltered condition, transitional phases to the more typical greenstone composed wholly of secondary minerals were found. The greenstone masses occur most commonly in the form of sills and dikes, though a few of them appear to be necks or stocks, and it is probable that some surface flows may be included among the more altered portions.

In a few places contact-metamorphic minerals have been developed around the margins of the intrusive bodies. On Newton Peak a veinlike deposit of chlorite and albite was observed in the contact of a greenstone mass with the schist. On the upper Casadepaga the limestone surrounding a greenstone mass is marmorized and contains brown hornblende phenocrysts. At the contact of a greenstone sill with the limestone on Tisuk Creek albite occurs in the limestone and tourmaline needles in the greenstone. The contact of a greenstone dike with schist on the upper Kougarok contains tourmaline in radiating groups of crystals and a pale-blue hornblende which has not been determined. The deep-blue hornblende glaucophane has been found in many of the greenstone masses.

In its schistose phases the greenstone seems to be composed entirely of secondary minerals, the most common being hornblende, chlorite, epidote, zoisite, garnet, secondary feldspar, sericite, quartz, and calcite Some chlorite-epidote schists that are widely distributed in small masses seem to represent extreme metamorphism of the greenstones and it is probable that many of the green schists made up of quartz and chlorite that have not been differentiated from the metamorphosed sediments may also be altered greenstones.

It will be seen from the geologic map that the greenstone intrusions occur in the undifferentiated schists of the Nome group, in the Port Clarence limestone, and in the slates in the York region, all of

---

[a] General geology of Harriman Alaska expedition, vol. 4, 1904, p. 44.

which are believed to be of pre-Carboniferous age. They have not been recognized in the Kigluaik group nor in the limestones exposed near Palazruk.

As many of the greenstone masses present schistose phases, it is evident that some and possibly the greater part of them were intruded before metamorphism of the Nome group. In some respects these greenstones resemble the extensive greenstone masses of the Devonian Rampart formation of the Yukon basin. The Rampart greenstones, however, are mainly surface flows and are interbedded with tuffs, limestones, and various other sediments.[a] They indicate that in parts of Alaska during Devonian time there was an enormous amount of volcanic activity, which probably continued into the early Carboniferous. If this volcanic disturbance extended northwestward as far as Seward Peninsula it was probably sufficient to account for the greenstone intrusives in the metamorphic series there. These intrusions evidently must have occurred after the close of the Silurian and most of them while the Silurian rocks were deeply buried.

BASALTS.

A second group of basic rocks is composed of unaltered basalts, which occur at many places in the northern and eastern parts of the peninsula. Typically they are dark-gray or nearly black lavas, usually very cellular or even spongy in appearance, but in some places compact and without amygdaloidal cavities. They are diabases and basalts, both rich in olivine. In the basalts especially olivine phenocrysts are abundant and are noticeable even in the hand specimen. Rocks of this type find their greatest development in the Fairhaven precinct, where Moffit[b] reports that there has been a succession of outbreaks of lava occurring through a considerable period of time. In the region about the head of Kuzitrin River the most recent outpouring of liquid rock occurred at no very distant date, for the ropy surface and irregular margin are still preserved. Caverns or tunnels produced by the cooling of the surface and the continued flow of the still liquid rock beneath are numerous. Flattened lenticular steam cavities 2 or 3 feet in diameter are exposed in some places and the upper surface of the sheet is marked here and there by smooth irregular elevations produced by the escape of steam and the welling up of the lava from below.

On Noxapaga River the writer found these later lavas overlying Pleistocene or late Pliocene gravels that were indurated near the

---

[a] Spurr, J. E., Geology of the Yukon gold district: Eighteenth Ann. Rept. U. S. Geol. Survey, pt. 3, 1898, p. 161. Collier, A. J., Coal resources of the Yukon: Bull. U. S. Geol. Survey No. 218, 1903, p. 15.

[b] Moffit, F. H., The Fairhaven gold placers, Seward Peninsula, Alaska: Bull. U. S. Geol. Survey No. 247, 1905, pp. 31-35.

contact and contain rounded pebbles of basalt derived from an older flow.[a] Along Koyuk and Tubutulik rivers Mendenhall [b] reports the occurrence of several masses of basalt of similar character. On the upper Koyuk the lava has a relation to the unconsolidated gravels similar to that seen by the writer on the Noxapaga. This is probably an outlying portion of the large lava field mapped by Moffit.

Basalts occur in small flows and volcanic necks at a number of localities north of Port Clarence and Grantley Habor. Mukacharni Mountain, the two small buttes to the east of this mountain, a small area on the east side of California River, and a small hill about 2 miles east of the forks of Don River are composed of unaltered basalts that exhibit both compact and vesicular phases and vary in color from gray to black. The vesicles are unfilled and are often confused in weathered specimens with the cavities left by olivine phenocrysts.

Mukacharni Mountain, the two buttes to the east of it, and the small hill to the east of Tozier Creek are volcanic necks cutting the schists and slates. The basalt areas west of Mukacharni Mountain and east of California River are remnants of surface flows or possibly sills which were erupted at the same time as the necks. The amount of erosion since these eruptions indicates that most of them are pre-Pleistocene in age, only the more recent occurring during the Pleistocene. The age of the older flows of the Fairhaven country and of the necks and flows north of Grantley Harbor, however, has not been satisfactorily determined. Moffit [c] states that the lowest possible age limit for the earlier lavas of the Fairhaven district is fixed by the coal-bearing formation of that district, which may be either Tertiary or Upper Cretaceous.

There are no criteria for age determination of the basalts north of Grantley Harbor except the amount of erosion which they have suffered. They antedate the formation of the Kougarok plateau, which is certainly older than most of the Pleistocene sediments. The lava flows of the lower Yukon and St. Michael Island, a region which lies about 100 miles southeast of Seward Peninsula, began, according to Spurr,[d] in the late Miocene and continued until the early Pleistocene. They are represented in part by dikes which cut Upper Cretaceous rocks, and as some of the eruptive masses were reduced to the base level of the Yukon plateau it is evident that the eruptions may have begun before the end of the Cretaceous.

---

[a] Reconnaissances in the northwestern portion of Seward Peninsula, Alaska, in 1901 : Prof. Paper U. S. Geol. Survey No. 2, 1902, p. 31.

[b] Mendenhall, W. C., Reconnaissances in the Cape Nome and Norton Bay regions, Alaska, in 1900, a special publication of the U. S. Geol. Survey, 1901, p. 206.

[c] Moffit, F. H., The Fairhaven gold placers, Seward Peninsula, Alaska : Bull. U. S. Geol. Survey No. 247, 1905, p. 34.

[d] Spurr, J. E., Geology of the Yukon gold district, Alaska : Eighteenth Ann. Rept. U. S. Geol. Survey, pt. 3, 1898, pp. 244 et seq.

St. Lawrence Island, which lies about 100 miles southwest of Seward Peninsula, is in part made up of tuffs and lavas that do not seem to be older than Pleistocene. In the central part of the island there are numerous volcanic cones that have suffered little erosion. On an extensive lava bed examined by the writer the original surface of the flow is still preserved in many places, presenting features similar to those of the most recent flow in the Kuzitrin basin. That volcanic activity began here prior to the Kenai (Eocene ?), however, is indicated by pebbles of lava contained in a conglomerate of that formation. If we assume that the lavas of St. Lawrence Island, Seward Peninsula, and the lower Yukon belong to the same period of eruption, the inference may be justified that the lavas and basalts of Seward Peninsula represent a series of eruptions which commenced in the early Eocene and continued into the Pleistocene.

None of the basalts or diabases included in this group have been affected by dynamic metamorphism and although in a few places they occur in the neigborhood of greenstone masses they are readily distinguished from the greenstones without the aid of the microscope. In the southern part of the peninsula, however, some small dikes of igneous rock have been found that greatly resemble the basalts but seem to present transitional phases to the typical greenstone. The diabase dikes of Fox River described on page 237 are of this character. Some basaltic bowlders which may belong either to the basalt or the greenstone group were found on Independence Creek, a tributary of Sinuk River, but they were not traced to their source in the bed rock.

GRANITIC ROCKS.

The siliceous igneous rocks of the peninsula are grouped together under "Granitic rocks," though many of them vary from typical granite both in texture and composition. The larger masses in the region examined by the writer are generally normal granites, but the smaller masses, some of which may be apophyses from the larger ones, present various pegmatitic and rhyolitic phases. The granites of the Kigluaik and Bendeleben mountains form a network of intricate intrusions in the rocks of the Kigluaik group and were probably intruded at intervals during a rather extended period. Some of these intrusions that were probably the earlier ones are gneissoid and cannot in all cases be distinguished from gneisses and biotite schists that are almost certainly altered sediments. Generally the gneisses occur as sills interbedded with the schists and limestones. The larger granite bodies, which occur in the heart of the range, are generally massive, but many of them are slightly gneissoid and cut by later intrusions that are massive. In the flanks of the ranges there are numerous sills and dikes of coarse-grained pegmatite. Several pegmatite sills of this character along Kuzitrin River were staked as quartz veins by the prospectors who went with the first rush to the Kougarok.

The granite in its massive phase is coarsely crystalline and consists essentially of quartz, orthoclase, and biotite, with plagioclase as an important accessory mineral. The gneissoid granites are similar in composition except that they contain a somewhat larger percentage of plagioclase. Their gneissoid structure is either original or due to entire recrystallization of the minerals, as the microscopic examinations afford little if any evidence of distortion or dynamic movement. The quartz and feldspar individuals are elongated without being crushed and the biotite plates are not deformed. The pegmatite phases, which seem to occur more commonly around the margins of the mountain areas, consist of quartz, orthoclase in large crystals reaching several inches in diameter, and muscovite, with biotite in smaller amounts if present. Tourmaline was observed as an important accessory mineral in one of the pegmatite sills near Kuzitrin River.

Batholiths of the character indicated by these observations regarding the granites of the Kigluaik and Bendeleben mountains would doubtless be competent to produce schists of the Kigluaik type from rocks that in neighboring localities might resemble the schists of the Nome group.[a] For this reason the detailed geologic survey of a portion of the Nome district that was commenced during the summer of 1905 may be expected to yield some important scientific results.

The granite masses that occur in the region north of the Kigluaik Mountains differ considerably in type from those just described. In general they are intruded bosses, few of which present gneissoid phases, but many of them are surrounded by a fringe of apophyses that differ from the central mass mainly in texture. Most commonly these offshoots are porphyritic dikes. The distribution of the granite masses of this type is of economic importance, as tin-bearing lodes occur rather commonly near their contact with the surrounding rocks. A typical granitic intrusion of this sort is found at Cape Mountain.[b] The large granite mass which composes this mountain is intruded through the Carboniferous limestone exposed near Palazruk. Around the margins some large masses of limestone are included in the granite, and many offshoots or apophyses from the main mass extend into the limestone. The largest of these differ only slightly from the granite of the main mass, while the smallest are apt to be aplitic. In the main granite mass there are some ill-defined veins of aplite (a granite rock containing none of the darker minerals) which have not

---

[a] See Van Hise, C. R., A treatise on metamorphism : Mon. U. S. Geol. Survey, vol. 47, 1904, pp. 707–728.

[b] Collier, A. J., Tin deposits of the York region, Alaska : Bull. U. S. Geol. Survey No. 229, 1904, p. 15.

yet been studied in detail. No gneissoid phases are exhibited, but a platy or sheet structure appears in many weathered outcrops. The granite of Cape Mountain is coarsely crystalline and somewhat porphyritic. It consists essentially of quartz, microcline, and biotite, but locally contains as accessory minerals albite, muscovite, apatite, tourmaline, pyrite, fluorite, and cassiterite. Lodes containing cassiterite in appreciable quantities occur on the mountain.

A smaller granite mass which is lithologically similar to that at Cape Mountain occurs in the Port Clarence limestone on Tin Creek, a tributary of Lost River. In this vicinity the limestone is cut by a number of porphyritic dikes that are probably connected with this or similar masses that are deeply buried beneath the limestone. One of these dikes has been studied in some detail, as it is intimately associated with a tin lode of economic value. Microscopic examination shows that the rock was originally a rhyolite or quartz porphyry containing quartz and feldspar phenocrysts in a finely crystalline groundmass. In its present condition, however, the phenocrysts and much of the groundmass are replaced by fluorite. Southwest of this locality there are several somewhat basic dikes that have been placed in the granitic group for the reason that what seem to be transitional phases between them and the typical granite porphyries have been found.

At Brooks Mountain, near the head of Lost River, occur granite masses and dikes, but their field relations have not been determined. The same is true of a mountain east of the forks of Don River, where granite porphyry dikes cut the slates on the periphery, and massive granite bowlders occur on the mountain slopes, but have not been traced to their source in the bed rock.

Ear Mountain is a granite boss of a type similar to Cape Mountain. The massive granite shows the same platy structure in weathered outcrops and the main body of granite is surrounded by a fringe of porphyritic dikes. The granites of Ear Mountain consist essentially of quartz, orthoclase, and biotite. A smaller body of pegmatite-granite is made up of quartz, orthoclase, and plagioclase, and a small dike in the same region consists essentially of quartz and orthoclase phenocrysts in a groundmass of quartz and feldspar, with muscovite, largely secondary, and a secondary growth of feldspar surrounding the larger orthoclase phenocrysts.

On the north side of the mountain several dikes have been extensively staked as tin veins. Apparently these rocks were originally rhyolite or quartz porphyry, but in thin sections they show considerable alteration. In one specimen the porphyritic texture of the rhyolite remains, but the minerals, especially the feldspar phenocrysts, are partly replaced by tourmaline and pyrrhotite, or magnetic pyrites. In another section the original texture is completely obliterated and the rock consists essentially of tourmaline in radiating groups of crystals surrounded by a groundmass made up principally of calcite with some quartz. These rocks, when carefully analyzed, were found to contain traces of tin, but no cassiterite was recognized in the thin sections. Specimens of tin ore that have been obtained at Ear Mountain indicate that, as at Cape Mountain, tin-bearing lodes are associated with the granite intrusives. Some rhyolite dikes that occur at several localities remote from any large granite mass, and some cassiterite veins in the slates of the York region probably indicate the presence of deep-seated granite magmas that have not yet been uncovered by erosion.

The granite from the large area near Midnight Mountain has not been examined microscopically, but seems to be of the same general type as that at Ear Mountain. The outcrops as a rule exhibit the platy structure noted in those localities. No fringing porphyritic dikes or contact minerals have been observed.

The granite masses in the eastern part of the peninsula, most of which lie outside the special area covered in this report, have been described by Mendenhall [a] and Moffit,[b] who investigated the geology of the Norton Bay region and of the Fairhaven district, respectively. These granites, which are confined to a belt extending northward from Norton Bay to Kotzebue Sound, seem to be of a distinct lithologic type from those described by the writer. Though in general appearance they resemble normal granites, they exhibit as a rule a dioritic phase that distinguishes them from the granites of the western part of the peninsula. According to Mendenhall, a broad belt of country, with a maximum width of about 12 miles, extending 55 miles northward from Cape Darby, in the Norton Bay region, is occupied by a great intrusive body of granite and granitoid rock that exhibits considerable variation in texture and mineralogical composition, but is regarded as belonging to one geologic body. At the south end of this belt the type rock is a diorite porphyry with large tabular phenocrysts of andesine or andesine-oligoclase, some colorless pyroxene, and abundant hornblende, in part, at least, secondary. Quartz is present but locally in very inconsiderable amounts, and titanite is a conspicuous accessory. Near the east edge of the northern part of the area the rock appears as a coarsely crystalline aggregate of pale-brownish orthoclase and smoky quartz, with a little biotite. A gneissoid phase of the same rock appears along the west side of its northern border.

A broad belt of schistose rocks and alluvium separates the granite mass of Cape Darby from the biotite-granite mass of the Bendeleben

[a] Mendenhall, W. C., Reconnaissances in the Cape Nome and Norton Bay regions, Alaska, in 1900, a special publication of the U. S. Geol. Survey, 1901, pp. 204–205.

[b] Moffit, F. H., The Fairhaven gold placers, Seward Peninsula, Alaska: Bull. U. S. Geol. Survey No. 247, 1905, pp. 27–30.

Mountains, which lies to the northwest. The intrusive nature of the granite of Cape Darby is indicated by masses of schist and limestone caught up in the intruding magma and by dikes from the granitic mass cutting the surrounding schists.

Moffit[a] reports that in the eastern part of the Fairhaven district the siliceous igneous rocks form a group comprising granites, quartz diorites, and intermediate forms that may be properly classed with the monzonites. These rocks find their greatest development in a high ridge that lies between Buckland and Kiwalik rivers and extends from Eschscholtz Bay southward to Koyuk River, forming the boundary between Seward Peninsula and the mainland. The granites form the core of this ridge and outcrop in a series of disconnected areas surrounded by andesites of later age. Hornblende is the prevailing dark mineral of the granites, but here and there biotite takes its place. By a decrease in the amount of quartz the granites approach syenites in composition, such phases being characterized by the abundance and large size of orthoclase crystals.

From a consideration of the above-stated facts it will be seen that the larger granite masses of Seward Peninsula fall into three groups according to the nature of their occurrence, their effects on the intruded rocks, and their lithology. They are all of intrusive origin, but the masses of the Kigluaik and Bendeleben mountains are batholiths formed at great depth in the earth's crust within the zone of flowage. The masses north of the Kigluaik uplift represent intrusions in a zone nearer the surface which has been called the zone of fracture, and the dioritic masses in the eastern part of the peninsula seem to be of intermediate character. In their lithology the rocks of the Kigluaik uplift resemble those of the northern part of the peninsula, the distinguishing features being abundant gneissoid phases in the former and porphyritic phases and surrounding dikes of rhyolite in the latter. The northern granite masses are further characterized by such accessory minerals as fluorite, cassiterite, and tourmaline in their peripheral portions.

There is no direct evidence in regard to the relative ages of these intrusions. Those of the Kigluaik region may have been and probably were intruded at intervals through a long period of time, as granites of several ages can be recognized. The same thing may be true of the other granite masses, though the evidence for it does not seem to be so clear. The granite mass of Cape Mountain, which is regarded as typical of the granites north of the Kigluaik Range, is intruded in Mississippian rocks and is therefore certainly younger than that horizon. Moffit[b] reports that the granites of the Fair-

---

[a] Moffit, F. H., The Fairhaven gold placers, Seward Peninsula, Alaska: Bull. U. S. Geol. Survey No. 247, 1905, pp. 27–30.

[b] Op. cit., p. 30.

haven district are evidently older than the coal-bearing beds overlying the schists, which, as has been shown, may be of late Mesozoic age. Mendenhall [a] also states that the period of intrusion seems older than that of the coal-bearing sediments. Neither of these writers has presented more definite evidence for this conclusion than the fact that the granites present gneissoid phases whereas the sediments are unaltered. The rocks have not been found in contact, nor have granite pebbles been found in the sediments.

The coal-bearing sediments examined by the writer in the Sinuk basin have been sheared and folded with the surrounding schists, but contain no pebbles of granite, though such pebbles are the main constituents of the Pleistocene deposits that overlie them.

The unaltered sediments of the Fairhaven district have the same north-south strike and high dips that are common in the highly metamorphic rocks, and those of the Norton Bay region are also highly folded, with approximately north-south strikes. If, as Moffit suggests, this structure was produced at the time of the granite intrusions, the sediments would seem to be older than the granite. A similar inference may be drawn from the unaltered sediments of the Sinuk basin, which are highly folded and strike parallel with the beds in the front of the Kigluaik Range.

In view of these facts any conclusion regarding the relative ages of the granites and the unaltered sediments seems hazardous. The granitic intrusions may have continued into the Tertiary, though they are certainly older than Pleistocene. If we assume that all the granitic rocks of the peninsula belong approximately to the same period of intrusion, which seems reasonable, then it may be stated that these intrusions occurred between the early Carboniferous and the end of the Tertiary. The larger granitic masses of the Alaska Range and the Pacific seaboard, whose age has been determined, were intruded during the latter part of Mesozoic time, and some of the intrusive granitic masses of the Kuskokwim Valley have been assigned to the early Tertiary.[b] On the assumption that the disturbance attendant on the intrusion of the granite masses of southern Alaska extended to Seward Peninsula also, the conclusion is reached that these masses were most probably intruded in late Mesozoic or early Tertiary time.

QUARTZ VEINS.

Veins of quartz and calcite occur in the metamorphic rocks of nearly all parts of the peninsula, but most abundantly in the schists of the Nome group near the localities rich in placer gold. Most of

---

[a] Mendenhall, W. C., Reconnaissances in the Cape Nome and Norton Bay regions, Alaska, in 1900, a special publication of the U. S. Geol. Survey, 1901, pp. 204–205.

[b] Spurr, J. E., Reconnaissance in southwestern Alaska: Twentieth Ann. Rept. U. S. Geol. Survey, pt. 7, 1898, pp. 229, 231.

these veins are small lenticular masses whose longer axes are parallel with the foliation. Here and there occur larger masses of the same type, which may be called "blanket veins." A few of these lenses and blanket veins are known to be connected with approximately vertical fissures and zones of fracture, from which they seem to be offshoots.

Larger veins filling fissures that cut across the foliation and bedding have been found in a number of localities, and some of these are of economic value. (See pp. 228–232.) So far as observed, however, all the veins in the vicinity of the gold placers are irregular, and even those of considerable size are apt within a short distance to divide up into many veinlets and stringers which can not be mined profitably. As a rule these veins do not follow any discoverable system. They are disseminated in an intricate network through the metamorphic rocks and are very abundant in certain areas that may be best described as zones or belts of mineralization. Some of these veins were formed previous to the development of the schistosity and are sheared with the inclosing rock. A few of them can be shown to belong to different periods of intrusion, as veins of quartz cutting veins of calcite, or vice versa, have been observed.

In the rich placer regions nearly all the veins have been found to carry traces of gold, and some of them contain rich ore. In places the gold is free, though it is more commonly associated with sulphide minerals. Occasionally veins of white quartz have been discovered in which a little free gold may be seen near one of the walls, though the mass of the rock contains no trace of any.

Quartz veins containing cassiterite as well as traces of gold occur in the western extremity of the peninsula,[a] and silver-lead deposits have also been discovered in several localities, the best known being at the Omalik mine, on the upper Fish River.[b]

As a general rule no causal connection between the quartz veins and the greenstone intrusives that are so common in the rocks of the Nome group can be demonstrated, and though greenstones are nearly everywhere present in the bed rock where there is placer gold and are in many places cut by quartz veins, they can not be regarded as indications of the presence of gold. The cassiterite-bearing veins of the western part of the peninsula, however, have been shown with a reasonable degree of certainty to be due to emanations from intrusive masses of granite and kindred rocks of an acidic nature. In the same region some lead-silver deposits have been found intimately associated with acidic dikes. Gold in small quantities is known to be associated

---

[a] Collier, A. J., Tin deposits of the York region, Alaska: Bull. U. S. Geol. Survey No. 225, 1904, pp. 154–167.

[b] Mendenhall, W. C., Reconnaissances in the Cape Nome and Norton Bay regions, Alaska, in 1900, a special publication of the U. S. Geol. Survey, 1901, pp. 213–214.

with the cassiterite and to occur in the same veins, and as cassiterite
in small amounts has been found associated with the placer gold at
many widely scattered localities it is not unreasonable to expect that
the influences producing the cassiterite veins have extended over a
considerable part of the peninsula.   Although, as stated above, some
of these veins were probably formed before the metamorphism of the
Nome group occurred, the majority of them were intruded at some
later period.   There is reason to believe that the tin-bearing veins
were deposited during the period of granite intrusion.   The rocks of
the whole peninsula were greatly disturbed at that time and the more
deeply seated ones were intensely altered.   It is therefore the writer's
belief that most of the unsheared quartz veins were formed during
the intrusion of the granites and are traceable either directly or indirectly to the same influence.

OUTLINE OF ECONOMIC GEOLOGY.

By Alfred H. Brooks.

GENERAL STATEMENT.

Though the general features of gold placers are so simple as to require no searching analysis, yet but few studies of even the more prominent characteristics have been made and there are many obscurities in regard to the occurrence, origin, and distribution of alluvial gold. This field of applied geology has been somewhat neglected, probably in the main because the data needed for the investigations must be gleaned during the mining operations or they are lost forever. It is therefore only in exceptional cases that the facts bearing on the occurrence of alluvial gold have been available to the geologist.

It is proposed here to discuss the gold placers of Seward Peninsula in their larger relations, and thus to present an introduction to the detailed descriptions by Mr. Collier and Mr. Hess which follow. Attempt will be made to formulate the laws governing the occurrence and origin of the alluvial gold, so far as the data at hand will permit. In the preceding pages Mr. Collier has presented in some detail the geography and geology of the peninsula, but brief comments will be made on those features that bear directly on the geology of placers.

A study of the maps (Pls. VIII and IX, in pocket) shows that this is essentially a region of mature topography, for with the exception of a few rugged mountain masses, the region is characterized by broad valleys and flat-topped interstream areas. As a rule the valleys have broad floors with gently sloping walls, but there are a few sharply incised canyons. The rock valley floors are usually gravel covered, and bed rock is exposed only in the smaller gulches. In most places the seaward bases of the uplands are buried in extensive gravel sheets that form coastal plains and are made up of elevated fluvial deposits. Near the sea many of these deposits break off in escarpments. The gravel plains mark former epochs of deposition, and evidences of periods of erosion and sedimentation are to be found in the many well-marked terraces which occur up to altitudes of 600 feet. Less well preserved benches occur here and there at still higher altitudes,[a] but most of the topographic evidence of such older epochs of erosion has been effaced. An exception must be made, however, of the erosion that produced the highest surface. This appears to have beveled the rocks of much the larger part of the peninsula, and evidence of the base-level thus formed is still preserved in a general accordance of summit levels of the interstream areas, a very noticeable topographic feature throughout the province.

All through the peninsula there is evidence of deep rock decay, for nearly everywhere, except in the higher mountain masses where there has been some recent glaciation, a heavy residual soil mantles the hills. The valley slopes are in general deeply buried in talus, and outcrops are abundant only along the crest lines of the ridges. This residual mantle is due to (1) absence of glaciation, (2) long stability of the land relative to sea level, and (3) slowness of erosion because of the thick mat of vegetation that covers the surface.

The data now at hand point to the following succession of events in post-Eocene time. (1) A long period of erosion which reduced much of the peninsula to a peneplain.[b] During this base-leveling, or subsequently, the rock surface became deeply weathered. (2) Uplift succeeded base-leveling and was probably of an intermittent character. With elevation erosion was revived. The sum of movement was upward, yet the elevations appear to have been interrupted by periods of depression. The succession of uplifts is recorded in the benches now found at various altitudes. When the high terraces near Nome were formed, the land mass must have remained stable sufficiently long to permit an extensive drainage system to be developed. Collier[c] has pointed out that at some later stages local base-levels were developed, but these do not appear to be recorded throughout the peninsula. The last extensive period of erosion is represented by the extensive gravel deposits found in the coastal plains and the interior basins. Later these gravels were elevated and dissected, and then the present creek deposits were laid down.

The physiographic conditions above outlined were favorable to the formation of placers, for the deep rock decay set free the gold in the bed rock and the various epochs of erosion and deposition brought about a concentration of the heavier materials in the gravels.

---

[a] These features have been fully discussed in Reconnaissances in the Cape Nome and Norton Bay regions, Alaska, in 1900, a special publication of the U. S. Geol. Survey, 1901, pp. 48–64.

[b] The age of this peneplain has not been determined, but it is probably the same as that of the Yukon province. See Brooks, A. H., Geography and geology of Alaska: Prof. Paper U. S. Geol. Survey No. 45, 1906, pp. 280–290.

[c] Collier, A. J., Reconnaissance of the northwestern portion of Seward Peninsula, Alaska: Prof. Paper U. S. Geol. Survey No. 2, 1902, pp. 34–44.

Furthermore, dissection of the older alluvial deposits, brought about by uplifts, resulted in a reconcentration of the gold, and thus some of the richest placers were formed.

Mr. Collier (pp. 65–83) has shown that the bed rock is chiefly metamorphic, and that it has been subjected to stresses producing folding, faulting, and jointing. He also notes that intrusives are plentiful and that stringers and veins of quartz, locally metalliferous, are not uncommon.

Though a number of stratigraphic units are recognized on the map, only the two oldest, the Kigluaik group and the Nome group, need here be considered, for they form the country rock practically throughout the placer districts. The Kigluaik falls into three subdivisions—a basal biotite gneiss, a middle member of heavily bedded limestone, and a younger member made up chiefly of biotite schists and thin limestones. The rocks are folded and faulted and carry some quartz veins, but these have not so far been found to be much mineralized. The Nome group includes a great thickness of quartz and calcite schists, greenstones, and greenstone schists, together with a massive limestone member which has been called the Port Clarence and which carries Silurian fossils. Schists occur both above and below the limestone. The placer gold thus far found has been chiefly associated with rocks belonging to the Nome group, which have a wide distribution throughout the peninsula. The schists of this terrane are locally much fractured and seamed with quartz and calcite stringers, and some of them carry larger veins in which metallic minerals are not uncommon. Igneous rocks have a wide distribution in the peninsula, the dominant types being granites, in the form of dikes, sills and stocks, and greenstones, occurring as sills and dikes. Some of the acidic intrusives have been found to be mineralized. (See p. 120.)

Mr. Collier has discussed the structure (pp. 60–63), and it will be necessary here only to draw attention to certain larger features. Though in detail the structure is probably exceedingly complex, the geologic maps indicate that there are certain areas of uplift which are marked by the distribution of the limestone. (See Pls. X and XI, in pocket.) These suggest structural domes of great irregularity, probably modified by folding and faulting.[a] As shown on page 122, these uplifts may have causal connection with the distribution of the gold, for it appears that the rocks have been most intensely deformed

---

[a] This type of structure was discussed by the writer in the previous report (Reconnaissances in the Cape Nome and Norton Bay regions, Alaska, in 1900, a special publication of the U. S. Geol. Survey, 1901, pp. 36–37) and, though by no means proved by the later work, yet appears to find some support. Furthermore, indications of a similar type of structure were found by Moffit in the Fairhaven district. (See Bull. U. S. Geol. Survey No. 247, 1905.)

along the margins of these uplifts and that in such zones of movement the most extensive veining occurs.

GENESIS AND CLASSIFICATION OF PLACERS.

Three conditions are usually operative in the formation of placers (1) the occurrence of gold in bed rock to which erosion has access; (2) the separation of gold from bed rock by weathering or abrasion; (3) the transportation, sorting, and deposition of the auriferous material derived by erosion. It is self-evident that unless gold occurs in bed rock and then is separated from it no placers can be formed. The third agency, namely, transportation, sorting, and deposition, though it has been operative in most regions of extensive auriferous alluvium, is, nevertheless, not absolutely essential to the formation of placers. In many parts of Europe, in the southern Appalachians, and in the Tropics workable placers have been found which were formed solely by the weathering in place of the auriferous bed rock. Rickard[a] has described residual deposits of this kind in Australia, where a surface concentration of gold has been brought about through the agency of wind, the lighter material having been removed.

The distribution and origin of the gold in bed rock, involving as it does the study of ore deposits, though of first importance to the study of placers, can here be only briefly discussed. Of equal importance and more closely related to the genesis of placers is the consideration of the agencies leading to the separation of gold from the bed rock and its subsequent transportation, sorting, and deposition.

That the changes brought about by secular decay of bed rock play an important part in the genesis of placers is a fact that has hardly been fully enough recognized.[b] In the text-books emphasis has usually been laid on the two types, the residual placer and the transported or true placers, without full recognition of the fact that the former often represents an intermediate stage between the bed-rock source of the gold and the true placer. It appears to the writer that, as a rule, the primary concentration of the auriferous detritus brought about by rock weathering is a necessary stage in the formation of rich placers.[c] As shown on page 112 this really amounts to a recognition of the geographic cycle as an important factor in the genesis of placers.

---

[a] Rickard, T. A., The alluvial deposits of western Australia: Trans. Am. Inst. Min. Eng., vol. 27, 1898, pp. 490–537.

[b] H. L. Smyth has drawn attention to this feature of the formation of placers. See The origin and classification of placers: Eng. and Min. Jour., vol. 79, 1905, pp. 1045–1046, 1179–1180, 1228–1230.

[c] J. S. Diller has made a similar suggestion for the California placers. See Revolution in the topography of the Pacific coast since the auriferous gravels period: Jour. Geol., vol. 2, 1894, pp. 48–53.

The transportation, sorting, and deposition of material furnished by the weathering of rocks, the most easily understood of geologic phenomena, are all important agencies in placer formation. Uplift may revive the forces of erosion, and in this way these agencies may be repeatedly effective, resulting in reconcentration of the alluvial gold.

A logical classification of the placers should be based, first, on genesis; second, on form.[a] The primary grouping, according to origin, would be "residual placers," "sorted placers," and "re-sorted placers." The residual placers are those in which there has been no water transportation, the concentration of the gold being due solely to rock weathering. The gold of the sorted placers is the result of transportation, sorting, and deposition by water. Placers of the third group are those in which the gold has passed through two or more cycles of erosion before its final deposition. Those of the first class are practically all of one type. The sorted and re-sorted placers embrace many subordinate types, named according to the form of the occurrence. The following list presents the larger groups and the more important of the subordinate types:

1. Residual placers.
2. Sorted placers.
    Hillside.
    Creek and gulch.
    River bar.
    Gravel plain.
    Bench.
    High bench.

3. Re-sorted placers.
    Creek and gulch.
    Beach.
    Elevated beach.

It is evident that in this, as in most other classifications, intermediate types are found which may belong to either one of two groups. Hillside placers, for example, are those that occur on hill slopes and do not occupy any well-defined channels. These, though usually water sorted to a certain extent, grade directly into deposits of a purely residual origin on the one hand and into stream or gulch deposits on the other. Again, creek and gulch placers occur both in material which has been sorted only once and in that which has passed through several cycles of erosion.

In the following pages an attempt will be made to trace the placer gold through the various stages, from its occurrence in the bed rock to its deposition in the reconcentrated placers.

---

[a] The subject of the classification of placers has been discussed in Reconnaissances in the Cape Nome and Norton Bay regions, Alaska, in 1900, a special publication of the U. S. Geol. Survey, 1901, pp. 144–150, where the same classification was suggested, though different names were applied.

GOLD IN BED ROCK.

INTRODUCTION.

The distribution of gold in the parent rock is of first importance to the mining industry; for it determines not only the occurrence of auriferous gravels, but also the possibility of lode mining. Unfortunately, the data at hand are entirely inadequate for a proper discussion of this important subject. Geologic research in Seward Peninsula has been of a reconnaissance character and only the larger features, with few details of the bed-rock structure and veining, have been determined. Detailed surveys of the Nome region proper have been begun by F. H. Moffit and F. L. Hess, but the results have not yet been worked up. The prospector, who might furnish additional facts, has assiduously devoted himself to the placers, which give promise of more immediate returns than the lodes.

Veins and impregnated zones are not uncommon in the placer districts; and though prospectors have not been backward in filing location notices, few excavations have been made and much of the search for auriferous quartz has been made by men either without experience or without capital. The Big Hurrah quartz mine (described by Mr. Collier on p. 228), one of the few examples of intelligently directed lode development, has attained a depth of 130 feet (1904). Elsewhere in the peninsula but few prospect shafts and tunnels have reached a depth of 20 feet (1904).

The search for lodes and the study of the bed-rock source of the gold are rendered difficult by the thick mat of vegetation which mantles much of the surface. A heavy talus covers the hill slopes, and the rock floors of the valleys are usually deeply buried in gravels. There is, therefore, relatively little opportunity to study the bed rock.

GROUPING OF DEPOSITS.

For this discussion a somewhat arbitrary subdivision of the ore bodies has been adopted, namely, (1) disseminated deposits and (2) concentrated deposits. In the first group falls all of the auriferous bed rock in which the gold does not occur along any well-marked individual fracture or zone of fracture, but is disseminated in larger rock masses. The second group embraces all the auriferous deposits which have been laid down along well-defined fractures or zones of fracture. All ore bodies having any prospective commercial value fall into the group of concentrated deposits, but it also embraces many in which the values are too low or too widely distributed to have any prospective value. It is evident that, as there is a gradation between the two types, some deposits might find place in either group.

Two other types of deposit are here briefly mentioned, though they have not been studied by the writer. These are the cassiterite-bearing lodes of the York region and the galena deposits of the Fish River basin.

In previous reports treating of the mineral resources of Seward Peninsula, emphasis has been laid on the fact that the gold was derived chiefly from disseminated deposits. The richness of the placers has been attributed rather to the erosion of extensive rock masses and subsequent concentration by repeated sorting than to the richness of the bed-rock deposits. Though this no doubt holds true for many deposits, yet the accumulation of more evidence both from placer-mining operations and from closer studies of the bed. rock suggests a far greater concentration of gold in the bed rock than was at first supposed. However this may be, the wide distribution of gold in the metamorphic terranes is attested by the occurrence of colors in the gravels of almost every creek whose gravels have been derived from metamorphic rocks.

The disseminated deposits appear to occur only in rocks having a secondary structure that is sufficiently well developed to permit the circulation of the mineral-bearing solutions. In general, however, they can be defined as deposits formed by mineral solutions that followed zones of fracturing not sufficiently intensified or localized to afford well-defined channels of circulation. These belts of mineralization, for such they are, probably have more or less zonal arrangement, as is indicated by the distribution of the placers, and appear to follow the major structural features. Some of the deposits carry little or no gangue, the evidence of mineralization being limited to the presence of pyrites or their oxidation products, together with some gold. In others lenticular masses of quartz and, less commonly, calcite, following the planes of foliation and carrying metalliferous minerals, are found. These lenses are discontinuous aggregates of gangue minerals, and are usually less than an inch in width. Locally the quartz and calcite occur merely as blebs a few inches in diameter, with no apparent connection.

The York region appears to offer an example of the dissemination of gold in the bed rock and consequent absence of rich placers, in spite of the extensive erosion that has taken place. In the metamorphic slates of that region (see p. 79) blebs and stringer veins of auriferous quartz are not uncommon in the bed rock. From a scientific standpoint, considerable areas of bed rock can be said to be mineralized, but these are disseminated through large masses, and up to the present time no well-defined impregnated zone has been recognized. This appears in a large measure to account for the absence of any considerable amount of alluvial gold, though certain conditions

of erosion,[a] discussed elsewhere, also have been adverse to the formation of placers.

CONCENTRATED DEPOSITS.

The concentrated ore bodies are of two general types—(1) impregnated zones of fracture and (2) fissure veins. The first are probably
the most abundant, but in many places the two grade into each other.

The impregnated zones of fracture appear to have been formed by
the shearing of rock masses along more or less well-defined zones and
the intrusion of the shattered rock masses by mineral-bearing solutions. The presence of included fragments of foliated rocks in the
shear zone indicates that the stresses that developed the secondary
structure antedated the shearing, which was accompanied by impreg-
nations.

Within the mineralized zone there is usually a system of intersect-
ing fractures that are followed by quartz or calcite veins and stringers,
the whole forming what Becker has termed a stringer lead. As a
rule, the entire zone strikes parallel to the dominant structures of the
country rock, but many of the fractures are discordant to the walls of
the deposit. The dips of the ore bodies vary greatly, but most commonly are nearly vertical. In most of the deposits which have come
to the notice of the writer the zone of mineralization is bounded by
slickensided walls. Here and there, however, the ore body merges
gradually into the wall rock. In some deposits gash veins penetrate
both walls.

At some localities two systems of veining were observed, one of
which, following the foliation, has been more or less deformed. The
second system is an infiltration along joints that intersect the lines of
foliation. Though both systems were found to carry sulphides, the
younger one appeared to be most heavily mineralized.

The dominant mineral in the vein filling is usually quartz, but in
some veins calcite is more abundant than the quartz, and they
very commonly occur together. Little attention has been given to the
ores themselves. Those that have come to the writer's attention are
chiefly pyrites, and among these, iron, arsenical, and antimonial pyrite
appear to dominate, but chalcopyrite has also been found. Other
metallic minerals that have been recognized are free gold, galena, and
stibnite. Scheelite is often found with the placer gold, but has not
been observed in place. Cinnabar occurs in abundance in the gravels
of Daniels Creek, and cassiterite associated with alluvial gold has
been found in some localities.

These deposits appear to be somewhat irregular, but no systematic
attempt has been made to prospect any of them. They vary in width

---

[a] See also Reconnaissances in the Cape Nome and Norton Bay regions, Alaska, in 1900,
a special publication of the U. S. Geol. Survey, 1901, pp. 137, 145.

from less than a foot to possibly 50 feet or more, but there are still wider ones which probably belong to the disseminated deposits. Most of them have not been traced farther than a few hundred feet, but in one case at least a zone of impregnation has been followed a mile or more along the strike.

Few of these ore bodies have been sampled and there is little information regarding values. As high as $3 to $10 a ton in gold has been reported to the writer as the result of careful sampling. Inexperienced men are often misled by the high values obtained from the assay of picked specimens. It appears that wherever gold is found some silver is present. It should be noted that many of these mineralized zones in which pyrite is present carry little or no free gold. Though some free gold has been found, it appears that most deposits of this type carry base ores.

The fissure veins, the second type of concentrated deposits, are relatively rare compared to the impregnated zones. This is at least true of those which are ore bearing, but large barren quartz veins are not uncommon in the metamorphic rocks of the peninsula.

The fissure veins are well defined and usually cut the foliation of the country rock. Like the impregnated zones, many of them are sharply differentiated from the wall rocks by slickensides. Some of them, however, send off gash veins into the country rock which follow the foliation. Quartz veins 2 to 15 feet wide have been found, but these appear to be barren and have been traced no great distance. The ore-bearing veins probably range from less than a foot to 8 feet in width. In the Solomon River Valley veins have been traced by surface croppings for a distance of more than half a mile. From the evidence in hand it would appear that the fissure veins were more persistent than the impregnated zones.

The vein filling is chiefly quartz, but in some veins calcite is practically the only gangue mineral. Gold, where present, is usually free. Iron pyrite is almost invariably present, and some galena and chalcopyrite has been found. It is reported that one chalcopyrite-bearing lode has been found near the head of Nome River and another (see p. 298) in the Kougarok region. Two types of vein filling have been recognized. In one the quartz is ribboned and little calcite is present. This is the characteristic deposit of the Big Hurrah Creek region, described by Mr. Collier (p. 228). In the other type the gangue is nearly all calcite, through which the gold occurs in a free state, together with some iron pyrite. No workable deposits of this type have been discovered, though many of the calcite veins yield high assay values. In general it may be said that the fissure veins thus far prospected, except those in the Solomon River region, have not been proved to be workable ore bodies.

Though the fissure veins and impregnated zones have been described as different types of deposits, yet it is not intended to imply that they represent distinct epochs of mineralization, for they may, in part at least, belong to the same period of intrusion. In many cases the formation of the impregnated zones appears to have been an identical process with that of the veins, except that in the fissure the schists appear to have been crushed without any notable open spacing.

CASSITERITE- AND GALENA-BEARING LODES.

Besides the auriferous deposits above described, two other types of ore bodies have been found on the peninsula, namely, the cassiterite-bearing dikes of the York region and the galena-bearing lodes of the Fish River region.

The tin-bearing lodes have been described by Collier [a] as mineralized granite dikes which appear to be apophyses from larger granite masses. Besides these, Hess [b] has described tin-bearing impregnation veins in the granite area at Cape Mountain. In some places the mineralization appears to have occurred along the peripheries of the granite stocks. A lode on Cassiterite Creek is described by Collier and Hess as essentially a mineralized porphyry dike, more or less altered, through which crystals of cassiterite are scattered. Besides the cassiterite the ore carries galena and some gold and a great variety of secondary minerals. Hess has also described a gold-bearing aplite dike which occurs on Cape Mountain.

Mendenhall [c] has described the Omalik silver deposit as a galena-bearing lode, occurring at the contact of a schistose intrusive and a white crystalline limestone. He notes the presence of pyrite and stibnite. His examination was confined to the outcrop, for at the time of his visit the mine workings had caved in. Analysis showed the presence of gold. It appears that this deposit is unlike those described from which the alluvial gold has been derived. Similar ore bodies have been reported by prospectors from other parts of the peninsula, but have not been examined by geologists.

These two types of ore bodies, occurring near the eastern and western extremities of the peninsula, indicate a wide distribution of the mineralization. They are also of interest to the miner, because they appear to be examples of types of deposits different from those which furnished the placer gold.

---

[a] Collier, A. J., The tin deposits of the York region, Alaska : Bull. U. S. Geol. Survey No. 229, 1904, pp. 16–29.

[b] Hess, F. L., The York tin region : Bull. U. S. Geol. Survey No. 284, 1906, pp. 145–157.

[c] Reconnaissances in the Cape Nome and Norton Bay regions, Alaska, in 1900, a special publication of the U. S. Geol. Survey, 1901, pp. 213–214.

DISTRIBUTION OF MINERALIZATION.

An adequate discussion of the distribution of the ore deposits must await a more complete knowledge of the geology. Some facts have, however, developed in course of the investigation that appear to give support to certain conclusions. These theories, being still largely hypothetical, are here presented with some hesitancy, because the determination of the loci of gold in bed rock is of such obvious importance to the mining industry. False deductions, if accepted by the miner, may lead to misdirected efforts in the search for auriferous placers and lodes.

As yet almost the only criterion of the occurrence of gold in bed rock is the distribution of the placers. This evidence must be used with caution, because (1) placers may be found a long distance from the bed-rock source of the gold, as in the Nome beach deposits, and (2) gold may occur in bed rock where the conditions are unfavorable to the formation of placers. It is a well-established fact,[a] however, that the gold of the gulch and creek placers is for the most part derived from bed rock of the immediate vicinity, and this fact makes it certain that the areal distribution of this type of placers is probably almost identical with the areal distribution of gold in bed rock.

The gold placers of Seward Peninsula fall into two broad belts, separated by the Kigluaik and Bendeleben mountains. (See Pls. X and XI, in pocket.) In the southern belt, at present the largest producer of gold, are included the Nome, Solomon, and Ophir regions and some smaller mining districts. The Bluestone, Kougarok, and Fairhaven placers lie in the northern zone. The absence of placer gold in intervening regions suggests the absence of gold from the bed rock, but may also be explained by the fact that glaciation and other conditions have not been favorable to the accumulation of auriferous gravels.

Within the two belts the placers are so irregularly distributed as to make it certain that some other cause must have operated in determining their occurrence than the accidents of erosion and deposition. On some creeks there are rich placers, while at near-by localities, where the character of the alluvial deposits is identical, gold may be almost entirely absent. (See maps, Pls. X and XI.) A natural inference is that this irregularity is due to the irregularity of the distribution of the gold in bed rock.

It has already been pointed out [b] that there appears to be a connection between structure and the distribution of the auriferous gravel. This view is borne out by the evidence of the geologic maps, which indi-

[a] Reconnaissances in the Cape Nome and Norton Bay regions, Alaska, in 1900, a special publication of the U. S. Geol. Survey, 1901, p. 141.

[b] Op. cit., p. 144.

cate that most of the workable placers occur along or close to the
contacts of limestones and schists. These contacts have in many
places been exposed to erosion as a result of the domal uplifts already
mentioned. This, however, is not everywhere the case, for some of
the limestone-schist contacts are simply the margins of lenses of limestone included in the schist. Be the relation of the limestone to the
schist what it may, it appears to be established that the bed-rock
source of the gold in most deposits is traceable to a limestone-schist
contact. Furthermore, these contacts appear to have been loci of the
greatest mineralization, either as impregnated zones or as fissure
veins. These contacts have only in part been studied and mapped,
and if the distribution of all the placers (Pls. X and XI, in pocket)
does not follow this law, this may only appear so because of the lack
of complete data on the bed-rock geology.

The close association of the placers with the contacts between limestone and schist is well illustrated by the Ophir Creek region, the
geology of which is indicated on the accompanying sketch (fig. 6)
in greater detail than on the general map (Pl. X). Several limestone areas are mapped which may belong to an older formation or
may be lenses in the schists which surround them. The contacts between the two types of rocks where examined give evidence of intense deformation, resulting in shattering and shearing. It is in these
zones of disturbance that the presence of quartz veins, many of them
ore bearing, is most noticeable. In the diagrammatic section (fig. 6)
the relation of the mineralized zones both to the limestones and
schists and to the placers is brought out. For the purpose of
this discussion it is immaterial whether the limestone is a distinct
lower member of the succession that has been infolded or whether it
is a series of lenses in the schist.[a] The point is that along the contact of the limestone and the schist shearing has taken place, and also
intrusion of mineral-bearing solutions. This may be due solely to
the mechanical influence of a massive bed in its contact with a schist
during the deformation, or also to the chemical composition of the
limestone, which may have had some influence on the deposition of
mineral veins.

Though the evidence of the placers makes it appear that the loci
of mineralization are usually confined to the contacts of the limestones and schists, yet there are known to be ore bodies which occur
in other associations. Mention has been made of the cassiterite-bearing dikes and the galena lodes occurring in crystalline limestone. It
is also probable that much of the placer gold may have been derived
from other than limestone and schist contact deposits. Detailed
studies by Mr. Moffit and Mr. Hess of the area covered by the Nome

---

[a] It should be noted that in one place, at least, there is evidence that the schist may
be an altered intrusive. See description of Bluff region, p. 285.

special map, though not completed, appear to indicate that the placer gold finds its source in a zone of intricately folded and faulted rocks, and not necessarily along limestone-schist contacts.

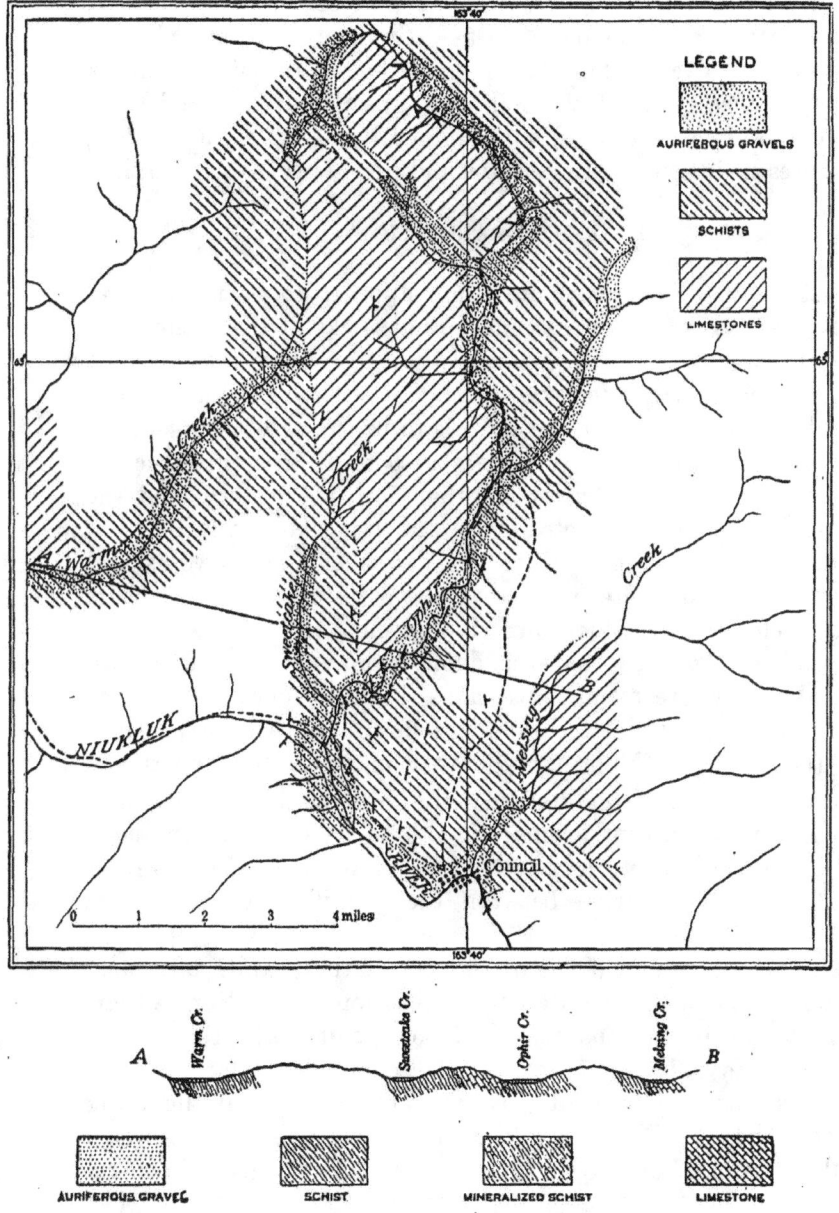

FIG. 6.—Geologic sketch map and section of Ophir Creek region.

All the deposits referred to in the foregoing discussion occur in the Nome group, a rather heterogeneous succession of limestones and calcareous, graphitic, and quartz schists, together with many greenstones. Most of the placer deposits occur in the rocks of this group and hence it would appear to afford the most likely field for ore bodies. It must be remembered, however, that the areas in which the Kigluaik group is the country rock had a physiographic history unfavorable to the formation of placers. The absence of placers, therefore, may not be taken as definite proof of the absence of lodes. As a matter of fact, large quartz veins are not uncommon in the upper formation of the Kigluaik group, but most of these have been found to be only slightly mineralized. A little gold is, however, present in some of these veins and they deserve examination.

AGE OF MINERALIZATION.

The last epoch of extensive intrusion appears to be that represented by the massive granites occurring as stocks or dikes throughout the peninsula. There is apparently some evidence that the last intrusion of the quartz veins was coincident with the granite injection. Collier and Hess have shown this to be the case with some of the cassiterite-bearing veins so closely associated with granitic intrusives. In any event it is certain that both the granite and the quartz veins have been injected since the last extensive epoch of deformation. The fact that there are deformed quartz veins and granites does not militate against this conclusion, for it appears that these belong to an older intrusive epoch. In a former report the writer has suggested that the dome structure may have been brought about by greenstone intrusions, but this does not appear to be borne out by the later investigations. In the northeastern part of the peninsula Moffit[a] found that many of the domes had a granitic core, and he suggested that the dome structure may have been caused by granite intrusions. If this is the case and the margins of the domes prove to be the loci of ore deposits, there is probably a genetic relation between the granite and the quartz veins.

There is little direct evidence of the age of the granite intrusion, but it appears to have antedated the deposition of the lignite-bearing beds found in various parts of the peninsula, probably of Eocene,[b] but possibly of Upper Cretaceous age. Spurr[c] found granitic rocks cutting beds of Lower Cretaceous age in the lower Kuskokwim basin. These are the nearest granites to Seward Peninsula of whose age anything is known. As no granitic intru-

---

[a] Moffit, F. H., The Fairhaven gold placers, Seward Peninsula, Alaska : Bull. U. S. Geol. Survey No. 247, 1905, p. 23.

[b] Brooks, A. H., Geography and geology of Alaska : Prof. Paper U. S. Geol. Survey No. 45, 1906, p. 242.

[c] Spurr, J. E., A reconnaissance in southwestern Alaska in 1898 : Twentieth Ann. Rept. U. S. Geol. Survey, pt. 7, 1900, p. 159.

sives have been found in any of the younger terranes of Alaska it seems probable that those of the peninsula were intruded in early Cretaceous time.[a]

SEPARATION OF GOLD FROM BED ROCK.

The separation of gold from bed rock is brought about by processes of erosion, of which secular decay is the most effective, but corrasion also plays a part. Though the long accumulation of weathered bed rock in place is the most favorable condition for the formation of placers, yet it does not follow that rich alluvial deposits may not be formed by rapid erosion of bed rock when the detritus is removed as fast as it accumulates. The ordinary processes of erosion, by which through the action of rain, frost, ice, vegetable life, and other agencies, hard rocks are converted into loose material, are too well known to require description. Of greater importance to this discussion is the rock decay that is not accompanied by any considerable mechanical removal of material.

Hard rocks by disintegration and decay are transformed into loose material. During these processes the soluble rock constituents are removed through the agency of water, and there remains a residual mass made up of relatively insoluble matter. The breaking down of the rock and the accompanying chemical changes of the constituent materials set free the gold, one of the relatively indestructible minerals, and this becomes intermingled with the other insoluble material. Clay dominates in the residual mass, but if the parent rock contained quartz, this, too, usually remains, being probably the most refractory of all the common minerals toward purely chemical agencies. Mineralized vein quartz very commonly carries easily decomposed minerals, such as pyrites, and is therefore readily broken up, allowing the insoluble ingredients of the ore body, such as gold, to be set free. This process is hastened by purely physical agencies, such as frost and changes of temperature, which break up the insoluble rock constituents.

Russell,[b] in describing residual rocks of the southern Appalachian region, says:

The quartz veins so common throughout the crystalline areas of Virginia, the Carolinas, etc., are more durable than the rocks inclosing them, and retain their integrity long after the associated schists have changed to plastic clays. Bowlders and pebbles derived from the breaking up of quartz veins cover the surface with a continuous sheet of débris over large areas, forming a residual deposit,

---

[a] The age of the mineralization in southeastern Alaska and the Yukon basin has been briefly discussed by the writer in Some recent publications on Alaska and adjacent regions: Economic Geology, vol. 1, 1906, pp. 342–355.

[b] Russell, I. C., Subaerial decay of rocks and origin of the red color of certain formations: Bull. U. S. Geol. Survey No. 52, 1889, p. 14.

the origin of which is revealed at a glance. The quartz itself, although opposing great resistance to disintegration, is not indestructible, as is shown by the brown, iron-stained lines that traverse it in every direction, marking the course of incipient fractures. The continual breaking of the quartz into smaller and smaller fragments produces an angular sand which everywhere forms an ingredient of the red-clay soil.

Hayes[a] has shown that under certain conditions even quartz is readily soluble by surface waters. It is evident, therefore, that the gold included in quartz veins, will, in course of the process of weathering, eventually be set free.

As a rule, the changes in a rock mass brought about by weathering result in a very material reduction in its bulk.[b] The loss of material by weathering among siliceous crystalline rocks, according to Merrill,[c] amounts to more than 50 per cent, and in the purer forms of limestone it may reach as high as 99 per cent. Pumpelly[d] has estimated that in the limestone areas of the Ozark Mountains the residual material represents only from 2 to 9 per cent of the original rock mass. Such reductions in volume necessarily result in more or less concentration of any insoluble material that may have been disseminated in the parent rock. This concentration will be materially greater in the case of substances of high specific gravity, such as gold, than in that of the lighter minerals, for the former will have a constant tendency to settle to the bottom of the loose material. On declivities gravity will accelerate the process and help to sort the material, producing in some places a rough stratification.[e] This is a secular process and will proceed as long as the rocks continue to disintegrate. Placers may be formed in this manner without mechanical transportation. Becker[f] has described such types of placers in the southern Appalachian region, where they at one time formed an important source of gold. To the same class probably belongs also the extraordinarily rich placer known as the Caribou Bill claim near the divide between Anvil and Dexter creeks on Seward Peninsula. (See p. 200.)

This process is here set forth in some detail because it is believed that the first concentration of gold after being set free from bed rock, by the contraction of the residual mass due to disintegration

[a] Hayes, C. W., Solution of silica under atmospheric conditions: Bull. Geol. Soc. America, vol. 8, 1897, pp. 213–220.

[b] Merrill has shown that in certain changes by hydration there is an increase in bulk. He estimated that in the conversion of granite into soil (District of Columbia) there had been an increase in volume amounting to 88 per cent. Compare Merrill, G. P., Principles of rock weathering: Jour. Geol., vol. 4, 1896, p. 718.

[c] Merrill, G. P., Rocks, rock weathering, and soils, New York, 1897, p. 234.

[d] Pumpelly, Raphael, The relation of secular rock disintegration to loess, glacial drift, and rock basins: Am. Jour. Sci., 3d ser., vol. 17, 1879, p. 136.

[e] Kerr, W. C., The gold gravels of North Carolina, their structure and origin: Trans. Am. Inst. Min. Eng., vol. 8, 1879–80, pp. 461–462.

[f] Becker, G. F., Gold fields of the southern Appalachians: Sixteenth Ann. Rept. U. S. Geol. Survey, pt. 3, 1895, p. 289.

and decay, has been an important factor in the formation of rich placers. It is not intended to imply that these residual placers are themselves of great value, except in certain places, but rather that they represent an intermediate stage between the gold in bed rock and the sorted placers.

If it is granted that the deep decay of bed rock is a favorable preliminary stage to the formation of placers, it is pertinent to consider the conditions under which such decay takes place. The agencies that bring about weathering are primarily solution, changes of temperature, beating of rains, gravity, and vegetation. All these agencies have played their part in the reduction of the rocks of Seward Peninsula. Though it is probably true that rock weathering is more rapid in warm than in cold countries, yet the sweeping assertions that have been made in regard to the absence of residual material in high latitudes have not been borne out by the writer's observations. Many of the placer-mining operations in this region have disclosed a considerable thickness of residual material, but the data are as yet too incomplete for a detailed discussion. The great irregularity in the distribution of the permanent ground frost—in some places reaching to depths of 100 feet or more, in others entirely lacking—would appear to favor the breaking down of the hard rocks. Throughout Seward Peninsula, except in the Kigluaik Mountains, which have been glaciated, the lower hill slopes are buried in a thick mantle of talus and residual material, and bed rock is usually exposed only at the crest lines. In the gulches and stream valleys residual clays 8 to 15 feet thick are not uncommon. All of this goes to prove that decay is not confined to the region of warmer climate and excessive rainfall.

It is evident that the effectiveness of all these agencies is proportional to the length of time in which they are operative. A land mass must remain stable, relative to sea level, for a long period of time to permit the accumulation of any considerable amount of residual material. Uplifts bring about renewed activities of the watercourses, and the residual mantle is quickly removed by erosion. It is evident that the conditions that are most favorable to the accumulation of residual material are those in which the land mass is at or near base-level, when erosion is reduced to a minimum.

Attention has been called to the topographic evidence of a long epoch of erosion, which planated much of the peninsula. If the facts are interpreted correctly, a peneplain was formed over nearly the entire peninsula since Eocene time. This indicates a long period of time during which the conditions would be favorable for disintegration and decay. The thesis that the writer would advance, then, is that the peneplanation of Seward Peninsula, previous to the present geographic cycle, has been an important factor in the concentration of the placer gold.

TRANSPORTATION AND DEPOSITION OF PLACER GOLD.

The third important agency in the formation of placers is transportation, sorting, and deposition by water. Material set free from the bed rock has a constant tendency to move toward base-level. This movement is brought about by gravity aided by the transporting power of water. The transporting power of a stream is dependent on its velocity, which is a variant determined by the gradient, volume, and load. When a stream is overloaded with sediment, the excess is dropped. When it is underloaded, it erodes. When equilibrium has been established, neither erosion nor deposition takes place. Gradient, volume, and load usually vary in the same stream, so that deposition may be going on in one part of its valley and erosion in another. When a stream is eroding, the material within reach of its activity is constantly moved in a downstream direction. All movements of this kind are accompanied by more or less sorting and make for the concentration of the heavier particles.

Deposition takes place in a stream when the velocity is decreased, either by the periodic changes in volume or by a change of gradient. Where there is a change of grade, resulting in diminished velocity, the gold is laid down with the other sediments. It must be remembered, however, that placer gold may find lodgment in inequalities of the bed-rock surface where no considerable deposition of detrital matter has taken place, though extensive placers are, as a rule, not formed because of irregularities in the bed-rock surface alone. The concentration of gold in river bars in analogous to its deposition in stream beds, for it is dropped where the velocity of the current is checked by the formation of eddies, due to the inequalities of the river floor.

CONCENTRATION OF GOLD ON BED ROCK.

The occurrence of the richest placers on or near bed rock is a normal condition wherever auriferous alluvium has been mined. Various theories have been advanced for this concentration on bed rock, but they appear to be inadequate to explain many occurrences. It is self-evident that a placer with an overburden not exceeding the depth to which alluvium is disturbed and sorted during one period of deposition would have its heavy ingredients concentrated at the bottom merely by the action of gravity. Gravity alone would appear insufficient to account for the concentration of gold in gravels having a depth of 20 to 200 feet, for such gravels were evidently not laid down at one time. The highest strata must have been deposited without disturbing the lower strata. If the gold were deposited at the bottom solely by the action of gravity during the process of deposition there should be a pay streak for each period of deposition. It is exceptional, however, to find more than one pay streak.

Posepney [a] has argued that the particles of gold were concentrated on bed rock by working their way down through the unconsolidated material below solely through the action of gravity and after the gravels had been deposited. Smyth [b] has, however, shown that such action would be inversely proportional to the size of the particles, and hence the finest gold should be at the bottom of the deposits. As a matter of fact the contrary is true. Moreover, many of the Alaskan placers are frozen and have probably been so for the most part since their original deposition.

Spencer [c] has described the placers on Gold Creek, near Juneau, showing that here deposition began on bed rock where there was a change in the gradient of the stream and that this action continued progressively toward the headwaters, and this appears to be a general law. In such a manner a series of placers would be formed, lying on or close to bed rock. It would seem, however, that this can not be advanced as a general law governing the formation of placers, unless the supply of gold furnished by erosion ceased when all the bed rock had been covered. If the gold from the parent rock had been concentrated by secular decay, leaving it intermingled with clay and other light substances, not only would the conditions then be favorable to the formation of placers, but such placers would carry more gold than later ones unless a period of quiescence with accompanying decay intervened.

The gold set free by secular decay is intermingled chiefly with clay, and this would be quickly removed by erosion if uplift renewed the cutting power of the stream. By this means gold already concentrated by weathering would be deposited as the bottom layer of the sediments laid down during the new geographic cycle. The erosion of bed rock after the mantle of weathered material had been removed might continue to supply gold, yet this could be concentrated only by the ordinary processes of sorting and would be more or less disseminated in the overburden covering the first placers formed. It seems probable, too, that more rapid erosion and the abrasion accompanying the destruction of the hard rock during the new geographic cycle may account for the absence of any large percentage of coarse gold in the later alluvial deposits.

In the southern Appalachians there appears to be a field where the first condition for the formation of placers has been met. During

[a] Posepney, F., Zur Genesis der Metallseifen: Oesterreichische Zeitschr. f. Berg- und Hüttenwesen, vol. 35, 1887, p. 327.

[b] Smyth, H. L., The origin and classification of placers: Eng. and Min. Jour., vol. 79, 1906, p. 1228.

[c] Spencer, A. C., The Juneau gold belt: Bull. U. S. Geol. Survey No. 287, 1907, pp. 77–85.

two cycles of planation [a] deep rock decay has taken place, and hence residual placers have been formed. The absence of any considerable uplift, however, has prevented the dissection of these residual deposits, and as a result there has been no sorting. The gold, therefore, is found on bed rock only where gravity has aided the concentration, namely, on hill slopes.[b]

Deep secular decay, a concomitant of base-level conditions, has, in the opinion of the writer, in most of the important placer fields of the world, preceded the period of rapid deposition to which the formation of placers is usually assigned.

The general cycle of events favorable to the formation is, then, base-leveling, followed by rapid uplift. The simple conditions above set forth are probably rarely fulfilled. In Seward Peninsula a general base-leveling antedated the formation of the placers, yet since that planation there have been many uplifts and depressions, probably separated by intervals of stability long enough to permit considerable secular decay.

RECONCENTRATION OF PLACER GOLD.

In the foregoing pages the succession of events in the formation of placers has been traced from the time when the gold is set free from bed rock until it finds a resting place in the auriferous alluvium. What has been described is the history of events that take place during an uninterrupted geographic cycle. If, however, after the placer has been formed, an uplift which renews the erosive power of the streams occurs, a re-sorting of the auriferous material may be effected and new placers formed that may be much richer than the primary deposits. Such a secondary concentration has been very marked in some of the Seward Peninsula placers. The processes of erosion and deposition are those already described and need not be considered again.

Among the important re-sorted types are the beach placers, formed by the action of the surf on the uplifted stream gravels of the coastal plain. This process of wave concentration has yielded some extraordinarily rich deposits. (See description of Nome beach placers, p. 151, and of Topkok beach placers, p. 289.) A less evident but probably equally important type of reconcentrated placers comprises those enriched by the dissection of the high gravels in the Anvil Creek region (p. 186). Reconcentrated placers are also found in the beds of the streams which traverse the costal plain near Nome.

---

[a] Hayes, C. W., and Campbell, M. R., Geomorphology of the southern Appalachians: Nat. Geog. Mag., vol. 6, 1894, pp. 63–126.

[b] Kerr, W. C., The gold gravels of North Carolina, their structure and origin: Trans. Am. Inst. Min. Eng., vol. 8, 1880, pp. 462–468.

In some places the placer gold has probably existed as such through several geographic cycles and has been re-sorted a number of times since it was freed from bed rock. The processes of reconcentration and the various types of placers thus formed have been considered at some length by the writer[a] and need not here be further discussed.

HINTS TO PROSPECTORS.

INTRODUCTION.

The foregoing pages are intended to lead to a better understanding of the detailed descriptions that follow and also to elucidate some of the general laws that control the occurrence of gold in this province. An attempt will here be made to present briefly the application of these theories to the placer-mining industry. Had the investigation gone far enough to formulate definite laws on the source and distribution of the alluvial gold, it would, no doubt, have great practical value, but even the incomplete data and more or less theoretical conclusions will, it is hoped, be useful to the prospector. Prospectors and miners are sometimes prone to regard the results of the geologist's study as of purely scientific interest, but as the application of geology to mining becomes better known this fallacy is gradually being dispelled. It is hoped that the matter here presented, though intended primarily for the inexperienced prospector, may be not without interest to those who are unfamiliar with the conditions prevailing in Seward Peninsula.

A placer is an unconsolidated deposit accumulated by mechanical processes, carrying one or more minerals in commercial quantities. All placers are secondary deposits—that is, the material of which they are composed was originally derived by erosion of bed rock. Although it is undoubtedly true that under certain conditions nuggets of placer gold have been enlarged through chemical precipitation, yet this action is a negligible quantity in placers.[b] Placers may be derived solely by rock weathering without water sorting, but more commonly are the result of water transportation, sorting, and deposition. Some of the richest placers are those formed by the erosion of older placers and the reconcentration of their gold.

A theory not uncommon among miners that the placer gold reached its present position by glacial action is not sustained by any facts known to the writer. The forces of erosion now in operation are ample to bring about the formation of placers, and it is not neces-

---

[a] Reconnaissances in the Cape Nome and Norton Bay regions, Alaska, in 1900, a special publication of the U. S. Geol. Survey, 1901, pp. 144–151.

[b] A. Liversidge (On the crystalline structure of gold and platinum nuggets and gold ingots: Jour. Roy. Soc. New South Wales, 1897, pp. 70–79) has shown that gold nuggets studied by him have no internal structure indicating chemical growth.

sary to have recourse to the various theories of glacialism, volcanism, etc., current among prospectors. It should be noted, however, that erosion may have been accelerated or retarded in the past by the uplift or depression of the land. Evidences of uplift are common in Seward Peninsula, the most striking being the elevated beach placers near Nome and the high-bench gravels near the head of Anvil Creek. Elevation may bring about the destruction of former placers and the concentration of their gold in new placers.

GOLD IN BED ROCK.

As has been stated, the placer gold has its source in the bed rock and in this field none except some of the finest has been transported far from its source. The facts previously presented indicate that the lines of contact of the limestones and schists are most favorable for the occurrence of gold in bed rock. This is especially true if the contact gives evidence of having been a line of movement. An exceedingly broken up and schistose condition of the rock is an indication of such movements, which are often accompanied by the development of micaceous minerals, giving the rock the appearance of talc or soapstone. If such contacts are found they should be carefully examined for evidence of mineralization, which may result in the formation of well-defined quartz or calcite veins, or may be simply an impregnation of the bed rock by mineral-bearing solutions. If mineralization occurs it is usually accompanied by the deposition of some of the pyrite minerals and then decomposition gives a characteristic iron or copper stain.

The character of the ores has been described, but it should here be noted that the gold of many of the mineralized zones appears to be included in arsenical and iron pyrite. In this type of deposit freemilling ores would be limited to a shallow surface zone. In another type, to which the Big Hurrah quartz mine belongs, the gold occurs free in a quartz or calcite gangue.

The more detailed knowledge gained since the former report was prepared makes it probable that the gold was not so widely disseminated in the bed rock as was at first supposed. On the whole, therefore, the cumulative evidence rather favors the discovery of lode deposits. It is with some hesitancy that the writer makes this statement, for it may be interpreted as an unqualified indorsement of the lode-mining industry of the peninsula, which is far from being his purpose. The sole object of this statement is to draw attention to this region as a possible field for quartz mining.

It can not be too strongly urged that every quartz claim be thoroughly prospected before any elaborate equipment is purchased. Here, as in other mining districts, legitimate ventures have been brought into disrepute by the ill-advised, if not downright swindling, mining schemes that have been foisted on the public. Nonresidents should be particularly careful in investing in mining enterprises concerning the management of which they have no personal knowledge. As a general rule, it is safe to presume that a man who has to find his capital by alluring advertisements in communities where he is personally unknown has something to dispose of that is unmarketable among those who are most familiar with it.

Though the question of costs in lode mining has not yet been carefully studied in Seward Peninsula, they are higher than in many more accessible regions. It is perhaps safe to say that under the present conditions (1905) no lode deposit not yielding at least $10 to the ton by carefully taken commercial samples is worthy of further investigation. The cost of mining is, however, being gradually reduced. When the placers are worked out the water ditches can be turned into power for lode mining, and by that time, too, there will be ample transportation facilities.

GOLD IN PLACERS.

The search for alluvial gold should be guided by the fact that it is usually deposited where the current of a stream has been checked. For example, a broad basin above a steep-walled canyon is more likely to carry gold than the valley below the canyon, provided the bedrock source of the gold is above the basin. Moreover, coarse gold is more likely to be found at the head of a filled basin than near its outlet. The same holds true of a stream that debouches on a coastal plain, which will deposit the coarse gold it may carry near the head of its delta. Some special conditions of deposition are described by Mr. Collier and Mr. Hess (p. 198) in the high-bench gravels between Anvil and Dexter creeks. Here the gold deposited by a former drainage system has been apparently laid down in steep-walled valleys and stream deposition has alternated with the accumulation of talus material (slide rock). This broken rock, making its way down the valley slope in huge slabs, has settled together, presenting the semblance of bed rock and furnishing a floor upon which the gold has been deposited. The piercing of these old buried talus slopes and the finding of gravels underneath has led some to believe that gravels occur under solid bed rock. A consideration of the origin of these deposits shows the utter fallacy of this conclusion.

High-bench gravels have been found in many parts of the peninsula, but only near Nome have they become important gold producers. The high gravels of Anvil and Dexter creeks are of such an extent and thickness as to indicate that there was a well-developed drainage system, which, having been disturbed by uplift, has lost a part of its accumulated deposits by erosion. There is good reason to believe that these ancient gravels may be gold bearing in other placer-bearing parts of the peninsula, and they are worthy of careful prospecting.

In seeking high-bench deposits the prospector should be guided by the topographic evidence, but should remember that every flat-topped shoulder of a mountain is not necessarily underlain by gravels. The best evidence is the presence of well-rounded material. Distinction should be made between placers occurring on hill slopes, the accumulation of minor streams belonging to the present drainage system, and those belonging to an old drainage system. Though the first type may be very rich it is less likely to include extensive deposits than the second.

Mr. Collier describes elsewhere (p. 189) the influence of talus slopes in crowding streams to one side of their valleys, and thus influencing the position of the placers. This has been a very important factor in the formation of the placers on Anvil and Ophir creeks (pp. 189–242).

The reconcentration of gold by stream and wave action is considered at length elsewhere (p. 130). It is shown that the dissection of placers and the re-sorting of the gold have led to the formation of many of the richest diggings. These theories can be made of direct commercial value by seeking the localities where such concentration has taken place. These localities may be old or present beach lines, or streams that traverse the coastal plain or that have dissected high-bench gravels. It is important that the prospector bear in mind the facts in regard to reconcentration or he may be misled in regard to values. Thus the gold in the stream traversing the tundra back of Nome has been concentrated from the gravels underlying the tundra and is not necessarily a measure of the gold tenor of the parent placer.

GEOLOGIC MAPS.

All the known facts regarding the geology of Seward Peninsula have been brought together in graphic form on the geologic maps (Pls. X and XI, in pocket). The prospector will naturally turn first with interest to the areas marked as carrying placer gold. These areas are outlined from the best information available, but final determination of the distribution of gold must await more detailed surveys and the accumulation of more data by mining operators. The facts here presented may, however, serve as a general guide. Those using the maps should bear in mind, first, that not all these auriferous areas carry commercial placers and, second, that placers may be found in areas not indicated as being auriferous.

The bed-rock geology indicated on the map has the same value, and the prospector in course of his detailed examination will no doubt find that some of the boundaries between the formations are incorrectly drawn. Nevertheless the general facts of the distribution of the limestone and schist contacts are probably correct and may be of aid in locating mineralized zones. The distribution of the granite masses is of practical importance, as it is along their contacts that lode tin has been found.

THE FUTURE OF THE MINING INDUSTRY.

Those who are devoting their labor or capital to the development of Seward Peninsula have a right to expect in this report some statement of the probable life of the placer-mining industry. Were the entire gold-producing area surveyed and studied in detail it would be difficult enough to form an estimate of the gold contents of the gravels; but with only the present general knowledge of the factors necessary to such an estimate the task may seem well-nigh hopeless. For this reason undue weight should not be given to the figures here presented, and the reader should carefully consider the character of the evidence and make his own deductions as to the reliability of the conclusions. No doubt any presentation of the facts or theories bearing on this subject will be distorted and misquoted by promoters of worthless mining enterprises for their own use, but this can not be prevented, however guarded a geologist may be in presenting his conclusions.

It is self-evident that if the areal distribution of the placer gold indicated on the map were correct and the approximate thickness of the alluvium were known, it would be a mere matter of multiplication to determine the cubical contents of the auriferous gravels. Unfortunately, however, these factors of the equation can be only approximated. The average gold content of the gravels, forming the third element of the problem, is also an unknown quantity. The following estimates are based on certain assumptions and on the best information available, but are not now susceptible of proof. Mr. Moffit has carefully measured the areas marked on the maps (Pls. X and XI; also Bull. U. S. Geol. Survey No. 247, 1905, Pl. III) as underlain by gold-bearing gravels. Allowing for the exaggeration of the areas because of the small scale of the maps, he finds the total area of auriferous alluvium to be about 210 square miles, or 650,496,000 square yards. The pay streak is assumed to average 3 feet in depth, which is believed to be a conservative estimate. By this calculation Seward Peninsula is estimated to carry a total of 650,496,000 cubic yards of gold-bearing gravels.

Although it is true that the pay streaks thus far mined have a very high gold content, it is equally certain that this gold tenor is not maintained throughout the auriferous areas, for the gravels mapped as auriferous include considerable alluvium that probably carries little more than colors of gold. A valuation of 50 cents to the cubic yard is therefore probably not too conservative. If these assumptions are correct the gold contents of the auriferous gravels of the penin-

sula can be valued at $325,248,000. It should be borne in mind that
it does not follow that these values, if present, will necessarily be
recovered, for they may be in part so disseminated or so difficult of
access as not to permit profitable exploitation.

Another estimate by Mr. Moffit is based in part on the number of
linear miles of gold-bearing streams. The total as measured on the
maps is 750 miles. If the average width of pay streak is assumed to
be 15 yards and the depth 3 feet, the total bulk of the creek gravels
would be about 20,000,000 cubic yards, of which probably 600,000
cubic yards have been worked out. This leaves a total for unworked
creek gravels of 19,400,000 cubic yards. This computation takes into
account the well-defined pay streak and not the areas of auriferous
gravels, and hence the gold tenor will be much higher than that used
in the previous estimate. Mr. Moffit found the average gold contents
of the gravels of seventeen streams, on data given in this report by
Mr. Collier and Mr. Hess and based on statements by owners and
operators, to be $5.93 per cubic yard. He obtained a gold content of
$5.98 per yard by dividing the total gold production from the creeks
by the number of cubic yards of pay streak estimated to have been
mined. The practical identity of gold tenor of pay streaks in these
two calculations is suggestive, though it may be purely fortuitous.
In view of the fact that some of the richest creek gravels have been
mined out, it is not safe to accept the above gold tenor as representative of all the creeks of the district. It is probable, however, that a
3-foot pay streak may average $2.50 a cubic yard along the water-
courses. With this assumption the gold contents of the creek gravels
would in round numbers amount to $50,000,000.

In addition to the stream placers above considered, there are also
the auriferous deposits in the gravel plains, of which that of the Nome
tundra is the most extensive, and in the high-bench deposits. The
gravel plain stretching inland from the coast at Nome, usually called
the tundra, is the richest of this class of deposits thus far prospected,
but similar deposits, some of which are known to be auriferous, occur
in other parts of the peninsula. It will remain for the future to determine what percentage of this type of placers can be profitably
exploited, but in the opinion of the writer it constitutes the largest
gold reserve of the peninsula. In estimating the gold contents of
the gravel-plain placers values of 25 to 50 cents per cubic yard of
pay streak have been adopted as being conservative. As in previous
calculations the pay streak has been assumed to be 3 feet thick. The
pay streaks of the coastal-plain gravels, known to contain considerable
gold, have been assumed to carry 50 cents and the others 25 cents to
the cubic yard. Many persons will doubtless take exception to these

valuations per yard, because they are familiar with some of the extraordinarily rich deposits which have been mined. These higher values are compensated for by the fact that much gravel is probably included in the above measurements of the gold-bearing areas whose tenor is far below 50 cents. The placers of the high-bench gravels, so far found only near Anvil Creek, are almost an unknown quantity and their gold contents have been put in as a lump sum in the following estimate. By these estimates the total values of the gravel-plain (tundra) and of the high-bench placers are found to be $215,000,000. This sum added to the value of creek placers gives a total reserve of $265,000,000, as compared with $325,000,000 in the previous calculation. The difference lies in the fact that in the second calculation the stream gravels lying outside of the 15-yard pay streak are not considered as gold-bearing.

Of the 750 linear miles of gold-bearing creeks, only 172 miles represent creeks which have produced gold in commercial quantities, but it must be remembered that many have not been carefully prospected and that there are probably other creeks not so marked on the accompanying map which will be found to carry gold. Mr. Moffit's computations indicate that those parts of the creeks which have been worked out carried values averaging probably $500,000 to the mile. Some of the richest creeks have yielded more than double this amount, but the yield of others is very much below it. These values are low compared with those in the Klondike, where, according to an estimate by McConnell,[a] 50 miles of paying portions of creek carry gold aggregating $95,000,000, or $1,900,000 to the mile. According to Hammond,[b] some of the ancient river channels of California have yielded $2,000,000 to $3,000,000 per mile. The deep leads in Victoria, Australia, have yielded nearly $2,000,000 worth of gold to the mile.[c]

Optimistic mine operators will undoubtedly regard the totals here presented as entirely too low, whereas those who have lost money in ill-advised mining ventures in Seward Peninsula must naturally evince skepticism toward the estimate of so large a gold reserve. On the one hand it is evident that it would not take many creeks as rich as Anvil, Dexter, and Ophir, to change this estimate entirely, but on the other hand the estimate is based on the assumption that gravels of much lower grade will be mined in the future than at present. The total is small compared with the output of the Cali-

[a] McConnell, R. G., Preliminary report on the Klondike gold fields: Geol. Survey of Canada, Ottawa, 1900, p. 18.

[b] Hammond, J. H., The auriferous gravels of California; geology of their occurrence and methods of their exploitation: Ninth Ann. Rept. California State Mineralogist, year ending December, 1889, 1890, p. 114.

[c] Lindgren, Waldemar, The deep leads of Victoria: Eng. and Min. Jour., vol. 79, 1905, p. 315.

fornia placers since 1849, which amounts to about $1,142,000,000,[a] and much larger than the estimated original total gold content of the Klondike district, placed by McConnell[b] at $200,000,000. In the Klondike district the number of miles of creeks bearing placer gold is estimated to be but 50, whereas there are 750 on Seward Peninsula. Moreover, all parts of the Klondike are much more accessible than much of Seward Peninsula, and consequently have been more thoroughly prospected. For this reason it is not to be anticipated that many important discoveries will be made in future in the Klondike, but the results of each year's prospecting in Seward Peninsula prove that the limit of new fields there has not been reached. The cost of mining in Seward Peninsula is as a rule less than in the Klondike, making it possible to exploit gravels carrying lower values.

All the above facts being taken into consideration, it is believed that an estimate of $250,000,000 to $325,000,000 for the placer-gold reserves of Seward Peninsula is conservative. Although these speculations may be of interest in showing the possibilities and probable direction of future growth, the actual estimate of the gold reserve must be regarded as little more than a bold guess.

The future of lode mining is impossible to predict. Some facts have been presented indicating the favorable conditions for the occurrence of ore bodies, but until more of these bodies have actually been discovered and found to carry commercial values no definite statements should be made.

The data bearing on the placer-gold reserves having been discussed, it becomes pertinent to inquire at what annual rate these reserves may be mined out. Here again the dearth of exact information makes it impossible to arrive at definite conclusions, but those here presented may be at least suggestive. Even were the amount of water available for placer mining known, there would still remain a large indeterminable factor, for the mechanical methods of mining are relatively little developed in this district and admit of much expansion. Moreover, with the gradual decrease in cost of transportation and the concomitant reduction in cost of fuel, machinery driven by petroleum or coal-consuming engines will undoubtedly be more commonly introduced to exploit gravels that lie inconveniently for economic handling by the use of water under head. The dredges now extensively used in the Solomon River region (Pl. IV, p. 232) are examples of the development of this method of mining.

---

[a] Based on figures furnished by Waldemar Lindgren. Compare also Lindgren, Waldemar, Geological features of the gold production of North America: Trans. Am. Inst. Min. Eng., vol. 33, 1903, pp. 816–818.

[b] McConnell, R. G., Report on the Klondike gold fields: Ann. Rept. Geol. Survey of Canada, 1905, pt. B, p. 61B.

It is probable that not more than 60,000 or 80,000 miner's inches of water may eventually be made available for hydraulic mining in Seward Peninsula, but this estimate is again little more than a guess. Purington [a] has shown that because of the low stream gradients and for other reasons the duty of an inch is very low in Seward Peninsula. The average of his table is 2.76 cubic yards per twenty-four hours, equal to about 300 for a season. According to these figures, 16,000,000 to 20,000,000 cubic yards of material could be moved in a season, but certainly not more than 5,000,000 to 8,000,000 cubic yards would be pay streak. If double this amount of gravel were excavated by mechanical means, the total maximum volume of pay streak which could be exploited in a single season would be 15,000,000 to 24,000,000 cubic yards. If there are 650,000,000 cubic yards of gold-bearing gravels in the peninsula they could not be mined at this rate in less than twenty-five to forty years. As a matter of fact, however, this maximum volume could be reached only after years of preparation and may never be attained. This is emphasized by the fact that during the last six years the average amount of pay gravels handled has probably been less than 1,000,000 cubic yards annually.

These speculations as to the life of the placers of Seward Peninsula are too indefinite to have much value, but they suggest, at least, that the gold production will probably increase rather slowly and that the outlook is favorable for a long period of rather uniform output.

---

[a] Purington, C. W., Methods and costs of placer mining in Alaska : Bull. U. S. Geol. Survey No. 263, 1905, pp. 138–141.

DESCRIPTION OF PLACERS.

By Arthur J. Collier and Frank L. Hess.

INTRODUCTION

It has been shown that the value of the annual gold production of Seward Peninsula is now (1906) about $7,500,000 and that the total, including 1906, is nearly $40,000,000. This has practically all been derived from the placers, though a small amount has been obtained from auriferous quartz veins. All of the placer gold carries some silver, the present annual production of which is valued at about $25,000. Of other mineral wealth tin is the only metal of immediate interest. Some placer tin has been exported, and the development of both lode and placer deposits is being actively pushed, but up to the present the tin deposits can not be said to have reached a productive stage. There are a few small areas[a] of coal-bearing rocks on the peninsula, but in only one of these has a workable seam been found, and this appears to be of value only for the very local demand.[b] Besides the above minerals, graphite and bismuth deserve mention, though their commercial value in this field remains to be proved.

A general account of the placers will next be presented, followed by a detailed description of the gold deposits of the Nome, Solomon, Council, Port Clarence, and Goodhope districts. In later sections of the report an account of the Bluff and Kougarok regions will be presented by Mr. Brooks, and of Iron Creek by Mr. Smith.

DISTRIBUTION OF GOLD.

On the geologic maps (Pls. X and XI, in pocket) is indicated the distribution of the unconsolidated silts, sands, and gravels, under the general group name Quaternary. These deposits have already been discussed at some length in the section devoted to the geology of the peninsula (pp. 83–99). As these Quaternary beds carry the

---

[a] See pp. 83–85 of this report; also Mendenhall, W. C., Reconnaissances in the Cape Nome and Norton Bay regions, Alaska, in 1900, a special publication of the U. S. Geol. Survey, 1901, p. 214.

[b] Moffit, F. H., The Fairhaven gold placers, Seward Peninsula, Alaska: Bull. U. S. Geol. Survey No. 247, 1905, p. 67.

alluvial gold their distribution is of first importance. It would require far more detailed mapping than has yet been done to determine, with any exactness, the distribution of the auriferous gravels. On the map these facts are indicated so far as the data at hand will allow. The auriferous alluvium is divided into two classes—first, that from which gold has been recovered in commercial quantities, and, second, that which is known to be auriferous, but in which it has not been proved that workable placers exist.

It is evident that this division is arbitrary. From a scientific standpoint all the alluvium of the peninsula is probably auriferous. If the tests were made with sufficient minuteness, traces of gold would probably be found in practically all the unconsolidated deposits. If this were a province of cheap mining, like California, the auriferous gravels of commercial importance would cover much larger areas than those indicated on the maps. Moreover, gravel with a low gold tenor, which may be absolutely worthless under the present costs of extraction, may in the future prove of value.

These facts are well illustrated by the reductions in mining costs that have taken place during the last six years. In the first years of mining in the peninsula but few pay streaks exceeding 5 feet in depth could be profitably exploited, whereas profitable mining has now been carried to depths exceeding 200 feet. It is probable that at that time gravels whose tenor was less than $10 to the cubic yard could not be worked at a profit, but at present, under certain favorable conditions, the cost of mining has been reduced to less than 50 cents a cubic yard. The boundaries of the gold-bearing areas, therefore, are not only more or less arbitrary, but also only in part determined. Yet it seems justifiable to indicate on the map the areas that give most promise of yielding returns to the placer miner. There are no doubt placers in some of the Quaternary deposits that have not yet been tested, and it is equally true that the areas indicated as gold bearing do not by any means everywhere carry workable placers. It is believed that if these facts are borne in mind the maps may serve as useful guides in directing intelligent prospecting. It is unfortunately true that they will probably also be used by unscrupulous promoters for foisting on the public worthless mining stocks, but this can not be avoided.

It has already been pointed out that the gold is derived from quartz veins and impregnated zones in the metamorphic rocks, which have been grouped together under the name Nome group. The distribution of loci of mineralization has been considered by Mr. Brooks in the preceding pages and need not here be dwelt on. Many facts bearing on the source of the gold are presented in the detailed accounts of the different districts.

CLASSIFICATION OF THE PLACERS.

INTRODUCTION.

All the productive gold placers here to be considered occur in the gravels of alluvial origin and are deposits concentrated from the bed rock by stream or wave action. Elsewhere in this report is given a discussion of the underlying principles which govern the occurrence of the alluvial gold, and with it is discussed a scheme of classification based on genesis. For the purposes of the present description the writers will hold to a topographic grouping of the various types, which is almost identical with the classification of placers recently published by Purington.[a] This group is as follows:

Classification of placers in Seward Peninsula.

Creek placers: Gravel deposits in the beds and intermediate flood plains of small streams.

Bench placers: Gravel deposits in ancient stream channels and flood plains which stand from 50 to several hundred feet above the present streams.

Hillside placers: A group of gravel deposits intermediate between the creek and bench placers. Their bed rock is slightly above the creek bed and the surface topography shows no indication of benching.

River-bar placers: Placers on gravel flats in or adjacent to the beds of large streams.

Gravel-plain placers: Placers found in the gravels of the coastal or other lowland plains.

Sea-beach placers: Placers reconcentrated from the coastal-plain gravels by the waves along the seashore.

Ancient beach placers: Deposits found on the coastal plain along a line of elevated beaches.

These types of placers in occurrence and origin have close affinities with the present topographic forms with which they are associated. Genetically they fall into five groups. One embraces those which are found in the present stream channels, such as the creek, river-bar, and in part the hillside placers. In a second group fall the gravel-plain placers, also of fluvial origin, but laid down by streams that have since shifted their channels. The bench placers form a third group, which includes most of the hillside placers as well, and these were for the most part stream deposits that have been elevated and dissected. The present sea-beach placers constitute the fourth subdivision, and the elevated sea beaches the fifth.

All these land forms are the result of extensive periods of denudation through which the peninsula has passed. This region has probably been exposed to subaerial erosion since late Tertiary time, and in this long interval there have been many upward and downward movements of the land masses relative to sea level. These movements have been irregular and broken by long intervals of stability during which erosion went on, and all parts of the peninsula have not

---

[a] Purington, C. W., Methods and costs of gravel and placer mining in Alaska; Bull. U. S. Geol. Survey No. 263, 1905, p. 27.

been equally subjected to these influences, for the movements have been differential, so that while one part was elevated above sea level other parts were flooded. In other areas the land surface has been warped.[a] This irregularity of uplift may make a coastal plain in one part of the region of identical age with a high bench in another part. The possible economic importance of this fact will be readily recognized.

A logical discussion of the various types of placers demands that the oldest be taken up first, and this class comprises the high-bench deposits which formed before the development of the present topography. Next in order of genesis are the benches on the slopes of the present valleys; then the gravel-plain placers, together with the elevated beaches, and finally the placers in the beds of the existing watercourses as well as those of the present shore line.

BENCH PLACERS.

Bench placers may be classified as low gravel terraces, spur benches, pocket benches, and high benches, according to their topographic relations to the existing streams. The gravel terraces are wide, flat gravel benches whose surface is considerably above the high-water level of the stream but whose bed-rock floor is only slightly, if at all, higher, than the stream bed. Spur benches differ from the terraces in that the stream bed is intrenched in the bed rock below the gravel deposit; and where, as is usually the case, the stream meandered over the valley floor at the old level these meanders have been cut down into the bed rock, leaving the spaces between meanders projecting from the valley walls as flat-topped spurs. The pocket benches are mere remnants of old channels which usually hang high on the valley walls. They can not as a rule be traced in any definite system, for on many streams only one or two small deposits of this kind are discovered. The high-bench placers belong to stream channels of older drainage systems, since the development of which elevation has taken place. The deposits of these ancient watercourses are, as a rule, only in part preserved, having been for the most part removed by erosion. Rich placers of this type have thus far been discovered in only a few localities, but it is to be expected that others will be found.

HILLSIDE PLACERS.

The hillside placers constitute rather an indefinite subdivision of the bench type. They are in general not indicated by the surface

---

[a] The physiographic history of the peninsula has been only in part deciphered. Some account of it will be found in Reconnaissances in the Cape Nome and Norton Bay regions, Alaska, in 1900, a special publication of the U. S. Geol. Survey, 1901, pp. 48–64; Reconnaissance of the northwestern portion of Seward Peninsula, Alaska: Prof. Paper U. S. Geol. Survey No. 2, 1902, pp. 34–43; and The Fairhaven gold placers, Seward Peninsula, Alaska: Bull. U. S. Geol. Survey No. 247, 1905, pp. 42–47.

topography, nor do they rest upon definitely flattened surfaces of bed rock. Many of them may be accounted for by creeping movements of the soil and decomposed bed rock, which have destroyed higher-benches and distributed their auriferous gravels over the hillside below, or by the gradual sidewise shifting and coincident downward cutting of the stream, leaving its gold-bearing gravels behind as it moved. Hillside placers, therefore, are likely to occur on the slopes of any valley which has benches on its walls.

GRAVEL-PLAIN PLACERS.

Most of the gravel-plain placers that have been worked in Seward Peninsula lie in the coastal plain that intervenes between Nome and the foothills, and they are usually referred to as the " tundra placers " or " tundra mines." Similar deposits have been discovered in the gravels of the Kougarok basin. As has been noted, the gravels of the coastal plain were deposited during a period of submergence after the bed-rock floor had been formed and were contributed mainly as deltas by the rivers and streams from the highland portions of the peninsula. As the gravels, except those close to the hills, were laid down mainly by the larger streams, their placers may be expected to be of the nature of river-bar deposits. The gold as a rule is finer than that of the creek and bench placers, and where concentrated it is in irregular pay streaks, due to variations in river and stream currents. The bed-rock floor upon which these gravels rest was probably produced by subaerial erosion, or, in other words, by streams which had their sources in the upland. It is quite possible, therefore, that there may be old stream channels, as rich in coarse gold as the stream and bench placers, beneath the deep covering of gravel and silt. Most of the gold that has been won from the so-called " tundra " has been obtained from the beds of streams which flow across it, but some has come from concentration on layers of clay. The placers of these intrenched streams differ from ordinary creek placers in the fact that their gold has been reconcentrated from the coastal-plain gravels.

CREEK PLACERS.

The creek placers, from which, up to the present time, most of the gold has been taken, constitute the best-known type of deposits. Brooks [a] has described this form of placer as follows:

The pay streak in these deposits is usually on bed rock, though it sometimes is found on a clay which overlies the bed rock. Where no clay is present the gold is found not only on the bed rock, but also where the rock is broken the gold has worked its way down into the joints and crevices. Streams are often

---

[a] Brooks, A. H., Reconnaissances in the Cape Nome and Norton Bay regions, Alaska, in 1900, a special publication of the U. S. Geol. Survey, 1901, p. 146.

found to have a layer of clay on bed rock, which gradually thins out upstream and finally disappears entirely. The presence of the clay on bed rock usually indicates that no gold will be found in the weathered rock below, as the impervious layers prevent the gold from working its way down. On most of the creeks the gravel overlying the pay streak is shallow, and the creek placers usually afford what are known as "open" or "summer" diggings. Very little stripping is necessary, as a rule, after a foot or two of moss and muck have been removed. It is common to find the gravel containing sufficient gold from the surface down to warrant sluicing it all. The horizontal extension of the pay streak differs very much, and no general law in reference to it can be formulated. Sometimes it forms a narrow but uninterrupted layer, running parallel to the side of the valley, but more often the pay streaks are not continuous and are of irregular outline. They very frequently suggest the windings of old stream channels. The coarsest gold of the region is found in the creek and gulch placers, it being there nearest to its source in the parent rock.

RIVER-BAR PLACERS.

The river-bar placers are similar in general origin to the creek placers, the main points of difference being that the former are deposited by larger streams, and that the gold contained in them consists mainly of fine flake or flour gold, which can be transported by running water. The gold of these placers is usually distributed through the gravel, though it may be richer next to the bed rock. River-bar placers occur on many of the larger streams of the peninsula, but they are of low grade as compared with creek and bench deposits, and have not yet been notably productive.

BEACH PLACERS.

The beach placers are confined to the narrow strip of ground along the coast which is affected by the sea waves. As has been noted, the coastal plain is for the most part bounded on its seaward side by an escarpment from 10 to 20 feet high, which marks the inland edge of the beach. The waves are continually encroaching on the coastal plain, cutting back this escarpment and concentrating the gold from its face in the beach sands. Every year the streams probably bring down a small amount of fine gold, which is also caught up by the waves and added to the beach placer. This action has been continuous for the long period of time that has elapsed since any movement of land has taken place relative to the sea. The amount of gravel, therefore, thus affected must have been very great, but as the concentration occurs only in the strip subjected to wave action the resulting placer is confined within the same limits. Some fine gold is also found in the gently sloping floor of the sea, but since this is probably derived from the beach, it is more disseminated and finer than beach gold, and can not at present be regarded as forming a workable placer.

NOME PRECINCT.

INTRODUCTION.

An area of about 2,000 square miles, embracing the extreme south-western part of Seward Peninsula, forms a recording district under the name "Nome precinct." (See fig. 2, p. 42.) Its northern boundary is a sinuous line stretching eastward from Capt Douglas and following the watershed between Port Clarence on the north and Bering Sea on the south. This boundary maintains a generally easterly direction as far as the highlands east of Bluff, throwing all the southward-flowing streams of this part of the peninsula into the Nome precinct. Up to the present time this has been the most productive mining district of the peninsula, and to it must be credited at least four-fifths of the gold production. Its accessibility, lying as it does close to tide water, has led to the more rapid development of its placers than has occurred in the more isolated camps.

Three topographic provinces (Pl. VIII, in pocket) may be recognized in this region—the mountains, the upland, and the coastal plain. The first embraces the Kigluaik Mountains, which stretch as a rugged mass along the northern boundary of the precinct, forming a part of the range which begins near Cape Woolley on the west and ends at Cape Darby on the southeast. These mountains are pinnacled and are broken by steep-walled gorges which head in amphitheaters. On the south the mountains fall off abruptly to an upland 2,500 feet in height, which stretches toward the coast with constantly decreasing altitude, its summits near the sea being less than a thousand feet high. At three points along the coast the base of the upland is washed by tide water, namely, east of Sinuk River, at Cape Nome, and at Topkok Head, but in the main broad coastal plains of crescentic outline intervene between Bering Sea and the highlands. These plains are in places gently rolling, are broken by numerous benches, and form the typical tundras of the North. One plain stretches southeastward from the northern boundary of the precinct and includes the delta of the Sinuk. Another extends along the coast from Cripple River to Cape Nome, and a third from Cape Nome to Topkok Head. These coastal plains, as has been shown, are built up of gravels, sands, and silts, some of which are auriferous and near Nome, at least, form important placers.

The inland margin of the lowlands rises in many places by a series of benches to the uplands, on whose seaward flanks washed gravels have been found up to altitudes of 500 feet. These high gravel deposits, which are found filling flat divides, obviously mark former stages of erosion. Less evident is the interpretation of the more elevated degradational surfaces, the highest of which is found in the flat-topped summits of the upland. That these, too, mark an epoch of former base-leveling, or partial base-leveling, can not be doubted, but the detailed physiographic history of this region has not yet been worked out.

Most of the streams take a southerly course to the sea. Their valleys are broad and gravel floored, and have gentle slopes up to an altitude of 600 feet, above which steeper grades carry them to the upland level. The smaller tributaries are of the same general topographic type, though the grades are usually steeper.

GEOLOGY.

The bed rocks of this region fall into two classes—the Kigluaik group and the undifferentiated schists and Port Clarence limestone of the Nome group. The closely folded gneisses, schists, and limestone of the Kigluaik group form a broad east-west belt whose geographic distribution is coincident with the outline of the mountains from which it takes its name. (See geologic map, Pl. X, in pocket.) This series of rocks is broken by many intrusions of granite. No placer gold has been found in the area of Kigluaik rocks, and they need not here be further considered.

Of greater interest is the second group of rocks, which embraces the schists, limestones, and greenstones here called the Nome group. These rocks occur in a broad belt forming the uplands south of the mountains. They are highly schistose and much jointed and faulted, with very complex structures. Greenstones are everywhere associated with the schists of this series, and occur in both schistose and massive forms.

Quartz veins and stringers are very common in the Nome group, and many of them are sulphide bearing and not a few auriferous. Most of these veins thus far discovered are mere stringers and without commercial importance except as indicating the source of the gold. A few veins that carry commercial values have been found, and one has been opened up and worked. In some places the placer-mining operations have uncovered zones of mineralization, in which a series of stringer leads has impregnated a rock mass of considerable width with ore-bearing solutions. No such zones have been found which when fairly sampled would yield workable values, though some of the included individual veins carry a high gold content.

The Port Clarence limestone is a heavily bedded rock of Silurian age which forms a member of the Nome group. Its general unaltered character and its distribution relative to the placers indicate that it is not gold bearing. It is not uncommon, however, to find the margins of these limestone masses the loci of intense deformation accompanied by injection of quartz veins. The distribution of the limestone is thus not without economic significance.

A small area of coal-bearing rocks in the Sinuk River basin, probably of Tertiary age, deserves mention. The coal is, however, without value.

The unconsolidated deposits of this area can be grouped under three general heads—coastal-plain deposits, stream and bench gravels, and glacial deposits. All but the last carry placer gold. The coastal-plain deposits are the most extensive and in some places aggregate more than 100 feet in thickness. They are the elevated stream and delta deposits that were modified more or less by wave action. The stream gravels hardly need special description. The bench gravels include not only those of the present drainage system, but also some high terraces, which were laid down in drainage channels quite different from those now existing. Evidence of this older drainage system is found widely distributed, but detailed study of its deposits has been limited to those near Dexter Creek, which are known to be auriferous.

DISCOVERY AND DEVELOPMENT.

Nome, the recording office of the district and the supply point for the whole peninsula, lies at the mouth of Snake River, on the edge of the tundra which skirts Bering Sea. A railroad leads from the beach across the coastal plain to the mines about Anvil Creek, and has recently been extended up Nome River, across the divide and down the Kruzgamepa to Lanes Landing. Other important settlements in the district are Dickson and Solomon, which lie 30 miles east of Nome and from which a railroad has been built up the Solomon Valley and across the divide into the Casadepaga basin; and Bluff, about 50 miles east of Nome.

The first find of gold in the precinct was made on Anvil Creek in July, 1898,[a] since which time developments have been rapid. The greater part of the gold production was taken from deposits within 15 miles of Nome, but there are many other mining centers in the precinct that have produced smaller amounts. In addition to the region contiguous to Nome the precinct includes the Topkok, Solomon River, Eldorado River, Cripple River, and Sinuk River regions, all of which have produced more or less placer gold. The present beach and some of the creek-bed deposits have been exhausted, but new discoveries in the elevated beaches and the gravel-plain, stream, and bench placers are continually being made and the construction of ditches tapping the water reservoirs of the high mountain valleys to the north, together with improvements in transportation facilities and in mining methods, is constantly extending the field of operations.

---

[a] Schrader, F. C., and Brooks, A. H.: Preliminary report on the Cape Nome gold region, Alaska, a special publication of the U. S. Geol. Survey, 1900, p. 31.

Placer-mining operations in the Nome precinct include summer work in open cuts, where many kinds of mechanical devices are used, and winter work by shaft and drifting methods. Up to 1903 most of the operated placers were best adapted to summer work, and the gold obtained from winter dumps was only a small part of the output. Since that time the winter production has increased enormously, so that it now (1906) amounts to about 50 per cent of the total.

In the following account the region immediately adjacent to Nome will be described first, both because of its great importance and also because there has been more mining done here, which enables the geologist to gather more facts regarding the mode of occurrence of the gold than in the less well-developed regions.

NOME REGION.

GENERAL OUTLINE.

In the immediate vicinity of Nome there is a region containing many rich placer deposits of great variety, including the present and ancient beaches, gravel plains, and creek and bench placers of various kinds. (See fig. 7.) In places these forms merge with one another and the deposits extend across the divides, making a complete subdivision of the field by drainage basins impossible. For this reason it seems desirable in the following descriptions to disregard drainage lines and to arrange the subject-matter according to the kinds of placers. The Nome region, as here defined, is a triangular area, with its base extending about 15 miles east and west along the shore of Bering Sea and its apex 25 miles inland near the head of Nome River. The richest placers are within a radius of 10 miles from Nome, which lies near the center of the base of the triangle. The coastal plain here is from 4 to 7 miles wide; back of it the upland is drained by Snake and Nome rivers.

The bed rock of the coastal plain is deeply covered with Quaternary gravels and sands. Back of the coastal plain the bed rock is made up of the schists and limestones of the Nome group, described on pages 70–82. The structure of the bed rock seems simple at first sight, but the developments made in the last few years show that the rocks are faulted in many places, and as they are generally covered deeply with surficial deposits, it has not yet been possible to work out the stratigraphic relations in detail. Mineralized quartz and calcite veins have been found in many places where the bed rock has been exposed by placer-mining operations. They occur both as fissure veins following fault planes and zones of fracture and as stringers spreading out between the structural planes from the fissures. In some places the bed rock contains a network of small veins, and it is possible that some masses of bed rock mineralized in this

way may be rich enough in gold to work as quartz mines. Of the
many samples taken by Government geologists from small veins
and stringers near Nome for assay, a few showed no trace of gold, but

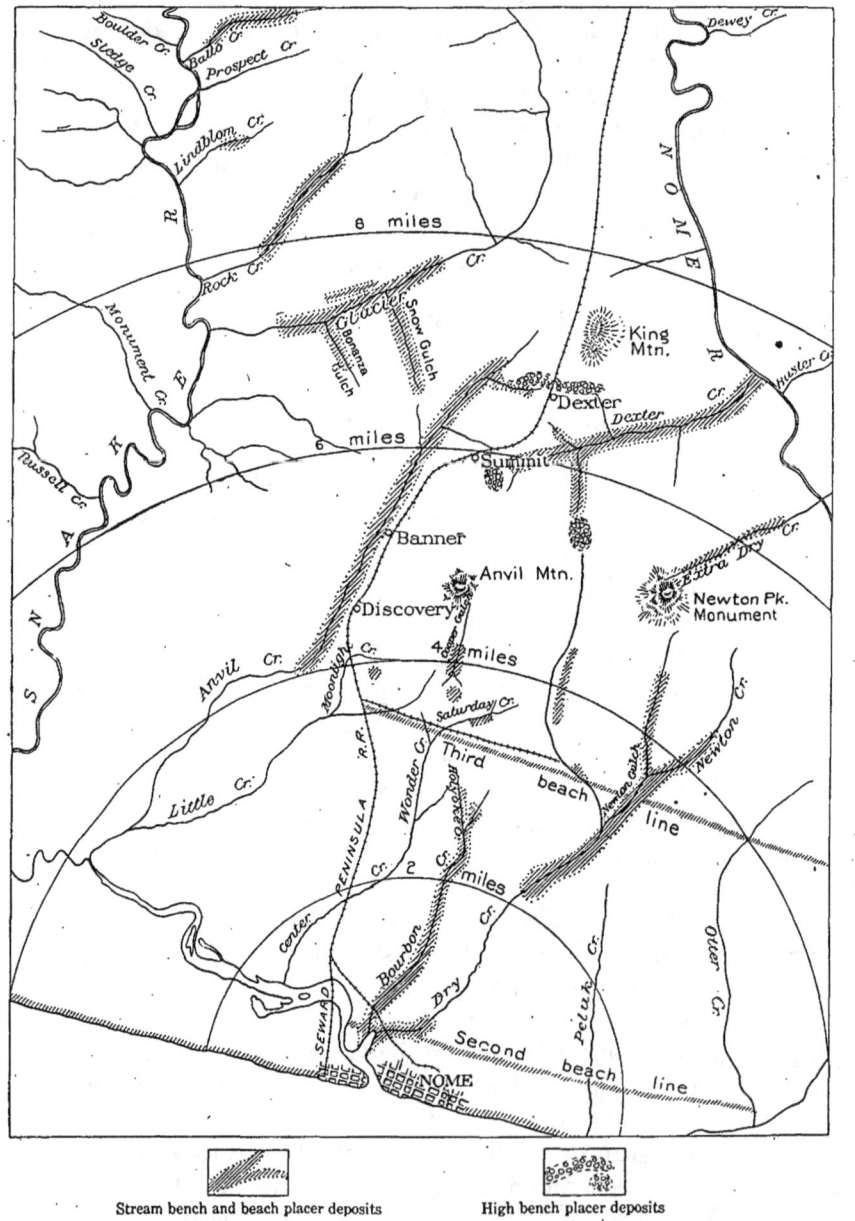

Stream bench and beach placer deposits   High bench placer deposits

FIG. 7.—Sketch map of Nome region, showing distribution of placers.

some were ore of high grade. Here and there free gold can be seen in
veins of nearly pure quartz and calcite, though as a rule the gold is
associated with pyrite, galena, and other sulphides. The placer gold

of the Nome region was undoubtedly derived from these quartz veins. Great masses of bed rock thus mineralized have been eroded away, leaving their gold content concentrated in the gravels. In newly cut valleys this concentration has been direct from the bed rock, but in the older, broader valleys and the coastal plains and sea beaches these gravels have themselves been eroded and again and again reconcentrated.

BEACH PLACERS.

The beach placers of the Nome region extended along the shore of Bering Sea for about 30 miles, practically from Cape Nome to Cape Rodney. They have been worked continuously every season since 1899, and though the greater part of their gold content was doubtless extracted before the end of the season of 1900 they are still producing small amounts of gold and will continue to do so for some years to come. The methods of mining do not differ materially from those used in 1900, though at the present time the placers are of little economic importance.

The total output of the Nome beach placers has probably been $2,000,000. Of this about half was taken out in 1899 and about $350,000 in 1900, since which time the production has been less each year. There are still a few people working with rockers, and in 1904 several sluicing and dredging plants were in operation.

The following account of the beach placers is quoted from Brooks,[a] who had exceptional opportunities to study these deposits when he visited Nome during the height of the excitement in 1899 and again in 1900:

It has been shown that the lowest bench of the coastal plain ends in an escarpment, 10 to 20 feet high on the seaward side. From the base of this escarpment the beach slopes to the sea at an angle of 4° to 5°, having a width of about 50 to 75 yards. Ordinarily the wave action is confined to the lower third, but during severe storms the surf sometimes rolls up the full width of the beach.

The surface material of the beach is usually sand with occasional shingle and gravel. The pebbles, which have the characteristic oblate spheroidal form of beach shingle, are composed largely of quartz, but also of the various types of country rock of the adjacent region. Pebbles of more than an inch or two in diameter are relatively rare, but occasionally small bowlders are found, which probably owe their position to the drifting of shore ice.

The beach sand consists largely of quartz, usually stained with iron, and mica and chlorite schist fragments. Reddish garnets form an important constituent, sometimes predominating over all the other minerals, and then constituting the " ruby sands " of the miners. Magnetite is always present, but usually forms less than 1 per cent by weight, though in the concentrated form

[a] Brooks, A. H., Richardson, G. B., and Collier, A. J., Reconnaissances in the Cape Nome and Norton Bay regions, Alaska, in 1900, a special publication of the U. S. Geol. Survey, 1901, pp. 85–91.

found in the pay streaks it may run as high as 10 per cent. Such percentage, where the material is fine, gives the sand a dark color, and it is then termed "iron sand" or "black sand" by the miners. Mr. Collier made a separation, by sifting, of a typical beach sand taken from a prospect hole near Nome, with the view of determining approximately the relative proportion of the different-sized particles by weight. A sample of sand was put through sieves of different meshes, with the following results:

38 per cent coarser than 20-mesh sieve.
42 per cent coarser than 40-mesh sieve.
16 per cent coarser than 60-mesh sieve.
3 per cent coarser than 80-mesh sieve.
1 per cent finer than 80-mesh sieve.

The coarse material was chiefly quartz and schistose rock fragments. The garnet percentage was about 4 or 5 by weight and the magnetite less than 1. Mica is plentiful, but forms a relatively low percentage by weight.

An average cross section of the beach sands, as determined by an examination of the pits made during the mining operations, was as follows: Near the edge of the tundra a blue-clay bed is found close to the surface and seems to slope seaward. Halfway down the beach toward the water this clay bed can usually be recognized at a depth of 5 to 7 feet. These statements in regard to the clay bed have a general application, but the writer does not wish to imply that this one bed can be traced the entire length of the beach. In some cases several seams can be recognized and in others the clay seems to be entirely lacking. It is true, however, that in most of the workings a clay bed has been found a few feet below the surface which slopes toward the sea. The thickness of this clay stratum has not been determined, but it must aggregate several feet. Immediately above the clay are usually fine sands, though sometimes coarse gravels. This layer of sand, which includes the gold of the beach placers, contains as a rule a larger percentage of the heavier minerals than the beds above. The higher beds include coarser and finer material, such as sands and gravels, with occasional thin seams of clay. Sometimes fragments of wood and other vegetable matter are found in the beach sands, but these are relatively rare. When traced horizontally the various layers which make up the beach sands above the clay bed are found to be thin lenses, which rapidly thin out and seldom can be traced a hundred feet. The coarse gravels are usually found in pockets, often but a few feet in lateral extent.

The beach sand, from Cape Nome to Rodney Creek, a distance of about 30 miles, nearly everywhere carries colors of gold. Broadly speaking, the richest diggings have been confined to about 20 miles of the shore in the central part of this stretch. In this belt colors are obtainable nearly everywhere at the surface, but the beach proper, as a rule, carries values only some feet below and on clay beds. The pay streak measures from 6 inches to 3 feet, and rests on a clay bed at nearly every locality that was examined. The thickness of the productive sand layer at any one locality is a variant, depending on the refinement of methods used in mining and extracting. Where the ordinary cradle is used, from 6 inches to 2 feet of sand are washed. With the employment of more elaborate methods, where larger amounts of material are handled, all of the sand and gravel from the surface down to the clay bed is often sluiced. Sometimes two pay streaks are found, one on the clay and one a foot or two above, separated by intervening layers of nearly barren sand. The gold-bearing sand slopes with the clay bed from the coastal-plain escarpment toward the sea. Near the margin of the tundra it is often found close to the surface, but near the sea it is from 3 to 6 feet below. Irregularities in the occurrence are introduced by the presence of two pay streaks, each resting on a minor clay bed. The accompanying photograph,[a] taken by Mr. Schrader, and reproduced from the report already cited, gives a typical section of the beach sand near Nome. The lowest pay streak seen in the picture, which is said to run $1 to the pan, is among the richest that have been found in the beach. The richest part of the pay streak in its horizontal extension, similar to the layers of sand and gravel, is found to be of lenticular form, thinning out toward the margins. The sand of the pay streak differs in no way from that of the beach, except that it is usually finer and carries a larger percentage of the heavy minerals. There is often a progressive increase of the heavier minerals toward the pay streak. While in the surface sands the percentage by weight of the magnetite is often less than 1, near the clay bed rock it may run up to 8 or 10 per cent. In any given section the amount of garnet also increases toward the bottom of the sand layers.

A sample of gold-bearing "ruby" sand from a locality east of Nome River, 5 feet below the surface, was sifted by Mr. Collier, and the following percentages by weight were obtained:

3 per cent larger than 20-mesh sieve.
29 per cent larger than 40-mesh sieve.
53 per cent larger than 60-mesh sieve.
11 per cent larger than 80-mesh sieve.
3 per cent larger than 100-mesh sieve.
1 per cent smaller than 100-mesh sieve.

A rough mineralogical separation of the same sample resulted as follows:

38 per cent quartz stained with iron, with some fragments of schistose rock.
61 per cent garnet.
1 per cent magnetite.

Another sample of auriferous sand gave the following results:

65 per cent chiefly quartz and mica, with some fragments of schistose rock.
34 per cent garnet.
1 per cent magnetite.

A third sample from Nome beach, 7 feet below the surface, was constituted as follows:

96 per cent quartz and schist fragments, mica, etc.
3½ per cent garnet, magnetite, etc.

The above separations show that the minerals of beach sand can not be regarded as having anything like a fixed ratio.

Time has been lacking to make a complete mineralogical examination of the auriferous sands. The following minerals, however, have been noted in abundance: Quartz, muscovite, chlorite, and garnet; and as accessories magnetite, ilmenite, and pyrite.

Large flakes of gold are occasionally found, but these are relatively rare. Some small nuggets have been reported, running up to the value of $1, but are very exceptional. In purity it is about 0.890, has a bright-yellow color, and amalgamates readily.

The gold is flaky, and the average grains as saved by panning will run from 70 to 80 to the cent, valuing the gold at $14 an ounce. A comparison of it with beach gold from other localities shows it to be rather coarser. The beach gold from Randolph, Oreg., averages about 110 colors to 1 cent. That from the Sixes mine near Denmark, Oreg., averaged about 600 colors to the cent. The river-bar gold from the Snake River, Idaho, which is mined at a profit, runs from

---

[a] Not reproduced in this volume.

about 900 to 1,000 colors to the cent.[a] In making these comparisons it should be remembered that much of the fine gold of the Nome deposit is lost during the process of separation, and that this will increase the average size of the colors that are saved.

The source of this beach gold has been the subject of much discussion among the miners and prospectors of Nome. Theories as to its glacial origin are common, and hypotheses which assign its source to volcanic or meteoric agencies are not without supporters. The more careful students of the question, be they prospectors who have gained their knowledge by practical experience or mining engineers who have had theoretical as well as practical training, are generally agreed that the beach gold has the same source as the creek gold, and are at variance only on the question of its mode of distribution.

There can be no doubt that the beach gold found its source in the bed rock of the adjacent region and that it was transported to its present position by water. A theory which is often advocated assigns its distribution to the action of shore currents—that is, that the gold was brought down by the rivers and distributed along the shore by currents. In support of this theory attention is drawn to the fact that the richest deposits occur near the mouths of rivers which have their source in the gold-bearing area. The objections to this mode of distribution are: First, we have no evidence that such currents exist along the coast, and to suppose that the beach gold was distributed by currents at Nome we must believe that the currents run in both directions from the mouths of the rivers. For instance, beach gold is found along the shore in both directions from Snake River. If this river contributed the gold, it is difficult to understand why it should be found in both directions from the mouth. Second, objection is made that the river bars do not contain as much gold as we should expect if they had been the immediate source of the beach gold. Third, the gold is too coarse to be carried by ocean currents. Fourth, its distribution in the beach sands shows that it must have been concentrated by wave action. If we are to suppose that the gold was brought alongshore by ocean currents, the waves must have washed it up on the shore, and subsequently given it its peculiar form of concentration. If the gold was borne by ocean currents, we should expect to find it in the present ocean bottom of the same degree of coarseness as it is found on the beach. As a matter of fact, the little gold that has been obtained off the coast of Nome has been much finer than the beach gold. These objections to the theory of current distribution seem to be insuperable.

Attention has already been directed to the fact that the sand and gravels of the coastal plain are auriferous. Evidence has been given that points toward the conclusion that these deposits were laid down at the mouths of streams when the land stood at a lower elevation relatively to the sea than it does now. The streams and rivers gradually encroached on the sea floor by the building out of their deltas. In these delta deposits gold was laid down with the other material, and was more or less disseminated.

In the "Preliminary report" Mr. Schrader and the writer advanced the theory that the gold was concentrated from the coastal-plain gravels by the wave action which cut the seaward escarpment of the plain. The diagrammatic sketch [fig. 8] which was published to illustrate this point is here reproduced. In this sketch the edge of the coastal-plain escarpment is marked and the position of the beach placers in reference to it is shown. During high storms the waves even now reach the margin of the escarpment and cut away the base of the bluff. The materials which go to make up the coastal plain are sorted by

---

[a] Washburn, W. H., Gold in the Snake River gravel bars: Min. and Eng. Press, December 29, 1900.

this wave action and the heavier particles, such as the gold, sink into the sand. It is a well-known fact that this fine gold will make its way through sand for some distance, provided its passage is not interrupted by any impervious layer. This downward movement is brought about near the surface of the beach by the constant motion which is given the sand by the waves. The percolating waters will also help to carry the grains of gold downward. It should be noted here that the beach sand, being well drained, is not frozen in summer. In the course of its downward passage the gold finally reaches a clay layer, and here becomes concentrated as it is found. There is another factor which may have accelerated the downward movements of the heavier particles. In the spring months southerly storms frequently pile up the ice on the Nome beach to a great height. When large floes are driven ashore on the shelving beach they must cause considerable movement in the underlying sands, and this motion is probably transmitted to a depth of several feet and causes a disturbance among the grains of sand. A sort of sifting process would take place by which the heavier particles would always descend and eventually be concentrated beyond the line of movement or on impervious layers which they could not penetrate.

The explanation by wave sorting may perhaps account for other beach placers. Mr. J. S. Diller informed the writer that the richest beach diggings on the Oregon coast are at points where waves have access to high bluffs. In fact, the Oregon and California beach placers have many things in common with those of Nome,

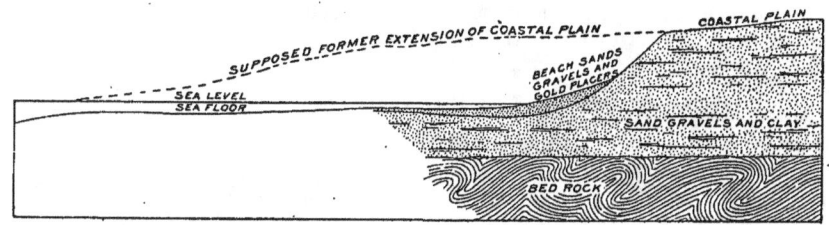

FIG. 8.—Diagrammatic section of beach placers.

but their gold seems to be much finer. In the early developments, however, of these southern beach deposits some diggings were mined which were more comparable to those of the Nome region in their richness and the coarseness of their gold; but these were long ago exhausted, and the operations are now confined to localities where careful and economic mining and separating methods enable the operators to work deposits which are of comparatively low grade.

A question of interest is whether there may not be old beach placers in this region similar to those that are found in Oregon. It has been assumed that this part of Seward Peninsula has in comparatively recent times been elevated to a height varying from 600 to 1,400 feet. Attention has been drawn to benches which are believed to represent former shore lines. It is a logical conclusion that during this period of uplift similar beach placers have been formed in the region, and a search for them should be a fruitful field of investigation. It has been suggested elsewhere that some of the escarpments in the coastal plain may mark former shore lines.

Mining operations on the Nome beach have all been carried on in a very haphazard way. During the first year the methods used were those of the hand rocker, and it was everybody's object to secure a rich patch of ground which would pay for the mining by this crude method. There was then ample room for the men who took part in the beach diggings. No claims were staked and no serious quarrels arose because of trespass. In the second year of the operations a vast deal of mining machinery was imported and many attempts were made to handle the beach sands at small cost. These enterprises were, on the whole, not paying ventures. The high freight rates in the Nome region and the great cost of fuel, as well as high wages, made the expense of running these plants very great. It was impossible for the operators to obtain exclusive right to any part of the beach, and the enterprises were constantly interfering with one another. Moreover, during the previous year the richest part of the beach diggings had been gutted, and, as the evidence of the working had been entirely destroyed by the ice of the winter, it was impossible to pick out fresh portions of the beach for mining operations. The choice of location had to be made at haphazard. The large number of men who were engaged in beach mining made intelligent prospecting previous to location entirely impossible. In August, moreover, a heavy storm destroyed a large part of the machinery along the Nome beach and put an effectual stop to many of the enterprises. In view of these facts it is not surprising that there were more beach-mining companies that lost money than there were that made money. A large amount was nevertheless taken out of the beach, which Doctor Whitehead estimates at $350,000.

As to possibilities of future mining developments, it can be prophesied that no large enterprises can be successfully carried out unless in some way a title for a given stretch of the beach can be obtained. A large percentage of the beach sand is probably far richer than that which is mined on the Oregon and California coast at a profit. Indeed, the fine gold, such as that found in the Oregon beaches, is not saved at all at the Nome beach.

PLACERS OF THE COASTAL PLAIN.

INTRODUCTION.

The auriferous gravels of the coastal plain, or tundra, as it is usually called, have not yet been very large producers. Many of the rich spots have been worked, especially near the base of the hills, but no means have yet been devised for handling the lower-grade auriferous gravels, which form very extensive deposits. Plans have been formulated for handling some of this material with steam shovels, and it has also been proposed to use hydraulic means, although this would necessitate some provision for the removal of the tailings, as much of the bed-rock surface underneath the alluvium is probably close to sea level, if not below it. The Brooks party studied this district in considerable detail in 1900, and not much can be added to the general description then published, though many additional detailed observations have since been made. Brooks's general account of these deposits is here quoted in full: [a]

The crescent-shaped coastal plain which intervenes between the highlands and the sea in the vicinity of Nome has already been described. The sandy beach slopes up from the ocean at an angle of about 5° to an escarpment 10 to 20 feet high. This same slope is carried seaward at a lower angle and gives the shallow bottom which extends out about a half mile from shore. From this

· [a] Brooks, A. H., Richardson, G. B., and Collier, A. J., Reconnaissances in the Cape Nome and Norton Bay regions, Alaska, in 1900, a special publication of the U. S. Geol. Survey, 1901, pp. 80–82.

escarpment the plain extends inland to another escarpment, which marks the seaward limit of a second bench which is about 50 feet high. In some places this second bench lies within a quarter of a mile of the shore, while elsewhere it has receded much farther inland. A third bench, 100 feet or more in height, lies still farther from the coast, and from this the coastal belt merges with the upland in a succession of smaller and less-marked benches.

It has already been shown that the region has been gradually elevated during recent geological time. The coastal-plain terraces, like the high benches already described, are made up of material that was deposited along the margin of the shore while the land stood at lower elevation relative to the sea than it does now. The successive benches mark a series of interruptions in the uplift, when the land stood at a constant elevation long enough to permit the accumulation of the material of which the terraces are formed.

The sediments of the coastal plain are predominatingly of fine material, usually sand and sandy clay, with some coarser gravel layers and frequent beds of clay. Broadly speaking, these sediments increase in coarseness in an inland direction. They were contributed by the streams and rivers, which gradually built out their deltas. These coastal-plain deposits were laid down on a rock surface which had been more or less channeled by streams and which must have been very uneven, and the consequence is that there is probably great variation in their thickness. Near the town of Nome bed rock reaches close to the surface, and in close proximity pits 30 feet deep have failed to reach bed rock. Locally the thickness may be 100 feet or more. Therefore mining enterprises which depend on reaching bed rock must in every instance make a preliminary determination of thickness by boring or by sinking test pits, as generalizations regarding thickness are liable to be entirely in error.

The gravels and other sediments of the coastal plain have the same origin as those of the creeks and gulches which have already been described, and hence we should expect them to be auriferous, and such is the case. The gold of the coastal plain, having been carried farther from the parent rock than the creek and gulch gold, is much finer. In character it is more closely allied to the gold of the river bars.

The gold deposits of the coastal plain may be grouped under two headings. The first are the gravel-plain placers, in which the gold first accumulated. The gold is more or less disseminated through the gravels, sometimes concentrated on a clay bed, but characteristically in an unconcentrated form. The placers of this description have received but little attention. Where they are unconcentrated they can be worked only on a large scale, with improved machinery. The success of such mining would be dependent on handling a large quantity of material and on saving all the gold by refined methods rather than on the finding of the rich pay streaks.

As these coastal-plain deposits were laid down at the margin of the ocean, we should expect old sea beaches to be found in these gravels. If such beaches are found, they are likely to prove as rich as the present beach at Nome. It would therefore be well for the prospectors to examine carefully the seaward side of the different escarpments which mark the limits of the terraces. These bluffs are quite likely to mark an old sea beach. In such types of deposits we should expect their extension to be more or less parallel to the present coast line.

The second form of deposits is composed of the gulch and creek placers of the coastal plain. This class of deposit includes the placers of the smaller streams whose sources lie within the coastal plain. As these streams do not reach bed rock, their placers must have derived their gold from the gravels in which they flow. These placers are in fact reconcentrations of the auriferous gravels of the coastal plain. They form natural sluices in which the disseminated gold of the gravel-plain placers has been collected. This concentration has usually taken place on a stratum of clay, or "false bed rock," as the miners term it. As these layers of clay are liable to be encountered at different horizons in the gravels, there may be a number of pay streaks, one above the other, separated by sands and clays.

Most of the development that has been made so far is in that part of the coastal plain which lies between Nome and Snake rivers. The probabilities are that this area includes the richer portion of the coastal-plain deposits. This corresponds in a general way with the distribution of the richer placers of the Nome beach, which, as will be shown, derived their gold from the coastal-plain deposits. In the detailed descriptions which follow, the heterogeneity of the coastal-plain deposits is well illustrated. Test pits but a few yards apart show very different sections. In the coarser gravels cross-bedding such as would be expected in deltas is not uncommon. The surface layers in all the sections consist of a foot or two of moss, other vegetable growth, and muck. Below this mantle the ground is usually found to be frozen, in summer as well as in winter. The few sections that were obtained, as will be seen, show a pay streak consisting of rather coarse gravel, 1 to 3 feet in thickness, and always overlying a clay bed. The pay streaks which have thus far been mined were found from 1 to 10 feet below the surface. The sands and gravels are usually found to carry gold from the surface down, but are never concentrated except on a bed of clay, The creek and gulch placers of the coastal plain at Nome are estimated to have produced about $150,000 during last season.

Since the above was written these auriferous deposits have received a constantly increasing amount of attention, and during the season of 1903 large areas were systematically prospected. Along the terrace east of Nome, already described, an old beach containing fine gold in large amounts has been uncovered in shallow prospect holes, verifying a prediction made in the previous report. Fossil mollusks and Bryozoa identified as Recent by William H. Dall have been found in the old beach sands uncovered by the mining operations.

OLD BEACH LINES.

One of the most definite placer deposits of the coastal plain lies about half a mile back of the present beach and is traceable for 9 miles from Hastings Creek to Dry Creek, being marked by the escarpment rising from 10 to 40 feet above the tundra already referred to. That this is a beach deposit is evident not only from the presence of the escarpment but also from the nature of the materials, which consist of characteristic well-rounded pebbles and white quartz sands containing rounded fragments of sea shells. The same beach line is probably continued westward, but is not marked by an escarpment, as it here formed a barrier between a lagoon and Bering Sea. It seems likely that the lower Snake River, whose course for 5 or 6 miles is parallel to the beach, follows the shore of the old lagoon. (See topographic map, Pl. VIII, in pocket.) The evidence for this supposition lies in a ridge formed of beach sand and gravel containing rounded fragments of sea shells, which extends along the south side of this part of the Snake River trench, parallel with the present beach. Ancient beach deposits may be recognized by the character of the material, even where no scarp or lagoon is preserved. Perfectly rounded pebbles and shingle with white sands, composed mainly of quartz grains but containing rounded fragments of sea shells, are infallible indications of beach material. The peculiar bedding of the beach sands can usually also be recognized. By these criteria ancient beaches now deeply buried below the tundra have been identified and reported by prospectors at a greater distance from the coast.

The east end of the ancient beach above described is being mined about one-fourth mile from the present shore near Hastings Creek, which enters Bering Sea 10 miles east of Nome. Hastings Creek is about 6 miles long, and its basin is entirely within the coastal plain. It has a low gradient, but carries sufficient water for sluicing. About 2 miles from the coast a prospect hole reached bed rock at 32 feet.

Gold was discovered near the mouth of the creek in 1901, and since that time more than $6,000 has been produced. The section exposed by the mining is as follows:

Section near mouth of Hastings Creek.

	Feet.
Moss and muck	2
Beach sand	10
Coarse gravel and clay.	

The gold occurs in three or four thin layers of " ruby " sand near the bottom of the beach-sand stratum and is of the same character as beach gold. This pay streak has been traced about 400 yards to the west, and also occurs on the east bank of the creek, where a few hundred dollars have been produced. When Mr. Hess visited the locality in June, 1903, a dump of about 1,000 cubic yards had been mined out and a gasoline pump was being installed on the creek to provide water for sluicing. A ditch to bring water from Flambeau River for hydraulicking has been surveyed.

There appears to be an extension of this deposit about half a mile east of Hastings Creek, where there is a well-marked bench about 40 feet above the sea. At this place several pits have been dug, and it is said that several thousand dollars in gold has been taken out. The excavations show sands and gravels with some seams of clay. Gold has been found in a layer of " ruby " and black sand about 8 feet below the surface, where the placers are said to have yielded 10 cents to the pan.

The same beach deposit is found again east of Nome River, where it has been traced to Peluk Creek. This elevated beach has been prospected for about 1½ miles, and one claim is reported to have produced $20,000 during the winter of 1902–3.

Similar deposits have been mined to a limited extent about half a mile east of Nome near Dry Creek, and also at the mouth of Bourbon Creek. Here the gold-bearing stratum lies near the surface of the tundra, about 20 feet above the bed of the stream. The excavations show a few inches of moss and muck overlying beach sands that contain well-rounded fragments of sea shells. The fine gold is mostly concentrated in thin layers of "ruby" sand interbedded with white sand. From some of the layers of this "ruby" or garnet sand Collier obtained 15 to 20 fine colors of gold in a shovelful of the material. In one place the deposit was sluiced early in the season with water collected by a dam on the tundra. In another place horses were used and the pay dirt was scraped down the hill to sluice boxes near the creek bed. Dredges have also been used to handle this material.

The latest authentic information on the elevated beaches is contained in a report by Moffit, which includes an account of the so-called third-beach line discovered in the fall of 1904. The following is quoted from this report: [a]

Our knowledge of the tundra has been greatly increased during the last two years by developments on the buried beaches. Two well-defined ancient gravel deposits of this sort have now been explored through a part of their length. [See fig. 9.] One of these, near Nome, lies about three-fourths of a mile north of the present coast line and extends eastward to a point within a short distance of Cape Nome. At Hastings Creek it is about a fourth of a mile distant from tide water, but east of that locality its position is not known, and it appears to have been removed through erosion. To the west it is probably represented by the beach deposits of Jess Creek. Its elevation above sea level is 37 feet and its location is in most places indicated by a steep, moss-covered gravel bank, at whose foot it lies. The other beach line is definitely located from the place where it is crossed by the railroad tracks at Little Creek to McDonald Creek, a distance slightly more than 5 miles. Its elevation above sea level, according to reliable information obtained at Nome, is 79 feet. It extends in a nearly straight or slightly curved line between the points mentioned, yet shows slight undulations and is interrupted by the valleys of Nome River, Anvil Creek, and Snake River, these streams lying below its level. These two ancient coast lines are generally known as the second and third beaches, the present one being regarded as the first. Mention of others lying between is frequently heard, but although this is not only possible but even probable, no other continuous beach has yet been traced.

A generalized section of the deposits exposed along the third beach would show gravel or sandy gravel, with coarse bowlders resting either on schist bed rock in which are a few limestone beds or, as is the case at the east end of the beach, on fine sands which in turn rest on schist bed rock. Above this is a considerable body of gravel overlain by "muck" and the surface vegetable matter. This general section is, however, subject to wide variations. The thickness of the muck varies from 1 to 2 feet to 20 feet or more. Underlying the muck in several shafts a blue clay was found. In places a heavy wash

---

[a] Moffit, F. H., The Nome region, Alaska: Bull. U. S. Geol. Survey No. 314, 1907, pp. 134–144.

occurs near the surface. Here and there the gravels are slightly cemented by the deposition of lime or iron oxide. Marine gravels are interbedded with creek wash. The character of the material varies both in composition and in coarseness—in fact, the deposits where first exposed on Little Creek were so varied in appearance and manner of deposition as to cause doubt whether any of them were of marine origin.

The gold-bearing gravels or pay streaks vary in width from 25 to perhaps 100 feet and have a fairly constant southerly slope of about 1 foot in 10. They rest in some places on bed rock, in some places on other gravel, and toward the east, as has been stated, on fine sand.

Only a part of the shafts on the second beach have been sunk to bed rock, as the pay streak usually lies on a false bed rock of clay or sandy clay and

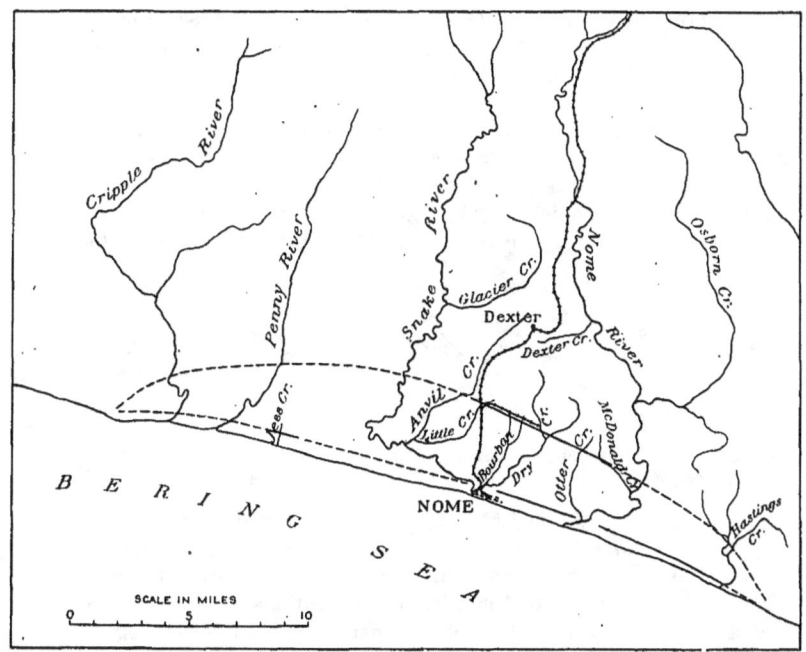

FIG. 9.—Sketch map showing the known parts of the second and third beaches (full lines) at Nome and their hypothetical continuations (dashed lines). The upper line also shows approximately the area of the Nome tundra.

gravel. There are few data, then, on which a complete generalized section could be based, but it appears that, though coarse, angular material is by no means lacking, it is not as abundant as on the third-beach line. Furthermore, the quantity of garnet, or " ruby sand," is far greater on the second beach, and the proportion of other sand and fine gravel is also greater. This is probably accounted for by the fact that much of the material of the third beach has traveled a shorter distance from its source and has been less subject to stream and wave grinding.

In the third beach, then, irrespective of any greater differences which may have occurred in the meantime, we have definite proof that the land now stands not less than 79 feet higher than it did when the beach formed the coast line. Further evidence of elevation, though of lesser amount, is furnished by marine shells taken from various shafts between the second and third beaches. Such shells in an almost perfect state of preservation are found on Center Creek and suggest that in that locality they accumulated in comparatively quiet water. They occur in gravels 32 feet below the surface and at an elevation of about 20 feet above sea level. Numerous marine shells from Otter Creek were obtained at approximately the same height above sea but at a depth of 50 feet below the surface. They are in a good state of preservation.

As a rule the deposits of the beaches and of the tundra in general are frozen from top to bottom, but there are places where this is not the case. One such area is located near the intersection of the third beach and Holyoke Creek and has caused difficulty in working the Bessie Bench claim because of the large amount of water circulating through the gravel. The boundary between the thawed and frozen ground was here located by drilling, and care was taken not to bring the workings too close. Thawed ground is in some places overlain by frozen ground and here and there is underlain by it also. The reason for the presence of unfrozen areas is not entirely understood, but they are probably due in part at least to the circulation of water through the gravel.

* * * * * * *

To those who year by year have followed the development of mining in the region adjacent to Nome it is a noticeable fact that during the summer of 1906 the attention of mining men was largely given to operating within the area of the Nome tundra. This is a condition which may probably continue for some years, until the gold of the beaches begins to fail or until all the available

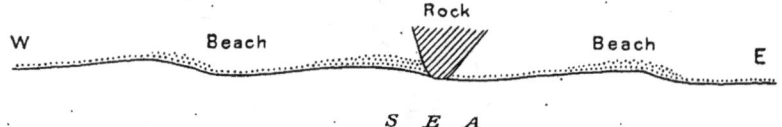

FIG. 10.—Diagram showing the manner in which gold is concentrated north of Nome in shallow depressions or on sides of cusps of the third beach.

ground is opened up. If, in addition to the operations on the tundra, those of Glacier Creek, Anvil Creek, and Grass Gulch are included, all the most important workings will have been taken into account, although elsewhere within the area shown on the Nome and Grand Central maps minor operations were conducted on a few scattered streams, probably the most extensive being on Buster Creek.

At present the buried beaches of the Nome tundra occupy the center of the mining stage, and the efforts of all operators have been given to the exploitation of the old placers or the search for new. Litigation touching the rights of property, however, has seriously obstructed the development of much of the most valuable ground, and may be expected to continue to do so as long as the present methods of acquiring and holding mining property are in force.

Mining is carried on most actively just now along the two ancient shore lines whose locations and principal features have already been described, but there has been more or less work on different streams, such as Dry and Bourbon creeks, and extensive drilling in various other parts of the tundra. On the first or present beach practically all work was suspended. Along the second beach many of the old properties were worked and some very good new ground was discovered in the vicinity of Otter Creek. The third beach is the principal producer of the region.

The third beach was discovered in the late fall or early winter of 1904, but when the summer closed in 1905 operations were confined to the immediate vicinity of Little Creek. An account of this locality has been published elsewhere.[a] During the winter of 1905–6 the eastward continuation of the beach was located to a point within a short distance of Nome River. Between Moonlight Creek or the railroad and McDonald Creek it is traced continuously. To the west and east of these localities it is not definitely known. Remarkably rich ground was exploited near Little Creek and between Holyoke and Bourbon creeks, and nearly all the claims between Little and Dry creeks have shown good values in gold. East of Dry Creek less work has been done, but most of the shafts have struck good pay. Nevertheless, some claims or parts of claims are of little value with the present cost of mining, for while the beach gravel deposits are continuous the pay in them is not so. There are intervals along the line where gold is present in small amount or is almost lacking. These places are sometimes referred to as "blanks." Further, the gold is not evenly distributed through any of the gravel. The writer was informed that in the shallow depressions such as occur at intervals along the beach a much greater concentration of gold took place on the east ends [see fig. 10], and that in one or two places where low ridges or rolls of bed rock reached the surface and projected slightly beyond the ancient beach exceptional amounts of gold were found on their west sides, a position of maximum concentration corresponding to that in the indentations just mentioned. This would indicate that the distribution of gold in the gravels was largely influenced by the prevailing direction of the ocean waves and currents. It is probable also that very rich deposits, such as occur at Little Creek and Bessie Bench, are due to their nearness to the source of gold or to streams which brought it to the sea. The character of both the gold and the gravel accumulations would indicate the same thing. On the western part of the beach the gold is, on the average, much coarser than at the east end, where it resembles in appearance and approaches in fineness that of the present beach. The gravels of the west end are more variable in character and exhibit a larger amount of coarse, angular stream wash than those toward the east, showing that the conditions under which the western gravels accumulated were less uniform—at one time stream deposits, at another sea deposits, being laid down.

Some ideas concerning the eastward and westward continuations of the third beach are suggested by an examination of the topographic map. The shore line must formerly have extended from the hills west of Cripple River to Cape Nome, and if one may judge by the portion now known it had the form of a broad arc of fairly uniform curvature, like the present beach, but with smaller radius. It is the belief of not a few miners at Nome that the third beach did not, like the first and second beaches, keep to the seaward of Cape Nome, but that it passed to the north through the broad, low depression between Saunders Creek and Flambeau River, thus forming an island of the Cape Nome granite area. The elevation of the depression between Cape Nome and Army Peak is only 115 feet, and the possibility of the cape being an island at the time when the third beach was formed can not be refuted by any evidence now at hand, although it appears improbable. Bed rock is traced northwestward from Cape Nome for a distance of nearly 5 miles, and in the low rounded hill between Hastings and Saunders creeks has an elevation of 297 feet. Between this point and the Army Peak schist mass, still farther to the northwest, is an interval of about 3 miles across a broad, low saddle where no rocks are exposed. As stated, the elevation of this flat at its lowest point is about 115 feet, but the depth of gravel is unknown. If it has a thickness of 40 or 50 feet it is possible that the third beach passes through. It seems far more probable,

[a] Moffit, F. H., Gold Mining on Seward Peninsula: Bull. U. S. Geol. Survey No. 284, 1906, p. 134.

however, that the controlling influence in determining the coast line here was exercised by Cape Nome, since it must have been a factor in directing the ocean currents and consequently the accumulation of sands and gravel. To judge from present conditions, it appears more likely that the sea would have built a connecting bar between Cape Nome and Army Peak rather than wash between them. At any rate, the force of the waves due to southerly and southeasterly winds would have been greatly diminished through the protection offered by Cape Nome, and the amount of concentration would have been thereby decreased.

Another idea which is maintained by some and may lead prospectors astray is that wherever bed rock can be found at an elevation of 79 feet above sea level the third beach will be present. The fallacy of this idea is immediately apparent when it is remembered that, so far as known, the old beaches were not laid down on a cleanly swept, somewhat uneven rock floor, but were formed over a surface whose inequalities had already been reduced by a filling of gravel and sand. This is shown by the fact that in many places they do not rest on bed rock, but are underlain by a variable thickness of loose deposits.

Further evidence of a somewhat negative character is afforded by the fact that neither the third beach nor either of the others is known to have formed reentrants at such places as their intersections with the valleys of Snake and Nome rivers, but rather that in each place where the river valleys lie below the beaches at such intersections the beaches end abruptly, for the present valleys through the loose deposits have been cut since the beaches were formed.

Another fact which must not be lost sight of in prospecting for the third beach is the possibility of recent warping. When formed, a beach is at sea level, and if raised uniformly throughout its length all parts will continue to have like relations to the sea. But changes of level do not always nor even usually take place in a uniform manner throughout all portions of an affected area. One part may be raised or lowered more than another, or one part may even be raised while another is sinking. It does not follow, therefore, that because one point of the beach has an elevation of 79 feet all other parts will have the same elevation, although in a small area such as this it is probable that they will not differ greatly.

Since, however, we now have no evidence that warping of any consequence has taken place, and since so far as we know it the third beach does maintain a fairly constant level, it is not to be expected that it will be found in any locality whose surface elevation is less than 75 or 80 feet above tide, even if such an area lies directly between or in line with points where it is known to be present.

Some probabilities concerning the distribution of gold in the known beaches or in others which may be found are gained from a consideration of its distribution in the gravels so far exploited. The richest gold-bearing gravels mined in the Nome tundra have been found in that portion of it which lies between Nome and Snake rivers. This corresponds also with the richest part of the first or present beach, whose greatest values were taken from the neighborhood of the mouth of Snake River and from sands to the east toward the mouth of Nome River. This area lies directly south of the mineralized area from which it is believed that the gold has been chiefly derived, and it is the locality toward which weathered material from the near-by hills may properly be expected to migrate, since it lies immediately between them and the sea. There appears, therefore, to be warrant for the assumption that in the future, as in the past, the most valuable placers will be found within the limits given. One apparent exception is to be noted in the old beach placers of Hastings Creek. There is a possibility, however, that these may be the result of more than one concentration, that they may contain the gold of several old beach lines converging toward

Cape Nome, and that the gravels from which they were derived may not originally have carried any very notable amount of gold.

UNCONCENTRATED GRAVEL-PLAIN PLACERS.

Under this heading will be described a number of localities at which gold has been mined from deposits not directly traceable to wave action or to reconcentration in stream beds. Though there is undoubtedly some gold disseminated throughout the gravels of the coastal plain, only a few small areas have been prospected and mined and the extent and nature of the workable deposits are not known.

Mining is in progress in a locality not visited by members of the Geological Survey, on the east side of Nome River about 2 miles below Osborn Creek. A ditch 4 or 5 miles in length has been constructed from Osborn Creek for hydraulicking. No details are known, but the pay streak is believed to belong to the class of unconcentrated gravel-plain deposits.

It is reported that $2,000 was taken out near the head of Peluk Creek about 2 miles from the coast during the winter of 1901–2, since which time only grubstakes have been obtained. The pay gravel consists of angular schist pebbles and clay resting on a clay seam 2 to 4 inches thick. Mr. Hess measured the following section:

Section near head of Peluk Creek.

	Ft.	In.
Gravel and clay	1	6
Sandy clay and muck with some angular gravel	5–6	
Angular gravel with some clay	4	
Clay seam	2	4

Mining was by shafts and drifting without timbers. The gold is coarse, rough, and iron stained, and the largest nugget was worth $1.25. It is reported that 15 prospect holes, sunk with an 8-inch churn drill, reached gravel at 6 to 10 feet below the surface and bed rock at 72 to 82 feet. The bed rock is gray mica schist and has a nearly level surface about 30 feet above the sea. The owners of the property report that the lower 40 feet of gravel carry from $2 to $8 in gold per cubic yard, which appears to be a surprisingly high tenor.

On Holyoke Creek, a small tributary of Bourbon Creek, at a point about 3 miles from the coast, a shaft 118 feet deep was sunk, in which it is reported that a layer of beach sand carrying 10 cents to the pan in fine gold and containing fragments of sea shells was encountered.

In 1903 a company attempted, without success, to mine the gravel plain with a bucket dredge at a point 3 miles from Nome, south of Cooper Gulch and half a mile from the foothills. The mine is about 100 feet above sea level, but a shaft 65 feet deep did not reach bed rock. The deposit from the surface down carries some gold, but the parts that have been mined consist of small pay streaks resting upon layers of clay. It is reported that the clay layers are of small extent and irregularly distributed, like the clay seams of the beach placers described on pages 152–153. The pay streak, as mined, is said to have carried from $2.50 to $3 per cubic yard.

In 1902 and 1903 numerous drill holes to bed rock were sunk between Anvil and Dry creeks. The depth to bed rock ranges from 80 to 120 feet, and although no well-marked stream channels were located, broad depressions or hollows were found under some of the present stream channels, and nearly everywhere the bed rock was found to be above sea level. The method of prospecting used was to sample the gravel systematically by taking pans every 2 feet. These samples were panned down and the concentrate, consisting mainly of black sand, garnet sand, and fine gold, was assayed. It is claimed that a large area, believed to lie along an ancient beach line, was located, in which the whole deposit from the surface down to bed rock contained enough gold to pay if an economic plan of working on a large scale could be devised.

From other reliable sources the existence of an extensive body of rather low grade auriferous gravel near the foot of the hills south of Anvil and Newton peaks is reported at an average elevation of about 100 feet above the sea. In the practical working of such deposits some unique problems are presented to the mining engineer on account of their great thickness, frozen character, and poor drainage. One plan proposes to place a large dredge of unusually strong construction on Snake River, and, after testing the river bars and the deposits in the river bed, to dig out a channel into original gravels under the tundra. It will be necessary to make the dredge strong enough to excavate the frozen ground. A dredge was built on Snake River in 1904, but the results obtained are not yet known. Frozen ground, however, will yield to hydraulic processes, and, if dredging should fail, some combination of the hydraulic giant and dredge or hydraulic elevator may ultimately be devised, which will render these deposits productive. An ample supply of water with sufficient head can be obtained from Nome River.

RECONCENTRATED GRAVEL-PLAIN PLACERS.

Nearly all the streams which have trenched their beds across the coastal plain contain some gold derived from the original deposits by reconcentration. On the smaller streams, which lie wholly within the coastal plain, all the placers are obviously of this origin, whereas the larger streams, which rise in the upland region, may contain gold derived directly from the bed rock. Bourbon and Saturday creeks, which rise on the tundra, may be taken as examples of the small streams whose placers are entirely derived by concentration from the coastal-plain gravels. The amount of concentrated gold which their beds contain, when compared with the cubical dimensions of their channels, affords a basis for estimating the amount of gold contained in portions of the gravel plain.

A small amount of gold has been mined about 4 miles from the coast on Hastings Creek. The locality, which is comparatively unimportant, was not visited and the details regarding the occurrence of the gold are not known, but the reports indicate that it is of reconcentrated type.

Bourbon Creek, whose valley lies within the coastal plain, is about 3 miles long and joins Dry Creek near Nome. It carries from 20 to 100 miner's inches of water. Two miles above its mouth it forks, the west fork being called Holyoke Creek. Its trench probably will average 15 feet deep and 500 feet wide. Mining began on Bourbon Creek in 1900, and small amounts of gold have been taken out along nearly the entire length of the creek.[a] Its output during 1900 has been estimated at $5,000; in 1903 the output of Bourbon Creek and its tributary, Holyoke Creek, was probably not far from $40,000. The total output up to 1903 was probably about $100,000.

Near the head of the creek two sections of the gold-bearing deposit at points half a mile apart have been noted. One about a mile from the head is as follows:

Section 1 mile below head of Bourbon Creek.

	Feet.
Moss and muck	2
Rock and blue clay	6
Gravel	1
Blue clay	½
Schist gravel, pay streak	1
Clay (of unknown thickness).	

About half a mile from the head of the creek another section was measured, as follows:

Section one-half mile below head of Bourbon Creek.

	Feet.
Blue and red clay	4
Gravel; the pay streak	2
? Sand (of unknown thickness).	

On a claim near the head of the east branch a dump containing 25 cubic yards of gravel mined in the winter of 1901-2 yielded $300, or $12 per cubic yard. On a claim about one-fourth mile farther down the creek a pay streak 10 or 12 feet wide and 5 feet thick has been sluiced, yielding about $4 to the cubic yard. These

[a] Brooks, A. H., Richardson, G. B., and Collier, A. J., Reconnaissances in the Cape Nome and Norton Bay regions, Alaska, in 1900, a special publication of the U. S. Geol. Survey, 1901, pp. 69, 83–84.

figures show that some of the gravels of Bourbon Creek are
rich enough to work by shoveling into sluice boxes. Owing
to the low gradient and small flow of water, however, machinery is required to work the deposits at a profit. In 1903 three
bucket dredges, each with an estimated capacity of 250 yards in ten
hours, were installed. These machines were not successfully operated
because they were too small and light. Late in the fall of 1903 one
of these dredges was placed near the mouth of the creek, where it was
successfully used for removing the tailings from a small hydraulicking plant.

Saturday Creek is the name given to the upper part of Center
Creek, which flows into Snake River about a mile above Nome, and
Wonder Creek is the middle part of the same stream. This diversity
of names is due to the fact that under the local rules, established by
the prospectors, one claim only was allowed to each man on a creek.
This rule was often circumvented by changing the name of a creek at
each tributary and locating a claim with each change of name. Only
a small amount of work has been done on Saturday Creek since 1900,
and it was not reexamined in 1903. In 1900 one claim was being operated. The section showed 2 or 3 feet of muck, below which was 3 feet
of schist gravel containing the gold. This was generally bright yellow, though some was dark, and all was rather coarse. The largest
nugget found here was worth $14.50. Nuggets worth from 25 cents
to a few dollars were common. Some of the gold was smooth and
some rough. Quartz was attached to much of the gold. The concentrates consist of magnetite and garnets, with some scheelite.

Dry Creek and Newton Gulch, where they are intrenched across
the gravel plain, resemble the streams just described, but as their
headwaters are in the upland it is evident that their gravels may
carry gold brought directly from the bed rock. It is probable, however, that most of their placer gold has been reconcentrated from the
gravel plain.

Newton Gulch, an eastern branch of Dry Creek, 3 miles from the
coast, is about 1½ miles long and carries less than a sluice head of
water during the summer months. Its source is on the south slope
of Newton Peak, but it flows across an elevated portion of the gravel
plain through most of its length. Considerable gold is reported to
have been taken out of the creek, but it appears that the rich gravels
were of no great extent and that those which could be worked at a
profit by primitive methods have been exhausted.

A section exposed on Newton Gulch just above Left Fork is as
follows:

Section on Newton Gulch above Left Fork.

	Ft.	In.
Sandy clay	3	6
Schist gravel, somewhat cross-bedded		6
Sand	2	6
Gravel		7
Sand and gravel		3
Pay streak, consisting of schist gravel		3
Clay (thickness not known).		

Dry Creek heads between Anvil and Newton Peaks and flows southward about 6 miles to Snake River. Its upper 2 miles lie within a narrow valley cut in schists and limestones, but below it flows in a shallow trench across the coastal plain. At its mouth it carries from 50 to 150 miner's inches of water. Mining in the reconcentrated placers of Dry Creek began in 1900,[a] and the reported production for that year was $2,500. In 1903 the creek bed was mined at several places within a few miles of the foothills, but the whole production did not exceed a few thousand dollars. The gold is light colored and of varying coarseness, and nuggets $1 in value are the largest reported. The concentrates in the sluice boxes consist of much black sand and considerable scheelite.

The following section was measured on Dry Creek about 1½ miles from the foothills, an eighth of a mile above the mouth of Newton Gulch:

Section on Dry Creek above Newton Gulch.

	Feet.
Moss and muck	3
Sandy clay	3
Gravel composed of mica schist, limestone, etc. (the pay streak)	8
Clay (thickness unknown).	

The depth to bed rock at this place is not known. At other places the gravels are worked to a somewhat less depth. A prospect hole 1 mile from the hills has a depth of 92 feet, but does not reach bed rock. Three-fourths of a mile above the mouth of Newton Gulch a steam shovel has been used, but most of the mining has been done by drifting in the winter. Near the point where the valley of Dry Creek debouches on the coastal plain there is a series of productive bench claims situated along the course of an old channel on the east side of the creek. These are described in connection with the creek and bench placers of the upland region (p. 185).

Anvil Creek, which in its valley portions is one of the richest in Seward Peninsula, flows across the coastal plain for about 2 miles before it joins Snake River. In this distance little mining has been done, but in view of the rich placers found above (see p. 186) there

---

[a] Brooks, A. H., Richardson, G. B., and Collier, A. J., Reconnaisances in the Cape Nome and Norton Bay regions, Alaska, in 1906, a special publication of the U. S. Geol. Survey, 1901, pp. 69–82.

is good reason to expect workable deposits of fine gold in this part of its valley. As in all the streams of the tundra belt, its bed rock is deeply buried and it can not be mined profitably by the simple methods employed in the creek and bench placers.

Nome and Snake rivers carry large volumes of water and flow across the gravel plain in broad trenches. Their placers, which are of the river-bar type, have probably been derived as much from the upland as from the plains gravels.

Nome River is trenched across the coastal plain for about 5 miles. No attempts have been made to work the bars or river bed on a commercial scale. Should the dredging plants recently installed on other rivers of the peninsula prove successful, the bed of Nome River will doubtless soon receive attention.

The lower 6 or 7 miles of the Snake River Valley is intrenched in the coastal plain, and it here receives as tributaries Dry, Center, Anvil, and Glacier creeks, the gravels of all of which have yielded placer gold.[a] Snake River has also many productive tributaries north of the coastal plain. A small dredge was operated on Snake River near its mouth as early as 1900, but it was a primitive affair and the work was soon abandoned. A larger and stronger dredge was placed on the river in 1904, but the results obtained with it were not learned.

BAR, CREEK, AND BENCH PLACERS OF THE UPLAND.

The greater part of the Nome region, as here defined, lies north of the coastal plain and comprises uplands from 800 to 1,800 feet in height, bounded on the north by the Kigluaik Mountains, which reach elevations from 3,000 to 4,700 feet. Many of the lower hills are flattened on top, suggesting a dissected plateau, and their sides are benched at various levels. Near the seaward margin of the upland are systems of broad terraces about 500 feet in height, deeply buried in gravel, and these contain the high-bench placers described elsewhere. The upland is dissected by the valleys of Nome and Snake rivers and their many tributaries, and in these valleys lie the creek and low-bench placers described in the present section. Below the extensive high gravel-covered bench or plateau already mentioned the walls of the valleys show many minor benches, some of which are stream terraces and others products of differential weathering of the bed rock and of local creeping or sliding movements of the soil.

NOME RIVER BASIN.

Introduction.—Nome River has a broad valley for 20 miles above the coastal plain, in most places with one or two terraces not over 100 feet above the river bed. (See topographic map, Pl. VIII, in

---

[a] Since writing the above wonderfully rich gold deposits have been discovered on Little Creek, which flows into Anvil Creek within a short distance of Snake River.

pocket.) The floor of the valley is deeply gravel filled, and in some places the gravels rise in bluffs from 20 to 50 feet above the present flood plain. The flood plain ranges from 100 feet to a mile in width and at its lower end merges with the coastal plain. For 14 miles the gradient is about 10 feet to the mile and for 10 miles above this it is about 50 feet to the mile; the upper 5 or 6 miles are torrential.

Colors of gold have been found on the river bars for 20 miles from the mouth and are reported from prospect holes in the flood plains 10 or 12 miles from the coast, but as yet no attempts to extract the gold have been successful. Placers have been worked on many of the tributaries from both sides. The placers of the tributaries from the east, which are first described, are of less importance than those on the west, which include some of the richest diggings on the peninsula.

Osborn Creek.—Osborn Creek, the first large tributary of Nome River from the east, occupies a rather broad valley, the upper part of which is cut in siliceous chlorite schists and limestones of the Nome group. The mountain north of the mouth of Osborn Creek is composed almost entirely of greenstone, great blocks of which have rolled into the creek, forming bowlders of all sizes up to 10 feet in diameter. The gravels also contain many pebbles and bowlders of granite, which have probably been transported southward from the Kigluaik Mountains. (See pp. 94–99.) Similar granite bowlders are found up to an elevation of 800 feet on the mountain west of the creek. Although gold was discovered on this creek as early as 1900, very little development work has been done. It is reported that during the season of 1903 about 40 men worked on the creek. The workings are scattered along the creek for about $1\frac{1}{2}$ miles above a point 5 miles from the mouth.

Along the right bank of the creek there is a gravel bench from 8 to 20 feet high, which is probably a remnant of a more extensive deposit filling the valley. This bench had not been prospected in 1903. The gravel deposits of the creek bed average about 100 feet in width and have been excavated to a depth of 5 or 6 feet, where a clay seam forms a false bed rock; the real bed rock has not been reached. The gravels are rather coarse, pebbles 3 or 4 inches in diameter being common. Some gold is found from the surface down, but the richest pay lies just above the false bed rock. The gold is coarse, bright, and most of it well rounded, though some of the pieces are angular. A considerable portion is in nuggets worth from $1 to $20. When sluicing was in progress, a daily yield of $10 to $18 to the shovel was claimed, which, if true, would require an average gold content of $2.50 to $4.50 a cubic yard. Apparently there are on Osborn Creek extensive deposits of gravel containing low values, which can barely be worked at a profit by the primitive methods now in use. It is estimated that 1,000 miner's inches of water can be brought from New Eldorado Creek and the head of Osborn Creek to these diggings, with a head of 100 feet, for which 6 miles of ditch would be required. Should such a ditch be constructed, a large part of the gravels of the lower creek can probably be worked at a profit, but thorough prospecting of the ground to determine the gold tenor should precede any extensive ditch building.

Buster Creek.—Buster Creek, a tributary of Nome River from the east about 7 miles from the coast, is about 4 miles long. In the upper 3 miles of its course it flows through a narrow V-shaped canyon; below it is trenched across the floor of Nome River Valley. About 1¼ miles from its mouth the creek forks, the northern fork being called Lillian Creek.

Mining has been in progress here since 1899. The mines are easily reached, as the wagon road from Nome to Venetia and Iron creeks follows up Buster Creek from Nome River. The freight rate from Nome does not exceed 3 or 4 cents per pound.

The bed rocks are calcareous and chloritic schists and limestones of the Nome group. No intrusive greenstones have been found in place in the creek valley, though pebbles of such rock, which probably come from the greenstone on the south side of the creek, occur in the gravels. The gravels are from 2 to 3 feet thick and contain pebbles of schist, quartz, greenstone, and some granite. Some claims on this creek made good returns in 1899 and again in 1900. Much of the gold extracted was coarse, and the largest nugget found was worth $18. The assay value of the gold is said to be $18.60 to the ounce. In 1903 some of the operations included a reworking of the gravels that had been sluiced in 1900; others were in virgin ground. In the earlier operations the bed rock, some of it carrying the richest pay, was not cleaned up, so that it has been found profitable to strip the old tailings and to take out the bed rock to a depth of 1 to 2 feet. About 2 miles from the mouth of the creek some undisturbed ground was to be seen in a gravel deposit about 100 feet wide and 3½ feet thick, in which the richest pay was on the bed rock and not in it. A ditch half a mile long raised the water about 30 feet above the creek bed, and a hydraulic apparatus, consisting of a canvas hose and sheet-iron nozzle, throwing a 4-inch stream was used. The tailings were removed by a team of horses and scraper. The plant required three men to operate it and handled about 100 cubic yards per day of ten hours.

The bed rock and gravel of Lillian Creek, which joins Buster Creek 1 mile from its mouth, are similar to those of Buster Creek, but the gravel is more deeply covered by soil and muck. The gravels are from 3 to 10 feet thick and in places are covered by an overburden 5 feet thick. Only a few men were employed on the creek in 1903, and the output probably little more than paid wages. Some of the gold is so coated with iron oxide that the nuggets might easily be mistaken for small pebbles of hematite. About a mile from the mouth the pay was said to be evenly distributed through the gravel, which was here 3 to 4 feet thick, and the output was from $12 to $13 per man shoveling in.

Union Gulch, not over a mile long, joins Buster Creek from the north about a mile above Lillian Creek. Its bed rock is a gray schist broken up by many small faults. The gravel deposit is from 10 to 20 feet wide and about 30 inches thick and has been estimated to carry from $3 to $4 a cubic yard. The values are nearly all on bed rock. The gold is coarse and rusty, and the largest nugget found was worth $16.

Gold-bearing gravels have been discovered in a bench 100 feet above the creek bed on the south side of Buster Creek, about 2 miles from its mouth. This bench has been traced for about 700 feet and is crescent shaped, with the ends pointing toward the present creek bed. A section of this deposit from the surface down is as follows:

Section of bench 2 miles above mouth of Buster Creek.

	Feet.
Muck and slide rock	18
" Low-grade pay gravel "	10
Glacial mud (probably frozen slide deposit)	10
" High-grade pay gravel "	3
Iron-stained schist bed rock.	

The width of the upper pay streak is reported as 140 feet, and of the lower pay streak 25 feet. The lower pay streak is inclosed by well-defined bed-rock rims, and pay is found in the decomposed schist to a depth of 18 inches. At the west end of the deposit the rim on the creek side is wanting, and the gold-bearing gravels are spread over the hillside below the old channel, probably by local sliding. A section here, probably in the end of the old channel, shows 18 feet of ill-assorted gravel and muck, containing many bowlders of gneiss, granite, and greenstone, but a few feet down the hill the deposit is not more than 4 or 5 feet thick. Pans are said to average from 10 to 18 cents, and from a pit 75 by 30 feet and 18 feet deep $4,000 is said to have been taken. The gold resembles that of Buster Creek, and is reported to assay $18.97 to the ounce. It is coarse, but has few nuggets of any considerable size, and is mostly well worn, though some is rough. About half a mile east of this place a smaller deposit at approximately the same elevation is cut by Grace Gulch, a small tributary of Buster Creek. This was worked in the winter of 1902–3, yielding about 60 ounces of gold similar to that of Buster Creek. The working of these benches is difficult on account of the small water supply. A ditch along the hillside from Grace Gulch brings to the first-described locality about 50 miner's inches of water, which is mostly seepage from the hillside above the ditch.

From the above facts it seems evident that this deposit was formed in an old channel of Buster Creek. After the lower and older pay was formed the channel was partly filled with material sliding in from the hillsides and raising the creek bed. In the higher bed thus formed the thicker and wider upper pay streak was deposited. The creek then deserted this channel, which was afterwards covered by slides from the hills above. These slides at the upper and lower ends of the deserted channel have continued and either carried the gold-bearing gravel down to Buster Creek or scattered it in the form of a hillside placer.

Dewey Creek.—Dewey Creek is a small eastern tributary of Nome River about 11 miles from the coast. During the summer of 1903 very little mining was done on the creek, and it was not visited by members of the Geological Survey. The following description is quoted from Brooks: [a]

This stream rises in the limestone hills a mile and a half north of the head of Lillian Creek and flows into Nome River about 4 miles above Buster Creek. At its head Dewey Creek has cut a gorge, from which it emerges to flow through the gravels adjacent to Nome River. Some gold has been taken from this creek, but not very much. A claim at an elevation of 300 feet has probably yielded the most. The section in the creek bed, to which work has been confined, is 1 foot muck, 1 foot sandy blue clay, 3 feet gravel, with pebbles of schist. quartz, greenstone, limestone, and granite. The gravel carries gold. Below is a false bed rock of sandy clay. The gold is rather light in weight, the largest piece being worth only a dollar. Considerable scheelite and garnet are associated with the gold, and not much magnetite. It has been reported that a natural amalgam occurs in the gravels of Dewey Creek, but it was not found by us, nor did we find any cinnabar. Colors have been found in the high gravels, which run up to 800 feet in the neighborhood.

Basin Creek.—Basin Creek, an eastern tributary of Nome River 13 miles from the coast, is about 3 miles long and flows westward. On the east side of the valley of Nome River at this point the hills rise to an elevation of 1,400 to 1,775 feet above the sea. Basin Creek through the upper 2 miles of its course occupies a deep valley in these mountains; in its lower half mile it flows across the broad valley of Nome River. At the edge of this valley the creek flows in a sharply cut gorge for half a mile, but above the gorge the creek valley widens somewhat, making a sort of basin, from which it takes its name. Early in June, 1903, the volume of the stream was probably not far from 500 miner's inches. The gradient is comparatively high, exceeding 100 feet to the mile. Gold was discovered and some mining was done on this creek as early as 1900. The amount of gold produced on the creek before 1903 probably did not exceed $20,000 or

[a] Brooks, A. H., Richardson, G. B., and Collier, A. J.: Reconnaissances in the Cape Nome and Norton Bay regions, Alaska, in 1900, a special publication of the U. S. Geol. Survey, 1901, pp. 78–79.

$30,000. In June, 1903, a great many loads of freight, consisting of pipes for hydraulic work, were being hauled to Basin Creek over a road which follows the gravel bars of Glacier Creek nearly to its head, then crosses Banner Creek and Nome River and follows the bed of Basin Creek up to the camps.

In the bed rock of Basin Creek limestone predominates and is interbedded with some calcareous and chloritic schist. The strike is generally north and south and the dip to the east at an average angle of about 20°. These rocks certainly belong to the Nome group, and it is believed that the limestones may be correlated with the Port Clarence, and they are so represented on the geologic map. The gravels consist mainly of limestone and calcareous-schist pebbles, with a few pebbles of greenstone. An average section of the gravel where it has been exposed by mining operations is as follows:

Section on Basin Creek.

	Feet.
Soil	2
Rounded creek gravel	3
Coarse angular material and clay (pay streak)	12
Bed rock containing gold in crevices	$\frac{1}{2}$–$1\frac{1}{2}$

In the pay streak over 30 per cent of the pebbles are more than 10 inches in diameter. The pay streak is about 150 feet wide, and it is said to extend along the creek for about 2 miles. The gold is bright, rough, and coarse. Nuggets worth from $1.50 to $2 are found, and many of the pieces are crystalline. The concentrates contain ilmenite, hematite, and scheelite. It is not probable that the gravels are of very high grade, for although mining began here about 1901 only a small area was worked previous to 1903. It appears that the gravels are not of sufficient value to pay wages if worked by the primitive method of stripping by hand and shoveling into sluice boxes. The limit of value can probably not be far above $6 to the cubic yard. Since 1903 mining on the creek has been done with hydraulic nozzles and elevators. The gradient of the stream bed is so great that the tailings need not be elevated more than 10 or 15 feet. With the hydraulic plants installed it is probable that the whole gravel deposit within the valley of the creek can be worked out in one or two seasons of continuous running.

Extra Dry Creek.—Extra Dry Creek flows into Nome River from the hills on the west side, about 7 miles from the sea. At its head it has a narrow canyon, but for most of its course it is trenched across the comparatively flat valley floor of Nome River. The creek gravels were being mined in 1900, when the workings were examined by Richardson.[a] In 1903 one man was working a rocker on the creek

[a] Brooks, A. H., Richardson, G. B., and Collier, A. J., Reconnaissances in the Cape Nome and Norton Bay regions, Alaska, in 1900, a special publication of the U. S. Geol. Survey, 1901, pp. 76–77.

and probably making scant wages. The whole output of the creek has probably not exceeded $20,000.

A section in the creek bed about one-half mile from its head, where the richest placers were found, showed 1 foot of muck, 1 foot of sandy clay, and 6 feet of schist and quartz gravel on mica-schist bed rock lying nearly flat. Only the lower 2 feet of gravel were worked with rockers, the water being obtained from springs. The largest nugget found was worth $13, but there were several averaging from $3 to $6, some of which were attached to pieces of quartz. In the lower part of the creek the gold is fine and little work has been done.

Dexter Creek.—Dexter Creek is a western tributary of Nome River about 7½ miles from the coast. It forks about 2 miles from its mouth, the southern branch being called Left Fork and the west fork receiving the name Grass Gulch. (See sketch map, fig. 7, p. 150.) About 100 yards above the junction Grass Gulch forks again, the northern branch being called Deer Gulch. Another tributary mentioned in the economic descriptions is Grouse Gulch, which enters from the north about one-half mile below the mouth of Grass Gulch.

The mines on Dexter Creek are easily reached from Nome by a wagon road which follows Dry Creek to its head and, passing over the divide, comes down Left Fork. Freight may be hauled by wagon from Nome, and in 1903 was delivered at any point on the creek for 3 to 4 cents per pound. The mines are also easily reached from either Summit or Dexter station of the Seward Peninsula Railway, to which the freight rate from Nome is at the rate of 1 cent per pound. These two stations are both on the divide between the basins of Dexter and Anvil creeks. A considerable settlement, which includes several road houses and stores, is located at the forks of Dexter Creek, the oldest establishment there being known as the " Sour Dough " road house.

Through the 2 miles of its course below the principal forks the gradient of the stream is about 140 feet to the mile. Above this point, where the several branches converge, the gradient is higher. The amount of water in the creek during the dry season is less than a sluice head, but in the spring when the snow is melting and after the late summer rain there is sufficient water for sluicing in a small way. The creek valley has a V-shaped cross section and in its floor the creek flows in a shallow trench or gorge, leaving benches along its sides from 10 to 50 feet above the bed.

The bed rock of Dexter Creek consists of limestones and mica and chloritic schists of the Nome group. The strikes are in general nearly parallel with the course of the creek. Although schists predominate in the valley walls, limestone is more commonly found in the creek bed, but it is associated with some bands of schist. Rather extensive gravel deposits have been found along the bed of the creek from its mouth to the sources of its several tributaries, and these range from 2 to 20 feet or more in thickness. Where the bed rock is limestone the gravels gradually pass into the decomposed bed rock and extend downward in crevices and fissures. Gravel deposits also occur at a number of places in benches lying parallel with the creek bed, and several of the tributary streams head in high-bench gravels covering the divides.

Dexter Creek was staked and gold on it discovered as early as 1899. In 1900 mining was in progress through the whole length of this creek and on each of its tributaries, the total output being estimated at $300,000. In many places water which passed through the sluice boxes was pumped back and used again and again. Elsewhere the gravels were washed with rockers, to which water was carried by hand or was hauled from Nome River in barrels. A criterion of the richness of the gravels lies in the fact that they were worked with rockers, using water purchased at $2.50 a barrel. In 1901 and 1902 the creek was almost abandoned, as the richest pay streak had been exhausted. With the completion of the Miocene ditch in the summer of 1902 the work on Dexter Creek was revived, and in 1903 the creek, probably produced as much gold as in 1900. The water from the Miocene ditch was used over and over again as it passed from one claim to another along the creek. In general the work was done by shoveling the pay gravel into the sluice boxes, but in several places the gravel was hydraulicked. As a rule the head was only 20 or 30 feet, and canvas hose and a galvanized-iron nozzle were sufficiently strong to stand the pressure. The gradient of the creek throughout its length is such that tailings are readily disposed of. With the revival of mining along this creek a successful search for bench deposits has been carried on, and it is probable that the gravels of the creek will not be wholly worked out for several years.

At a point one-half mile from the mouth of the creek the workings in 1903 showed 6 feet of pay gravel containing many large bowlders mixed with finer sediments and clay. Most of the bowlders were of limestone and the clay is probably residual from the limestone and schist.

One-half mile farther up, on the right bank of the creek, pay gravel in the form of a bench about 150 feet long by 50 feet wide was found resting upon blocky limestone 10 feet above the present creek bed. The gold extended for some distance into crevices of the limestone, which was taken up and washed during the mining operations.

A short distance above, the pay streak in the creek bed consisted of broken limestone fragments that seem to have slid down from the hillside on the southern bank of the creek. This deposit is about 7 feet thick and about 30 feet wide.

On the left bank opposite this point bench gravels have been found about 75 feet above and 200 feet north of the creek bed. A pit to bed rock showed the washed gravel to be 30 feet thick. The bed rock at the bottom of this pit is reported to have a well-marked rim on the side toward Dexter Creek, indicating an old channel parallel with the present channel. Gold was found near bed rock, but the deposit had not been thoroughly prospected.

About 100 yards above the mouth of Grouse Gulch the workings on the left bank of the creek extend into a great mass of gravel, which may mark the upper end of the old channel just noted. The bed rock was here stripped for 100 feet or more, revealing a nearly level floor, beyond which it rises toward the hill. This bed rock is a decomposed schistose limestone with an irregular surface, due to the fact that some parts of it apparently disintegrated more rapidly than others. Above this bed rock lies a gravel bed which on the north side away from the creek bed has a thickness of nearly 20 feet. The section shows a layer of several feet of yellowish clay, below which there is nearly 15 feet of gravel consisting of badly sorted schist, limestone, and granite pebbles, mixed with a large amount of yellowish sandy clay.

Near the forks of Dexter Creek the gravel deposits are wider than those exposed farther downstream, but nearly all of the gravel worked in 1903 had been picked over in 1900 and 1901, when, owing to the scarcity of water, only the richest parts could be handled. The mining was done here by means of a 2-inch hydraulic nozzle under a head of 60 feet, with which the surface was first piped off almost to bed rock and carried through the sluice boxes, after which the remnant of the gravel and the loose bed rock were shoveled into the boxes by hand. The deposit worked covered an area about 60 feet wide and 250 feet long by 6 to 10 feet deep, and about half the bed rock was schist, the remainder being limestone. Gold was not found in the schist bed rock to a depth of more than a few inches, but in the fissures of the limestone the gold had penetrated to an indefinite depth. Some holes in the limestone have been found in which the gold-bearing gravels are mined to a depth of 20 or 30 feet. Though the bed rock rises on the south side of the deposit, the limits of the pay gravel had not been reached. The gold here is all waterworn, the pieces are smooth, and no corners or angles remain.

About a quarter of a mile above the forks, in Grass Gulch, is one of the richest of the Dexter Creek mines. Here a pit 150 feet wide by 300 or 400 feet long, mostly on the south side of the original creek bed, had been worked out in 1903. At the upper end of this pit a face of gravel about 25 feet high was exposed at the time of Mr. Collier's visit. (See Pl. II, A.) This gravel carried gold from a depth

A. ANGULAR GOLD-BEARING GRAVELS ON GRASS GULCH.

B. CREVICED LIMESTONE BED ROCK ON GRASS GULCH.

178A

within a few feet of the surface down to the bed rock. Above the gravel there was a strip of moss and soil not over 2 feet thick. The section from the surface down is as follows:

Section in Grass Gulch.

	Ft.	In.
Moss and soil	2	
Brown gravel of rounded pebbles and clay	6	
Dark streak, probably stained with manganese		3
Brown gravel of rounded pebbles and clay	9	
Fissued limestone bed rock.		

The gravels consist mainly of pebbles and bowlders of limestone and schist, but contain also some pebbles of granite. The pebbles are well rounded and range from one-half inch to 10 inches in diameter. They are not well sorted in layers, though the whole deposit appears to be water-laid, and is cemented by a cinnamon-colored, rather sticky clay sediment. Shingling is not very distinct, but seems to indicate a current flowing toward the east.

The bed rock exposed by mining is chiefly limestone that has been fissured, and the erosion along the fissures has left a very irregular surface. (See Pl. II, *B*.) The sticky brown sediment extends into fissures and crevices for an unknown distance. These fissures connect with deep-seated channels in the body of the limestone, as is shown by the fact that they can not be filled with water, and that in the deeper excavations the sound of running water can be heard. A thin bed of decomposed schist included in the limestone forms a reef across the floor of the mine, into which the placer gold has not penetrated more than a few inches. The gravel above the schist reef is not as rich as that resting upon the limestone. These deposits are only partly frozen, and after hydraulicking the frozen parts remained as pinnacles. The gravels are angular and gold occurs in them from the surface down (Pl. II, *A*) and in the fissures of the bed rock as far as the excavations have gone, but the greatest values occur in the gravel layer resting upon bed rock. Many of the residual bowlders have been found to be spangled with alluvial gold. The gold is comparatively coarse, but ranges from fine dust to large nuggets. The largest nugget found, worth about $412, was well worn and apparently consisted originally of a sheet of gold surrounded by quartz, but the quartz has worn away and the edge of the gold has been worn smooth.

The mine was worked with water from the Miocene ditch, which has an elevation above the mine floor not exceeding 20 feet. About 700 feet of sluice boxes, with riffles through the whole length, were used, and the pay streak was, for the most part, handled with shovels.

Above this mine the valley of Grass Gulch is not well defined. Extensive prospecting has been done here and the limestone bed rock, which underlies the valley, has been penetrated to a considerable depth, exposing a series of fissures and crevices probably connecting with underground channels similar to those just described, but no large body of gold-bearing gravel is known to have been discovered. The mine workings of the Summit bench, north of Anvil Mountain in the divide between Grass Gulch and Specimen Gulch, lie from one-fourth to one-half mile west of the highest workings on Grass Gulch and are described on page 206. From their position it seems quite possible that the upper gravels on Grass Gulch are connected with these bench deposits, but the bed rock on Grass Gulch appears to lie 50 or 60 feet lower than that of the bench.

On Deer Gulch, a northern tributary of Grass Gulch, a succession of schists and limestones is exposed and some gold has been obtained. The gravels were worked to a small extent in the early part of 1903, but operations had been abandoned before September.

The gold-bearing gravels of Dexter Creek also extend up Left Fork and probably connect with the gravels in the high-bench deposit on the divide between Dexter and Dry creeks. About one-fourth mile above the mouth of Left Fork a force of men were employed in the fall of 1903 exploiting the creek gravels with a hydraulic giant supplied with water from the pumping station on Nome River. The bed rock here consists of limestone and thinly bedded calcareous schists. The pay streak is not more than 50 feet in width and the gradient of the stream is such that the sluice boxes could be laid along the stream bed and the gravels above readily piped into them.

Grouse Gulch, a tributary of Dexter Creek from the north side, heads in the deep gravels of the Snow Flake and Sugar mines, from which it has a steep gradient southeastward to the creek. In 1901 and 1902 some mining was done along Grouse Gulch and a bed rock consisting principally of limestone but containing also some thin beds of schist was exposed. In the upper part of the gulch the bed rock is mainly schist. About one-fourth mile above the junction of Grouse Gulch and Dexter Creek a pit had been excavated on the left bank of the gulch, in which a face of gravel 10 feet thick was exposed, showing 6 inches of moss and muck; 4 feet of yellow sandy clay, containing a few rounded pebbles; 6 feet of stratified gravel and sand, consisting mostly of flat schist pebbles arranged parallel to the stratification, and yellowish-red and brown sand containing a great deal of fine mica. The bottom of this layer contains some large, flat pieces of rock, which show little if any rounding. It rests upon a very much decomposed schist bed rock. Near bed rock is a layer of vein-quartz pebbles. The bedding of the gravel dips toward Dexter Creek at an angle of 3° to 5°. Several pieces of granite were found in the dump from these workings, but a diligent search failed to show any granite in the face of the cut. About half a mile from Dexter Creek considerable gold has been obtained from a tunnel extended from the bed of Grouse Gulch into a gravel bed, which is described in connection with the high-bench deposits (p. 199).

Banner Creek.—In 1900 Brooks wrote as follows concerning Banner Creek: [a]

> This is a small tributary to Nome River from the west, about 3 miles above Dexter Creek. Its bed rock is schistose greenstones and limestones. Colors have been found on this stream, but no paying placers. It must be said, however, that but little prospecting has been done. The stream is a small one and during dry weather does not furnish a sluice head of water.

In 1903 the writers traversed the creek, but, though some evidences of old workings were seen, no miners were on the creek and it is doubtful if it has even produced grubstakes.

Hobson Creek.—Hobson Creek joins Nome River from the west, about 18 miles from the sea. It has a rather narrow valley and southerly course. The bed rock is massive limestone with greenstone schist near the head of the stream. Considerable prospecting was done in 1900, but very little gold has been produced. The creek, which carries about 1,500 miner's inches of water, is one of the feeders of the Miocene ditch.

Dorothy Creek.—Dorothy Creek, a western tributary of Nome River, 24 miles from the coast, flows through a small canyon having a general northerly course. Above the canyon the valley broadens out to a small basin having a gravel floor. The bed rock is limestone, tentatively correlated with the Port Clarence, and greenstone intrusives, which strike in a northeasterly direction and dip to the northwest. Mineralized quartz veins are very plentiful. The creek is reached by a wagon road from Nome, and the cost of hauling freight in 1903 was less than 10 cents a pound. Where it emerges from its canyon into the valley of the Nome, Dorothy Creek has a broad delta, in which the gravel is about 3 feet thick and overlies a greenstone-schist bed rock. The gravel deposits of the creek bed have been mined since 1900, and, according to T. A. Campion, who in 1903 owned most of the claims, have produced in all $44,000, but at small profit to the numerous operators. Most of this gold was obtained from claims within 1 mile of the mouth. The gold is light colored and said to assay between $14 and $15 to the ounce. In 1902 the Campion ditch was completed to this creek from Buffalo Creek, near the head of Nome River, and it was proposed to hydraulic the whole gravel deposit of the creek bed.

Above Dorothy Creek on the west side of Nome River some extensive gravel deposits lying in the divide between Nome River and the

---

[a] Brooks, A. H., Richardson, G. B., and Collier, A. J., Reconnaissances in the Cape Nome and Norton Bay regions, Alaska, in 1900, a special publication of the U. S. Geol. Survey, 1901, p. 79.

head of Stewart River have been developed in excavating for the Campion ditch. Boer Mountain, which lies west of Nome River between the heads of Sinuk and Stewart rivers, is composed of green quartz-chlorite schists containing many veins and stringers of clear glassy quartz. Some pieces of float quartz resembling that in the mountain, but containing large pieces of free gold in one side of the specimen while the other side is barren, were found along this ditch. One nugget was found which weighs three-fourths of a pound and which is nearly all gold, but includes some quartz similar to that from the mountain. It is proposed to hydraulic the benches along the west side of Nome River after the Dorothy Creek placers are exhausted.

Divide Creek.—Divide Creek rises in the broad divide at the head of Stewart River and flows eastward into Nome River. It is a small stream flowing over deep gravel deposits that may be in part of glacial origin. At the head of the creek these deposits are very deep. Some sluicing was done on the creek during 1903. It is proposed to hydraulic these gravel deposits with water from the Campion ditch.

Boer Creek.—Boer Creek is a small stream which rises in Boer Mountain, between the headwaters of Stewart and Sinuk rivers, and flows northward to Hudson Creek, which flows into Buffalo Creek, a tributary of Nome River. The distance from Nome is about 30 miles. It is reached by a wagon road, and freight rates in 1903 were not less than 10 cents a pound. About $2,800 was produced in 1901 and 1902, but the creek was not worked in 1903.

The bed rock is green chloritic schist containing many veins of clear glassy quartz, and the pay streak is reported to be narrow and confined to the creek bed. The thickness of the gravel ranges from 18 inches to 8 feet.[a] In the upper part of the creek the bed rock is covered with a deposit of conglomerate composed mainly of subangular schist pebbles firmly cemented by hydrous iron oxide. The nature and limits of this conglomerate have not been determined, but it is reported to carry values in placer gold and in this respect resembles the uncemented pay gravel of the creek. Two grades of gold are obtained—one light-colored gold assaying $16 to the ounce, the other assaying $18 to the ounce. Mining on the creek has been shut down pending the completion of the Campion ditch.

SNAKE RIVER BASIN.

Introduction.—There are many placer-bearing streams tributary to Snake River, which heads in the upland region about 20 miles north of Nome and flows nearly southward to Bering Sea (Pl. VIII, in pocket). The lower 7 miles of its course in the coastal-plain belt is described on page 170. Above this portion it occupies a valley 4 or 5

---

[a] This information is furnished by T. A. Campion, of Nome.

miles wide, and here the river has trenched the valley floor, leaving low terraces along the sides. The flood plain is about half a mile in width, beyond which the slopes rise gently 300 or 400 feet and then become very much steeper. It is a noteworthy feature that all bench gravels and placers thus far discovered occur below the level of these steeper slopes. The bed rock of the greater part of the Snake River basin is composed of undifferentiated schists and limestones of the Nome group, which are here locally more highly mineralized than the average rocks of this group, for gold-bearing quartz veins and stringers are of more abundant occurrence.

No attempts at mining the river bars have been made above the coastal-plain belt, but the bars are known to be more or less auriferous throughout the length of the stream.

The low benches that occur along both sides of the Snake River valley are in part rock cut and in part built up of gravel. At a locality on the east side of the valley about 10 miles from the coast, strippings of the sloping valley floor made in a search for quartz veins have revealed a rock surface covered to a depth of 4 or 5 feet with muck and residual schist fragments, with here and there some washed gravel. Here in some places gold believed to be derived directly from the disintegration of quartz veins has been found on bed rock. The average-sized pieces of this gold are worth not more than one-twentieth of a cent and are not in the least rounded or water worn. Some of these deposits doubtless carry placer-mining values, and they are well situated for ground sluicing or hydraulicking, but they have thus far been neglected.

Pioneer Gulch.—In 1903 a company was preparing to mine the gravels of Pioneer Gulch, near the mouth of North Fork of Snake River, about 16 miles from the coast. A 4½-mile ditch bringing about 500 miner's inches of water from Last Chance Creek to this company's property had been completed, and actual mining operations commenced in September, 1903, but the results obtained since that time have not been learned.

Newton Gulch.—Newton Gulch is a small valley with smooth, grassy walls incised in the south side of Newton Peak, whose slopes are gentle and moss covered up to an elevation of 400 feet, above which they are steep and bare. The stream is about 2 miles long and carries less than a sluice head of water into Dry Creek. Its head lies in a sharp canyon well up on the mountain, but most of its valley is only slightly trenched into the more gentle slopes at the edge of the coastal plain. The gravels are here from 30 to 150 feet wide and from 6 to 27 feet deep, but the deepest pay streak reported is 10 feet beneath the surface and rests upon a clay seam. At the mouth of the gulch, where the valley floor merges with the coastal plain, the bed rock is 27 feet below the surface.

Claims near the mouth of the gulch are reported to have produced $75,000 from 1901 to 1903; the pay streak was from 2 to 6 feet thick and averaged over 100 feet in width. The gold is coarse, mostly bright and rather rough. Estimates of the gold tenor of these deposits are rather indefinite, but a large part of the pay streak must have run over $15 to the cubic yard.

About three-fourths of a mile north of the mouth of Newton Gulch a pay streak has been found in a gravel deposit on the slope of the hill above the creek bed, and a large area has been worked to a depth of 10 feet without finding either bed rock or definite limits to the pay streak. An average section of the deposit is as follows:

Section near mouth of Newton Gulch.

	Feet.
Tundra vegetation and peat_____	2
Sandy subangular gravel (pay streak)_____	6–8
Blue clay (thickness not known).	

The pay streak contains a few bowlders as large as 18 inches in diameter. The gold in the pay streak is said to average above $2 to the cubic yard. It is bright and mostly smooth, but some of it is very rough. It is coarse and easily saved, though the largest nugget found was worth only $1.50.

In 1904 this mine was worked by hydraulicking with pumped water from a 2-inch nozzle under a head of about 100 feet and the tailings were dumped into Newton Gulch. Though the deposit was frozen it was easily worked in this way after the tundra growth had been removed. The origin of this deposit is difficult of explanation, with the meager data available regarding it. As noted, it lies upon a gentle slope which continues upward to an elevation of 400 feet. Similar gentle slopes rising to about the same elevation surround the bases of many of the hills and are features of the larger valleys of the Nome region. It seems likely that they represent remnants of an old eroded surface. If this be true, bench gravels may be expected to be found up to an elevation of 400 feet, and below these gravels hillside deposits would probably result from the destruction of such benches by the creeping movement of the soil and loose bed rock down the slopes.

Dry Creek.—Dry Creek heads in the flat divide separating it from Left Fork of Dexter Creek, and joins Snake River near Nome, after traversing the coastal plain. The lower part of the creek has already been described (p. 169). In its general form it much resembles Dexter Creek, as its source lies in a high gravel terrace, in which it has cut a basin and below which the valley contracts. Much of the valley is deeply filled with gravels that are probably not rich in placer gold. Although some claims along the creek bed and along Bear Creek, a western tributary, have been worked, they have not been very productive, and in 1903 all the creek placers were idle. Most of the gold produced has been obtained from an old channel on the east bank 1 to 2 miles above Newton Gulch, about 50 feet above the creek bed, which has been traced for about three-fourths of a mile. This channel is about 600 feet east of and parallel to the present creek channel. It has a grade of about 100 feet to the mile. The pay streak is from 20 to 60 feet wide and the gravel is shingled in the same direction as that of the present stream. It is covered by about 50 feet of gravel, slide rock, muck, and silt, and consists in general of slightly rounded schist pebbles, but contains a few pieces of greenstone and bowlders of granite. In some places the pay gravel is cemented with iron oxides, but most of it is uncemented. On the gently sloping hillside above there are many evidences of creeping movements in the surficial layer. At both the upper and lower ends of the portion mined prospecting has been done to locate the continuance of the channel, but so far without success. Near the north end of the channel the rim is wanting on the west side and the pay streak pinches out where the surface of the bed rock dips down toward the present creek bed. The bed rock, of schist, is 20 to 50 feet below the surface and the pay streak, 4 feet thick and 60 feet wide, rests upon it. It is reported that the pay dirt mined during the winter of 1902–3 yielded between $9 and $12 to the cubic yard. About one-fourth mile south of this locality the depth to bed rock is 50 feet, the section being as follows:

Section on Dry Creek about 1½ miles above Newton Gulch.

	Feet.
Muck and fine dark-colored silt	45
Fine sand	5
Pay gravel	½–2
Schist bed rock.	

The pay streak is here 20 feet wide and is defined by well-marked rock rims. Six thousand 8-pan buckets of pay dirt are reported to have yielded $16,000, which is approximately equivalent to $50 a cubic yard, 150 pans being considered equal to a cubic yard. The gold of this bench is not greatly waterworn and many of the small nuggets contain quartz. The gold is comparatively coarse, the average-sized pieces probably being worth about 1 cent, but it ranges from fine dust to coarse nuggets. Magnetite, garnet, ilmenite, scheelite, and quartz are associated with the gold in the concentrates. A sample of the concentrate from which the free gold had been separated by panning and treating with quicksilver was assayed, yielding 13.76 ounces of gold and 2.64 ounces of silver to the ton.[a] Some of this gold was undoubtedly contained in small quartz grains remaining in the concentrate, though a part must have been held by some other minerals. Mining is done by the shaft and drift method, and

[a] Assay by E. E. Burlingame & Co., Denver, Colo.

the dump is washed early in summer with water collected from melting snow on the hillside above.

Cooper Gulch.—Cooper Gulch, which in general characteristics much resembles Newton Gulch, is a small valley cut out of the south slope of Anvil Peak, at the base of which there is a well-marked bench between 325 and 400 feet above the sea. Cooper Gulch is trenched across this bench and below it spreads out in a fan-shaped delta, beyond which its waters are not again collected in a definite channel. In 1900 a few men worked here with rockers, making bare wages, but early in 1903 a pit about 100 by 300 feet was hydraulicked with water from the pumping plant on Snake River. In September of that year the place was found abandoned and the machinery removed. The bowlders in the tailing piles are principally limestone, but there are also some of granite and greenstone. The edges of the cut show from 2 to 4 feet of brown, sandy, stratified material overlying gravel of unknown depth.

Anvil Creek.—Anvil Creek, an eastern tributary of Snake River, has a length of about 6 miles. The upper 4 miles of its course is in a valley lying between Anvil Peak on the east and a lower flat-topped ridge on the west. This valley is comparatively broad and is not symmetrical in cross section, as the east side rises by a gentle grass-covered slope to an elevation of 400 or 500 feet above the sea, whereas the slope on the west side is much steeper. The creek bed is trenched to a depth of 20 or 30 feet in the valley floor, leaving a sloping bench on the east side, with no corresponding bench on the west side. (See fig. 11, p. 189.) This bench is more or less covered with gravel and slide material, and near its lower edge a series of old channels rich in placer gold have been discovered.

The bed rock of the creek is composed of various schists belonging to the Nome group. In general these rocks strike approximately parallel with the creek bed and dip at a high angle both east and west. They are faulted in many places, a majority of the fault planes striking a little east of north, parallel with the general course of the valley. At the lower end of the valley the bed rock for about half a mile is graphitic schist. Above this for half a mile it is a gray chloritic calcareous schist, and above this there are belts of both graphitic and quartz-chlorite schist. All these rocks contain small stringers and veinlets of quartz and calcite, some of which are mineralized with pyrite and other sulphide minerals, while others contain free gold. No veins of economic importance had been discovered up to 1903.

In the last five years the creek has produced over $5,000,000 in gold. The discoveries of gold which first attracted attention to the Nome region were made near the lower end of the Anvil Creek valley. Below the point where the creek emerges from the highlands its gravels have not produced placer gold in notable amounts, and this part of the valley had been only cursorily examined by geologists. Within the upland the flood plain of Anvil Creek ranges in width from 300 feet at its lower end to about 50 feet near its upper end. The gold of the valley-floor gravels has practically been exhausted, though it is proposed to rework some of the gravels by more refined methods and extract the small values that may be left. Present mining operations are confined to the gravels found in bench deposits along the east wall of the valley. These deposits have been described by Brooks[a] as follows:

The gravels of the creek floor have all been derived from within the basin and include the various types of bed rock which have been described. They have a thickness varying from 3 to 5 feet, and are usually covered by 2 to 3 feet of muck and clay. The gravels generally carry pay throughout their thickness and are all put through the sluice boxes. In the mining operations from 1 to 2 feet of the underlying weathered bed rock is usually found to carry values and is sluiced with the gravels. The richest deposits are found at and in the bed rock. Where the creek valley broadens out, about a mile above its mouth, bed rock has not been reached in the prospect holes. Some gold has been found on a layer of blue clay about 8 feet below the surface. This gold, however, is finer than that found on the bed rock on the stream above and the deposit is not nearly so rich.

Anvil Creek has produced gold throughout its length and all of the claims have paid profits. They are, however, not all of equal richness, and this is probably due to the fact that some of the gold has been contributed by the old terraces, which, being dissected by smaller tributaries, have contributed their gold to the main stream.

The bed of Anvil Creek for nearly 3 miles has been practically all worked over once during the last two years. Future developments will be directed toward the bench claims, which up to 1900 were practically untouched, and toward reworking the creek gravels with more refined methods of separation. In the lower course of the creek, dredging may also be found profitable.

Anvil Creek gold is both coarse and fine. Nuggets up to $300 in value have been found, and $2 and $3 nuggets are very common. It is irregular, generally chunky rather than flat, and has a dark-yellow, sometimes almost brown color. The fineness of the gold, according to Dr. Cabell Whitehead, is 0.890, or $18.33 per ounce. Quartz and fragments of a schistose rock are often found attached to nuggets. Among the heavier minerals in the concentrates magnetite and garnet are most common and occasionally scheelite is found.

More detailed examinations show that although nearly all the gravels of the creek bed have been derived from points within the basin and consist of the various types of rock that make up the Nome group, together with much vein quartz, they also include some granite pebbles and bowlders that are found in place only in the Kigluaik Mountains.

The irregularity in the distribution of the gold is no doubt in part due, as Brooks has indicated, to a reconcentration from older chan-

[a] Brooks, A. H., Richardson, G. B., and Collier, A. J., Reconnaissances in the Cape Nome and Norton Bay regions, Alaska, in 1900, a special publication of the U. S. Geol. Survey, 1901, pp. 73–74.

nels, but it is also a result of the inequalities of occurrence of the gold in the bed rock. The rich placers on the upper part of the creek, for example, can. be attributed partly to the reconcentration of high-bench gravel washed down by Nekula Gulch, and the richness of the placers near Discovery claim is in all probability due to veins in the local bed rock, for it is there that all the phenomenally large nuggets have been found, and these could not have been transported far. The facts necessary for estimating with any degree of exactness the gold tenor in the Anvil Creek placers are not at hand, but the average for all the gravels mined along the creek can not have been less than $5 a cubic yard. In the richer spots much of the gravel contained more than $50 to the cubic yard. The gravels of the benches worked at the present time are of lower grade and many of them could not be worked by the expensive methods employed in the first few years of mining.

The richest of the creek gravels were within half a mile both above and below the site of the original discovery, which is near the point where Anvil Creek emerges from its valley and enters the coastal plain. From Discovery claim the gold is progressively finer downstream, a condition to be expected, as no gold is here derived directly from the bed rock and the placer contains only that brought in by the creek.

Discovery claim was one of the richest found in Alaska. The gold was generally coarse and included two very large nuggets, one worth $1,500 and the other worth $1,700. Although many of the smaller pieces are rough and angular, these large nuggets are battered smooth. During the season of 1903 no mining was done immediately on the creek at this place, but there were extensive operations on a bench lying about 10 feet above the creek bed and extending 500 or 600 feet eastward to the foot of Anvil Peak. Gold was discovered on this bench several years ago, but sluicing was not undertaken, on account of the lack of water. In 1902 prospect holes were sunk to the bed rock and a small amount of gold was obtained with rockers, but the great richness of these gravel deposits was not manifest until the summer of 1903, when the water of Anvil Creek, no longer needed to work the creek claims, was brought to the bench by a ditch. An area of more than 1 acre was worked over in 1903. This led to the finding of the largest nugget so far discovered in Alaska, weighing about 170 ounces. If pure gold, this nugget would be valued at about $3,200, but it contains considerable quartz and its gold content is estimated at about $2,600.

On a part of this claim there is reported to be a covering of moss and yellow clay from 12 to 14 feet thick, below which the gravel also contains some lenses of yellow clay, but the average overburden is peaty muck about 3 feet thick. The pay gravel ranges in thickness from 4 inches to 4 or 5 feet, and the bed rock to a depth of 6 inches also contains placer gold. In the thin places the clay layer from the top penetrates the gravel, which there resembles an old channel filled with mud. The bed rock is graphitic schist, similar to that found in the creek bed. Although it has an irregular surface, it shows no well-marked rim or evidence of a definite channel limiting the pay streak. The limits of the gold-bearing gravels had not yet been reached in 1903. About three-fourths of a mile above Discovery claim the bed-rock rim of the present channel forms an abrupt escarpment about 50 feet high on the east side of the creek, and above this a bench rises in a gentle slope toward the foot of Anvil Peak. In the face of the escarpment black schist is overlain by gravel. About 300 feet back of the rim, at an elevation of at least 50 feet above the present creek bed, a deposit of gravel about 100 feet long has been excavated. When the gold was discovered here it was supposed that there was an extensive bed of gravels, but the whole body bounded by rock walls was worked out in three weeks. The prospectors describe the occurrence as a

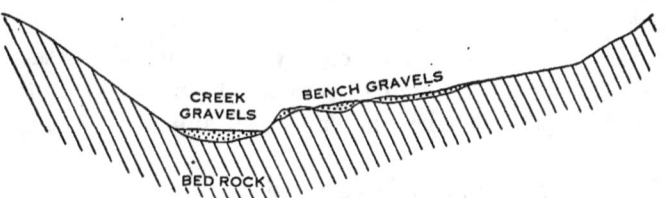

FIG. 11.—Diagrammatic section of Anvil Creek valley, showing bench gravels.

"pothole." This deposit (see fig. 11) was evidently the remnant of a more extensive channel of Anvil Creek formed at some earlier stage of the creek's development. It is similar to the other old channels described in the next paragraph, and as it lies at about the same elevation, 200 feet above sea level, it may represent the same period of erosion. This placer was very rich and in three weeks yielded sufficient gold to pay for the machinery used in extraction, consisting of an engine, pump, and pipe.

From a point about 1 mile above Discovery claim an old channel, or series of old channels, in the main less than 20 feet above the creek bed and parallel with it, has been traced for about a mile and a quarter. These, as has been stated, are old channels of Anvil Creek. They are characterized by well-defined bed-rock rims, but are everywhere narrower than the present creek bed. The overburden is for the most part slide material from the hillside above, and the surface contour gives no indication of the existence of the channels. At the lower end of this series of channels the rim on the creek side is wanting. The face of the workings shows about 10 feet of rather fine gravel containing more or less sand and clay, through which gold is somewhat uniformly distributed. Some of this gold is coarse, but the greater part is very fine. The section from the surface down is as follows:

Section in old channel of Anvil Creek.

	Feet.
Brown clay and soil containing some angular pebbles_____	3
Fine gravel containing sticky clay sediment_____	6
Coarse gravel and sand_____	3
Bed rock, not exposed.	

Mining was done here with a hydraulic monitor, having a head of only about 20 feet and supplied with water taken from Anvil Creek about a mile above. About 200 feet of 14-inch sluice boxes supplied with pole and Hungarian riffles throughout were used. The bed rock rises gently toward the hill on the east, and sluice boxes laid directly on the bed-rock floor have the proper grade. The tailings are dumped into the present creek bed.

Above this point the old channel is from 60 to 80 feet wide for about a quarter of a mile. Its east rim toward the hills is well marked and comparatively abrupt; the west rim between it and the present channel has been partially destroyed and is in places wholly wanting. The bed rock here differs from that at Discovery claim, as it consists of a light-gray calcareous schist, which is very much crumpled, but in a general way strikes parallel with the creek bed and dips to the northwest at a high angle.

About 100 yards above the upper end of these workings and one-half mile below Specimen Gulch an excavation about 300 feet east of the present channel of Anvil Creek exposes an old channel for a distance of 500 to 800 feet. This claim was worked by a hydraulic monitor; a hydraulic elevator raised the gravel about 20 feet to the sluice boxes, which dumped the tailings into the present bed of Anvil Creek. The face of the excavation showed about 6 feet of light-brown sand and angular gravel at the top; then a 1-inch seam of brown clay overlying 15 to 20 feet of gravel containing some large bowlders near the bottom and resting upon bed rock, in which there is a channel about 20 feet wide and 10 or 12 feet deep. The pay gravel extends over the rims of the bed-rock channel. A smaller channel about 8 feet wide and 7 feet deep runs into it from the side toward the present bed of Anvil Creek. What appears to be the same channel is exposed about one-fourth mile below Specimen Gulch, where it had been excavated for about 300 feet. It is here 50 feet wide by 20 feet deep and extends parallel to the creek. The bed rock here consists of graphitic schist striking north and south and dipping eastward at an angle of about 25° and contains many stringer veins of quartz, some of which are mineralized. No mining was in progress here in September, 1903.

Specimen Gulch, a small easterly tributary of Anvil Creek, cuts across the old channel and was probably the source of some of its gold. The floor of the gulch, made up of eastward-dipping graphitic schist, is about on a level with the floor of the old channel. The workings showed the following section:

Section in Specimen Gulch.

	Feet.
Moss and muck	1
Coarse gravel	8
Clay bed	1

The best pay lay on the clay bed, though the whole deposit carried some gold. The gold is bright and coarse, the average-sized pieces being worth not less than 5 cents. It is not greatly waterworn, though some of the edges and corners are rounded.

The old channel of Anvil Creek is also exposed north of Specimen Gulch, 400 feet east of the creek, in a pit 70 feet wide, 200 feet long, and 20 feet deep. The channel is here 70 to 120 feet wide and several feet lower than the bed rock in the workings on Specimen Gulch, just described. On the east side the rim is steep, but on the west side, facing the present channel of Anvil Creek, it rises very gently and is mantled by a gravel bed that probably extends all the way to the creek. The bed rock in the channel is a soft graphitic schist containing many small stringer veins of quartz. Overlying this is a gravel pay streak 7 feet thick, succeeded by about 12 feet of sandy clay. Neither the clay nor the gravel is frozen. The pay gravel contains pebbles and some large bowlders of mica schist, greenstone, and granite. All of the gravel is shingled, indicating a current running nearly south. A bowlder of galena coated over on the outside with a yellow alteration product and weighing about 40 pounds was found in the pay gravel. The gold has not worked deeply into the bed rock. It is fine, occurring chiefly in grains about the size of a pin head, though the largest nuggets were worth about $20. This mine was worked by a hydraulic monitor and elevator. The clay and gravel above the bed rock were piped down through a flume to a sump, where they were raised by a hydraulic elevator to the sluice boxes, which dumped into older excavations. The lowest gravel, together with the loose bed rock, was shoveled into the flume.

Between the north end of this pit and Anvil Creek a smaller channel about 150 feet east of the present creek, and probably more recent than that described above, has been opened. This channel, which is cut in decomposed chloritic schist, is about 20 feet wide and 12 feet deep, and in the workings of the mines makes a bend in the form of the letter **S**, which brings its lower end very near the present creek channel.

For half a mile above Specimen Gulch there were in 1903 no mines in operation on the bench east of Anvil Creek, and it was impossible to determine the course of these channels, but farther upstream a bench on the east side has been worked to a distance of about 300 feet back from the original creek bed, leaving an old channel exposed for nearly half a mile. At the outside limit of these workings a face of gravel about 15 feet high resting upon bed rock 10 or 12 feet higher than the old creek bed is exposed. A well-defined fault that strikes N. 50° E. is exposed in the workings; the bed rock northwest of it is gray chlorite-mica schist, dipping about 25° S. 20° E., while that southeast of the fault is graphitic schist, very much crumpled and twisted, striking parallel to the fault plane. The old channel here is about 60 feet wide, and lies from 200 to 300 feet east of the modern creek bed and several feet higher. Although gold occurs in the gravels all the way from this old channel to the present creek bed, the richest pay is confined to the old channel. In some places the rim between the two channels is low and the pay gravel extended across from one channel to the other without much diminution in value.

In 1903 no mining was in progress on Anvil Creek above this point, for the claims along the creek and its tributaries were mined out in 1900. Near Nekula Gulch some bench claims were still in operation, but these are described in connection with the high-bench deposits (pp. 198–207).

Glacier Creek.—Glacier Creek, an eastern tributary of Snake River, 8 miles from the coast, is about 6 miles long. It is reached by a wagon road from Banner station, on the Seward Peninsula Railway, and freight rates from Nome in 1903 did not exceed 4 cents a pound. For a mile from its mouth the creek crosses the floor of Snake River valley, which rises gently from a 20-foot scarp to an elevation of 400 feet at the base of the steeper valley walls. This slope is partially gravel covered, and the tributaries of Snake River, including Glacier Creek, have cut trenches across it. The flood plain of Glacier Creek is here 200 to 500 feet wide. The creek enters Snake River valley from a broad valley incised in an upland, with a flood plain from 100 to 300 feet in width. The valley slopes, like those of Snake River, rise gently to an elevation of 400 or 500 feet and then perceptibly steepen. These lower gentle slopes are locally gravel covered, and it is reported that washed gravel occurs as high as 300 or 400 feet, but it is not uniformly distributed and in many places bed rock is exposed at the surface. The cross section of the creek valley shows no evidence of either rock or gravel benches. Snow Gulch and Bonanza Gulch are the only tributaries of economic importance.

The bed rock throughout the Glacier Creek valley is nearly all chloritic mica schist, containing small beds of limestone and having some graphitic phases. Small veins of mineralized quartz are common. Opposite the mouth of Snow Gulch there is a mineralized zone in the schist, in which a number of stringers of highly mineralized quartz follow the schistosity of the bed rock. Some of these stringers are 6 inches thick and the schist between them is also impregnated with sulphides. A picked specimen of the ore yielded one-half ounce of gold to the ton. The source of the placer gold of Glacier Creek is probably in such veins and its local distribution is an indication of their relative number and richness in different parts of the bed rock.

The first placers found in this basin were those on Snow Gulch, a small southerly tributary about 2 miles from the mouth of Glacier Creek. This gulch carried some of the richest placers found on the peninsula, but they were practically exhausted before 1903. The bed of Glacier Creek for 3 miles from the mouth is floored with gravel, generally more than 100 feet wide and ranging from 5 to 20 feet or more in depth. The lower 2 miles of this flood plain are known to carry gold values and extensive operations have been inaugurated. No reliable estimate of the amount of gold to the cubic yard of gravel was obtained. Some of the deposits mined previous to 1903 must have contained more than $5 a cubic yard, though the gold tenor of the gravels worked in 1903 was probably much less. Hydraulic monitors and elevators supplied with water from the Miocene ditch were being operated at two places in 1903.

A mile from the mouth of the creek the gravel deposit had a thickness of about 20 feet and rested upon a bed rock consisting of chloritic schist. The exposure shows 2 or 3 feet of muck and soil resting upon 10 or 12 feet of gravel. The pay streak is said to be 300 feet wide and to overlie a nearly even rock floor. It carries gold through a thickness of 10 to 15 feet and extends downward in the crevices of the bed rock to a depth of 1 to 3 feet. The gold is fine, bright, and well rounded. Opposite the mouth of Snow Gulch the gravels of the creek bed have been worked since the summer of 1900 and have probably all been handled at least once. A hydraulic plant was installed in the summer of 1903 for the purpose of rewashing these gravels. (See Pl. III, A.)

In 1903 mining was also in progress about one-half mile above the mouth of Snow Gulch, where a force of men was shoveling the gravels into sluice boxes, to which water was supplied from the bed of Glacier Creek by a short ditch. The gravels were loose and probably had been already worked over once, but still carried sufficient values to warrant sluicing them again. Above this point, although prospect holes, short ditches, and crosscuts indicate that some mining has been done in the creek bed, all operations had ceased by 1903 and, so far as known to the writers, no values have been found in the gravels.

Hot Air Bench.—The " Hot Air Bench," situated on the right bank of Glacier Creek opposite the mouth of Snow Gulch, is an old channel deposit resembling those of the Anvil Creek valley. This old channel of Glacier Creek was probably abandoned on account of slides from the hill above. The workings are about 100 feet above the bed of the creek, to which they are nearly parallel for a distance of 400 or 500 feet. Although the surface contour gives little evidence of a bench at this place, the excavations have disclosed a well-defined channel which runs parallel to Glacier Creek and whose bed lies about 20 feet below the surface. The exposures show 4 to 5 feet of vegetable matter and clay resting upon 4 to 5 feet of gravel made up of schist and limestone, with a few pebbles of granite; this gravel overlies chloritic schist similar to that in the bed of the creek. The channel in the bed rock where best defined has a width of about 100 feet, with well-marked rims on both sides. Near the east end of the deposit the southwest rim is broken down and the pay streak mantles the slope toward Glacier Creek. Before the end of 1903 the greater part of the channel deposit had been worked out. A cut was made through the south rim and through this cut a long line of sluice boxes carried the gravel and tailings to Glacier Creek. A 12-inch sluice box was used and the gravel was washed into it by means of a hydraulic nozzle and canvas hose. Water was obtained through a ditch about 3 miles long from Divining Creek, a tributary of Snake River. During the melting of the snow in the spring this ditch carried about 150 miner's inches of water. The penstock was about 150 feet above the mine, but the amount of water in the ditch was not sufficient to maintain this head with a 2-inch nozzle, and the pressure actually used in the mine was probably not far from 30 feet. A part of the deposit was hydraulicked with water brought in from Glacier Creek with a pressure of not more than 50 feet. At the east end the boundary of the deposit has apparently been reached where the pay streak turns toward Glacier Creek, but at the west end the boundary is not so sharply defined, for the pay streak thins out. The gold is similar to that of Glacier Creek. The total production is generally thought to have been not less than $600,000, but no very definite figures have been obtained. If this estimate is anywhere near right, the pay streak must have averaged more than $50 to the cubic yard. Auriferous gravels are said to have been found on the hill slopes above the Hot Air Bench to an elevation of several hundred feet, which suggests that there may be higher channels.

Bonanza Gulch.—Bonanza Gulch is tributary to Glacier Creek from the south about a mile from its mouth. The stream is about half a mile long, has a fall of about 200 feet, and, except in the spring when the snow is melting, carries only a few inches of water. The bed rock is chloritic schist and where exposed by mining operations is much fractured and sheared. An area 150 feet long and 10 or 12 feet wide has been mined out. Here 2½ feet of auriferous gravel is overlain by 5 or 6 feet of sandy clay and a foot of moss and muck. The prospecting indicates that the pay streak extends for about a quarter of a mile along the gulch bed and that it probably connects at the lower end with the pay gravel of Glacier Creek. The gold is nearly all very fine and well rounded. This gulch has been worked in the early summer by the use of snow water, with which the over-burden is sluiced off, and the pay gravel is shoveled into sluice boxes. If a proposed plan to hydraulic the deposits with water from the Miocene ditch be carried out, the whole gulch bed could probably be easily mined out in one season.

Snow Gulch.—Snow Gulch joins Glacier Creek about 2 miles from Snake River. In 1900 Brooks [a] described this gulch as follows:

Snow Gulch is the only tributary of Glacier Creek which has thus far produced gold. It joins Glacier Creek about 3 [2] miles from Snake River. Its source lies in a shallow basin, which is gravel filled, while its lower course is through a rather sharply cut gorge; the total length is about a mile.

The bed rock near the mouth of Snow Gulch is chloritic schist, and the strikes are about at right angles to the course of the stream. Near its head some blue limestones are exposed. A quartz vein near the divide has been found to carry some gold. The gravels are 3 or 4 feet in thickness and are chiefly quartz and mica schist.

Its topographic form makes Snow Gulch almost ideal for the concentration of gold. In point of fact it has proved the richest stream of its size so far discovered, for in the two seasons that it has been worked it has produced upward of a million dollars. At present the stream gravels have been practically all worked over, so that the gulch has reached its maximum production.

Stringer veins of quartz are common in the bed rock, and near the head of the gulch some calcite veins showing free gold have been opened up. In a tunnel which pierces the divide between Snow Gulch and Anvil Creek the chloritic schist and limestone bed rock is faulted at numerous places and contains many small stringers and veins of mineralized quartz and calcite, assays of some of which showed a small gold content. Veins of this kind are believed to be the source of the placer gold of the gulch. As stated by Brooks, the bed rock has been swept clean of gravel, and no further placer mining is possible.

Rock Creek.—Rock Creek rises in Mount Brynteson, north of Glacier Creek, and flows southwestward to Snake River, in a narrow trench incised in the sloping valley floor. Its bed rock is gray mica

---

[a] Brooks, A. H., Richardson, G. B., and Collier, A. J.: Reconnaissances in the Cape Nome and Norton Bay regions, Alaska, in 1900, a special publication of the U. S. Geol. Survey, 1901, p. 75.

schist, with some interbedded limestone. A number of mineralized quartz veins have been found here, some carrying values. These veins strike a little north of east.

Gold was discovered and some mining was done on Rock Creek in 1900. During the season of 1903 five claims were being worked on the creek, all by one company. Though the gravels of this creek have been exploited since 1900, the total production has not been large, but no estimate can be given.

The gravels on Rock Creek are about 5 feet deep and range in width from 50 to 100 feet. The gold is distributed throughout the gravel. It is rather fine and is saved with mercury, but a few rough nuggets have been found. One nugget was seen which included a piece of black graphitic rock, and another held a mineral resembling scheelite. In the concentrates scheelite occurs in subangular pieces, and it is reported that a vein of this mineral, 3 or 4 inches wide, has been uncovered in the bed rock. In addition to the scheelite the concentrates contain magnetite, limonite, and garnet. The concentrates are saved and worked over in an arrastre with mercury.

Lindblom and Balto creeks.—Lindblom and Balto creeks are two small streams which, like Rock Creek, are trenched across the Snake River valley floor. The occurrence of the gold is similar to that on Rock Creek. In 1903 only one claim on each of these creeks was worked, and the total production was small. A ditch was then under construction to bring water from Grouse Creek for hydraulicking on Balto Creek.

Boulder Creek.—Boulder Creek, about 5 miles long, empties into Snake River from the west about 8½ miles in a direct line from the coast, and can be reached by wagon road crossing Glacier and Rock creeks. Its lower course for about a mile traverses the gently sloping valley floor of Snake River, above which it occupies a deep V-shaped valley cut in limestones and calcareous graphitic schists of the Nome group. It carries between 50 and 200 inches of water. Twin Mountain Creek, its most important tributary, joins it from the north 2 miles from Snake River. The gravels carry many bowlders of mica schist up to 3 feet in diameter. Several quartz veins, the largest 5 inches wide, were seen cutting a flaggy limestone that outcrops near the mouth of Boulder Creek. The best assays from them are said to have shown from $3 to $4 to the ton in gold. Attempts to develop quartz veins have been made at several places, but without success.

Colors of gold were found here in 1900, but in 1903 only a few men were at work, and they claimed to be making no more than wages. The gold is coarse and easily saved; the largest nuggets have a value up to $13. Besides the gold the concentrates contain hematite, magnetite, ilmenite, stibnite, and native bismuth.

On Twin Mountain Creek the gravel deposit is shallow and not more than a few feet wide. Mining here seems to have been unprofitable, as the miners left their claims during the season of 1903 to work for wages.

Last Chance Creek.—Last Chance Creek empties into North Fork of Snake River from the west side, about 16 miles from Nome in a direct line, but the distance by the wagon trail, which follows the bed of Snake River above Prospect Creek, is much greater. The creek is between 3 and 4 miles long, and occupies a V-shaped valley between hills about 2,200 feet high. The country rocks are graphitic, calcareous, and chloritic mica schists. Gold was discovered on the creek in 1900 and rather extensive prospecting was done in 1901 by a company that disposed of its holdings at the end of the season. In 1903 a small force of men worked on the creek during a part of the season. The small amount of gold obtained is coarse, rather rough, and bright.

Grub Gulch.—Grub Gulch, a small stream running into Snake River from the east, about 1½ miles above the mouth of North Fork, carries about 50 inches of water. It can be reached by a rough wagon road over a low divide from Nome River. The gulch has a narrow valley cut through mica schists, and its gravel consists of schist and vein-quartz pebbles. On a claim worked in 1903 the pay streak is said to have had an average width of about 40 feet, and a thickness of 5 to 6 feet, and to have yielded 2½ cents to the pan. The gold is rather rough and coarse, the largest nugget reported being worth about $1.75.

Goldbottom Creek.—Goldbottom Creek, which joins Grouse Creek about 16 miles from the coast to form Snake River, rises west of Mount Distin and flows in a southerly direction. It has not been examined by members of the United States Geological Survey, but it is reported that some claims were worked in a small way during the season of 1903, and that the results were not encouraging. A sample of the concentrates obtained from A. W. Lane contained a few pieces of stream tin.

Grouse Creek.—Grouse Creek rises on the east side of Mount Distin and flows southward to its junction with Goldbottom Creek. It carries a large volume of water and is the objective point of several proposed ditches. Its valley is V-shaped and has a grade of about 200 feet to the mile. The bed rock at the mouth is chloritic mica schist of the Nome group; its headwaters are in massive limestone. One claim, about a mile from the mouth, was worked during 1903. Here three men were employed and report taking out more than $2,000. The gold is mostly bright and rough, ranging from very fine dust to nuggets weighing 2 pennyweights. The pay streak, about 40 feet wide and from 1 to 3 feet thick, is made up of coarse gravel, chiefly limestone pebbles, and rests upon a limestone bed rock. The concentrates are composed mostly of hematite pebbles.

HIGH-BENCH PLACERS.

INTRODUCTION.

Under this heading will be described the placers occurring at altitudes ranging from 500 to 600 feet above the sea. (See fig. 7, p. 150.) These placers are the elevated remnants of deposits made by an older drainage system, which have been dissected, and, for the most part, removed. Although the workable placers of this kind so far found are limited to the Anvil and Dexter Creek region, the evidence of the epoch of erosion during which they were formed has been found in various parts of the peninsula, and there is a strong probability that other workable high-bench placers will yet be found.

The discovery of these high gravels must be credited to Schrader and Brooks, who also forecasted that they were likely to prove aurif-erous. The following passages are quoted from the first report on this region:[a]

Some 8 miles north of Nome, between the heads of Anvil and Dexter creeks, in the region of Mount King, some benching and gravel terraces, seemingly marine, were observed on the lower slopes of the mountains up to a height of about 1,000 feet [estimated]. They occur at irregular intervals and mark successive stages in an elevation of the land which is probably still in progress.

* * * * * * *

The gravels of the terraces on the mountain slopes which have already been described, have a similar origin to those of the coastal plain and are probably of a similar character. We can judge of this only by inference, as we unfortunately found no section of these gravels exposed.

* * * * * * *

As has been explained, these benches and terraces have a similar origin to that of the tundra plain and, as their material has a similar source, they are likely to contain gold.

The above was published before a single high bench had been prospected. Two claims were opened up in 1900 and yielded about $150,000. At the close of the second season's investigations, Brooks said:[b] "Comparatively little investigation has been made of these high gravels, but they will probably become important gold producers."

These predictions have been abundantly verified, for during the last five years these high-bench placers have produced probably

---

[a] Schrader, F. C., and Brooks, A. H., Preliminary report on the Cape Nome gold region, Alaska, a special publication of the U. S. Geol. Survey, 1900, pp. 12, 16, 20.

[b] Brooks, A. H., Richardson, G. B., and Collier, A. J., Reconnaissances in the Cape Nome and Norton Bay regions, Alaska, in 1900, a special publication of the U. S. Geol. Survey, 1901, p. 78.

$2,000,000 in gold. They have contributed especially to the prosperity of the district because they have afforded a field for winter work. Nearly all the mining operations on the high-bench deposits have been made by sinking shafts and drifting on the pay streak, with the aid of steam thawers. If water could be brought to this high level at not too great a cost, there is no question that all these deposits of gravel could be sluiced at a profit.

GENERAL DESCRIPTION.

The high-bench placers occur in alluvial deposits forming terraces and flat saddles at altitudes of 500 to 800 feet in the divides which separate Anvil and Dexter creeks and Dry and Dexter creeks. This same level of erosion and deposition has been recognized in other parts of the region, but the deposits elsewhere have not yet been proved to carry workable placers. They have, however, been but little prospected.

Some of the high benches are cut on bed rock; others are deeply buried in gravel, as on the Anvil-Dexter Creek divide. In the region lying west of Nome River rounded granite and gneiss bowlders from the Kigluaik Mountains are strewn over many of the benches up to this level. These bowlders have, in part at least, been brought in by floating ice during a period of submergence. In two of the localities under discussion such pebbles and bowlders of foreign origin are distributed throughout the gravels; in a third the foreign material is found only at the surface, and the greater part of the deposit, including the pay streaks, is of local origin. No fossil remains have been seen in any of these deposits, although marine fossils have been reported from one of the prospect holes. In one or two places a shingling in the gravel has been observed, which indicates deposition by stream action.

HIGH-BENCH PLACERS OF KING MOUNTAIN.

On the south slope of King Mountain a broad, sloping bench lying between 600 and 650 feet above the sea stretches from the head of Nekula Gulch S. 60° E. for about half a mile, crossing the Anvil-Dexter Creek divide at right angles. (See fig. 12 and topographic map, Pl. VIII, in pocket.) At the west end of this bench the surface slopes down to Nekula Gulch, a tributary of Anvil Creek; to the east the level surface breaks off to a slope which stretches to Nome River.

Gold-bearing gravels have been found in several deep shafts and open cuts along the line of this bench for a distance of about 3,000 feet. The productive mines of this belt are known as the Caribou Bill, Mattie, Madeline, Lena, Snowflake, Honey, and Sugar. During the summer of 1903 only the workings of the Caribou Bill, Mattie, and Snowflake were accessible. Although the surface of the bench varies slightly in elevation and at the west end is dissected by Nekula Gulch, the shafts penetrating to bed rock reach a pay streak at a nearly uniform elevation above the sea, thus indicating that the floor on which most of the deposits were laid was approximately horizontal and had an elevation of 650 to 700 feet.

The bed rock is principally much-decomposed chlorite-mica schist, with which are interbedded some comparatively massive beds of limestone whose relation to the schist has not been determined. (See pp. 71–73.) The upper few feet of the gravels of the bench contain scattered waterworn bowlders of granite, but lower down the gravels include only pebbles derived from the schist and the limestone of the adjacent bed rock. They are in general not greatly waterworn, a slight rounding of the edges and corners being the only evidence of water erosion. All the pebbles are deeply weathered and partially decomposed. The surface of the bed rock upon which the gravel deposits rest is more or less irregular and some gravel deposits have been found below masses of schist that appear at first sight to be in place, but are evidently slides from the hill slopes above. In some places the limits of the pay streak are well defined and lie in a bedrock channel; in others no definite channel rims have been found, and in one place at least the pay streak leaves the schist bed rock and mantles an older deposit of gravel. A sketch map and profile of this bench showing the relative elevations of the surface, pay streaks, and bed rock at the various openings, is given in fig. 12.

At the Caribou Bill mine, near Nekula Gulch, the bed rock of the bench crops out at an elevation of about 525 feet above the sea. It consists of limestone and decomposed calcareous schist. Gold-bearing gravels fill a great cavern or crevice in the bed rock. Here a pit about 30 by 50 feet and 20 feet deep has been excavated. The lateral dimensions of the deposit are well defined, but it extends nearly 90 feet in depth and its origin is difficult of interpretation from the evidence in hand. It seems probable, however, that the gold-bearing gravels have been carried into a cavern of the limestone and calcareous schist and deposited by a stream of water. The gravel consists of slightly rounded pebbles of schist and limestone cemented with yellow clay. This placer is of extraordinary richness, and $20 to the pan was not uncommon in the material excavated. Collier saw one pan of the material which yielded about $7. This sample consisted of fragments of schist bound together by yellow clay, and the colors of gold could be readily seen before panning. It is estimated that the excavated material carried about $100 in gold to the wheelbarrow load, which would probably be equivalent to about $1,000 to the cubic yard. The gold is coarse and angular. It seems likely that this material is nearly in place.

The lower workings of the Mattie claim lie about 100 yards south-east of the pit just described. At the west end the bed rock outcrops at an elevation of about 525 feet and the surface of the ground rises toward the east, so that the excavations, which followed the horizontal surface of the bed rock, exposed a face of gravel about 30 feet high at the east end. This pit is about 50 feet wide and 150 feet long.

The Mattie claim was the first of the high-bench claims to be developed in the Nome region. The lower end of the claim touches

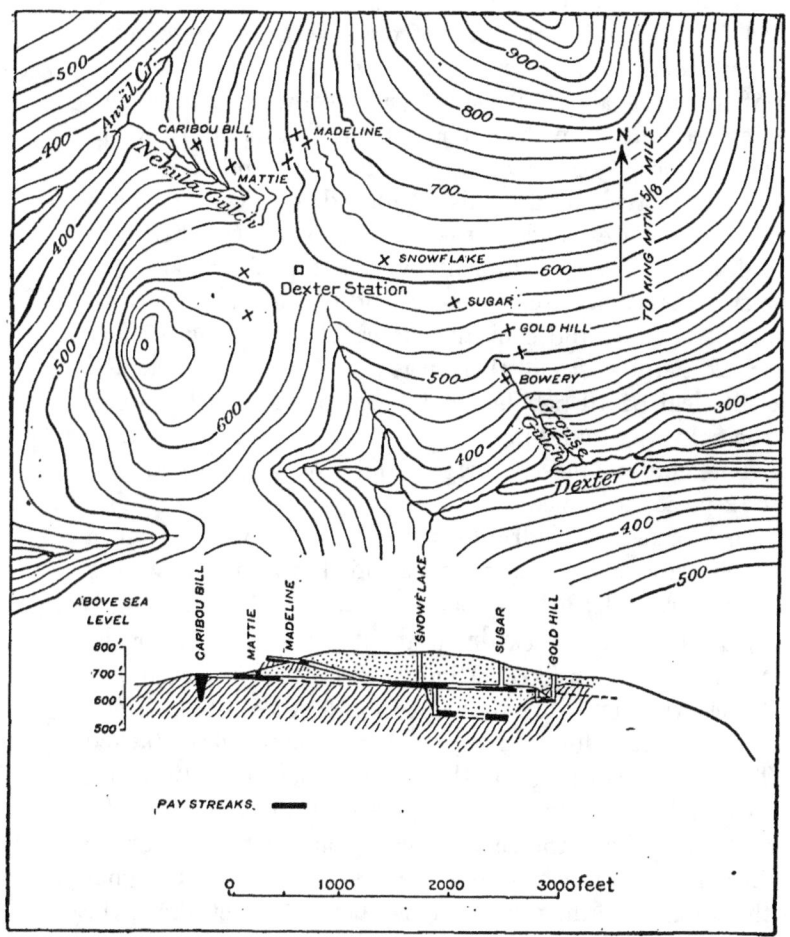

FIG. 12.—Sketch map and profile of high-bench gravels south of King Mountain.

Nekula Gulch, a small tributary of Anvil Creek, which was staked and prospected to some extent in 1899. Early in 1900 rich pay gravels were discovered on the Mattie claim and during that season about $90,000 in gold was taken out; the claim then sold for $100,000. One of the original owners of the claim informed Mr. Collier that the gravel mined in 1900 averaged about 35 cents to a 6-pan bucket, which is about equal to $9 per cubic yard.

The bed rock is faulted and otherwise so much disturbed that in the small area exposed by mining operations it varies greatly in dip and strike. It consists of alternating thin layers of mica or chlorite schist, graphitic schist, and calcareous schist, with some beds of limestone. There is little doubt that it is chiefly of sedimentary origin, but it may contain some igneous material. Many small veins of quartz are present, some of which cut across the bedding, following the fault lines, while others lie parallel to the bedding planes. The largest of these veins is a lenticular mass which thins out rapidly and does not extend far. Samples of quartz from several of these veins when assayed were found to contain traces of gold.[a]

In the face of this mine about 30 feet of gravels are exposed, showing the following sections from the surface down: Light sandy wash, containing some pebbles, 5 or 6 feet; gravel, 25 feet; and schist bed rock. Two kinds of gravel could be distinguished; one, comprising a body of gravel near the west end of the cut, was well stratified and consisted of ligh-brown clayey sediment, with flat pebbles arranged parallel to the stratification. Near the east end of the cut the gravel deposit was more irregular and contained more clay. About 10 feet above the bed rock at the east end there was a body of well-stratified sandy clay or silt clear of all pebbles. The rich pay streak lies upon bed rock and the gold was found in crevices of the schist to a depth of several feet. The gold is mostly bright and not waterworn, and averages 10 or 12 pieces to the cent.

This mine was worked by hydraulic methods, with water from a ditch about 100 feet above the rock floor, supplied from the pumping plant on Snake River. The gravel was piped into a 200-foot length of sluice box, placed in a bed-rock drain about 6 feet deep, near the lower end. After most of the gravel had been washed through this flume, the bed rock was picked up by hand to a depth of about 2 feet and, after washing in the sluice boxes, the larger slabs were stacked in the cut. The upper workings of this claim, opened in 1904, are about 300 yards northeast of the opening just described, at an elevation of about 625 feet above the sea, and are reached by a shaft 65 feet deep, indicating that the bed rock here is about 50 feet higher than that in the lower workings, where it is about 510 feet above the sea. This section is as follows:

Section at upper workings of Mattie claim, King Mountain.

	Feet.
Turf	1
Sand and gravel	60
Pay gravel	4–9

[a] Assay by E. E. Burlingame & Co., Denver, Colo.

The pay streak lies in a definite bed-rock channel from 15 to 150 feet wide. It has been traced for about half a mile and is identical with the pay channel of the Lena, Honey, and Snowflake mines.

The main shaft of the Snowflake mine is located on the south flank of King Mountain about 1,200 feet S. 60° E. from the workings of the Mattie and 625 feet above sea level. The surface is here nearly level and appears to be a bench. A shaft 130 feet deep reaches bed rock, which has an elevation of about 495 feet. The section from the surface down is as follows:

Section at Snowflake mine, King Mountain.

	Feet.
Muck and gravel containing granite and gneiss bowlders	5
Silts and gravels of local origin	125
Quartz-chlorite schist bed rock.	

The lower 3 to 9 feet of the gravel is gold bearing. The pay gravel is of local origin and consists of pebbles of chloritic schist, limestone, and vein quartz. Many of the pebbles, as well as the upper part of the bed rock, are much weathered. From the bottom of the shaft the mine workings extend about S. 60° E. and N. 60° W. toward the Mattie mine on the one hand and the Sugar mine on the other. The pay streak has an average width of about 100 feet and is bounded on the north and south by appreciable rises in the bed rock, regarded by the miners as the rims of an old channel. On the north side the rim merges with the slope of the hill; on the south side it rises only a foot or so above the level of the pay streak. About 70 feet east of the foot of the shaft the bed rock disappears below a layer of barren gravel, which is called by the miners "old run gravel," on account of its position underlying the pay streak and the more weathered condition of the pebbles. A winze has been sunk below the main pay streak 70 feet east of the main shaft. Bed rock, overlain by a second pay streak, was reached at a depth of 100 feet. This lower pay streak had not yet been developed nor its limits determined in 1903. The lower part of this winze descends for 15 or 20 feet along a wall of decomposed schist that marks the southwestern limit of the gravel deposit. Near the contact between this schist and the gravel the schist shows slickensiding, indicating a fault striking approximately northwest and southeast. This fault possibly antedates the deposited gravel, as the main pay streak has not been affected by it.

The gravels outside of these pay streaks are known to carry some gold, which is probably not very regularly distributed through them. The zone of perpetual frost extends from the surface downward to a depth of 90 feet. The workings of the mine on both the 130-foot and 230-foot levels are in unfrozen ground. At these depths the gravels are comparatively dry, and water penetrates only along the shaft. The gold is bright in color and comparatively coarse, the average-sized pieces being not far from 1 cent in value. The largest nugget thus far found weighed 9 ounces and 9 pennyweights. This gold shows little evidence of stream erosion; the points and corners of few pieces are battered, and many small octahedrons are preserved. A few of the nuggets still include pieces of quartz and calcite. The gold in nearly all the larger nuggets is porous and honeycombed, probably by cavities from which the associated minerals have been weathered out. Many of the large nuggets are so porous that they can be readily broken with the fingers.

In the workings of this mine a gangway was driven as nearly as possible along the center of the pay streak to the boundaries of the property, and crosscuts were then driven at intervals to the limits of the pay. The mining began at the remote end of the gangway, and as the pay gravel was removed the timbers were withdrawn. The pay gravel was moved to the foot of the shaft in cars and raised by a steam hoist. On the surface the cars were run on an elevated track to a high dump. This dump was sluiced early in spring with water obtained from melting snow. In order to procure sufficient snow water for this purpose fences were built on the hillside above the mine during the winter, causing drifts to form. By this means sufficient snow was accumulated to furnish water to wash the dump mined during the winter. It is probably impossible to mine gravels at less than $6 or $7 a cubic yard by this method, and only the richest part was taken out.

The main shaft of the Sugar mine lies about 600 feet southeast of the Snowflake shaft, at an elevation of 600 feet above the sea. During the summer of 1903 the deep workings of this mine were not accessible, for the mine had been developed in winter from a temporary shaft, which had caved in. At a depth of about 100 feet a 10-foot pay streak of what is reported to be rather low-grade gravel was encountered. This pay appears to rest upon an older deposit of gravel that is similar to that in the Snowflake mine and probably continues with it. The gravel below the main pay streak of the Sugar mine has been prospected by means of drill holes, which reached bed rock at a depth of 205 feet. Above this bed rock some gold-bearing gravel was found, but no details in regard to the deposit are known to the writers. Gold was discovered on the Sugar claim in 1901, and since then considerable auriferous gravel has been taken out and washed with snow water in the spring. Up to the time of visit (1903) only the pay streak on the 100-foot level had been worked.

Considerable prospecting has been done to determine the extension of the pay streak on other claims to the southeast of the Sugar mine. About 300 feet to the southeast, at an elevation of 575 feet, or 25 feet lower than the surface of the Sugar mine, a 100-foot shaft shows the following section:

Section in shaft 300 feet southeast of Sugar mine, King Mountain.

	Ft.	In.
Muck and silt	4	
Loose blue gravel containing a few colors	25	
Decomposed chlorite schist	51	
Washed gravel and schist		10
Schist	20	

At the 80-foot level the decomposed schist contains a 10-inch streak of washed gravel, below which the schist is very hard all the way to the bottom of the shaft. This gravel bed is about 495 feet above sea level and not far from the level of the upper pay streak in the Sugar and Honey mines. From the bottom of the shaft a tunnel was driven westward through hard schist for 15 feet, beyond which the rock shaded into a soft, decomposed schist containing some washed gravel. The drift was continued for a distance of 85 feet to the boundary of the property, where an upraise was made and at the 80-foot level a layer of washed gravel, 3 feet in thickness and containing one-half cent in gold to the pan, was found. Above this gravel there were 10 inches of decomposed schist underlying 3 feet more of gravel, above which the hard schist was again found. A tunnel was also driven for 35 feet north from the bottom of the shaft, encountering nothing but schist. An upraise from the end of this tunnel revealed no gravel. From the 80-foot level a tunnel was driven northward following the thin layer of gravel mentioned above, but this bed pinched out in 10 feet. A tunnel was also driven southward for 60 feet, but here the washed gravel also pinched out in a distance of 20 feet, and at the end of the drift only decomposed schist was found.

About 100 feet southwest of this shaft a prospect hole has been sunk to a depth of 40 feet in washed gravel without reaching the bed rock. The gravels thrown out consist of waterworn material mixed with sand. The section reported is as follows: Brown gravel, 20 feet; gravel and blue clay, 4 feet; gravel and yellow sandy clay, 16 feet. No pay streak was located.

A tunnel has also been driven northwestward, starting on bed rock in Grouse Gulch, 480 feet above the sea, toward the workings of the Sugar mine, which are about 500 feet away. It is reported that the bed rock here showed some irregularities, and the gravels carried values in the depressions, but also carried some colors of gold on the raises. On the dump from this tunnel, which has caved in, were found gravels made up of schist, limestone, and vein quartz. Some of the schist pebbles carry small veins of quartz.

To summarize, auriferous gravels have been found more or less continuously from Nekula Gulch westward to the head of Grouse

Gulch, a distance of about half a mile, but have not been traced beyond. An upper pay streak, in part confined to a definite bed-rock channel and in part resting upon an older gravel bed, has been traced almost continuously throughout this distance. This pay streak has an elevation of 568 feet at its west end and 490 feet at the east end. A lower pay streak near the west end of the bench lies at an elevation of about 525 feet, and one near the east end at about 400 feet, but the two localities are half a mile apart and their continuity has not been established. The lower deposit at the east end is, of course, older than that overlying it, but there is at present no evidence available to show the relative ages of the upper and lower pay streaks near the west end. The overburden of these deposits is distributed over a much larger area than that covered by the pay gravel. It is made up of material which was probably contributed mainly by slides from the surrounding hills and only to a slight extent rolled and sorted by water action, but its upper part contains alluvium transported from a distance. The pay gravel is all of local origin and the gold has probably been derived from small mineralized veins in the adjacent bed rock.

HIGH-BENCH PLACERS NORTH OF ANVIL PEAK.

A second high-bench placer lies on a divide 525 feet above the sea, between Grass Gulch, a tributary of Dexter Creek, and Specimen Gulch, which discharges into Anvil Creek. The development here consists of several shafts located along an east-west line for 600 or 700 feet. Prospecting was begun on this bench in the summer of 1900, but pay gravel was not discovered until 1901 or 1902. About 300 feet of the pay streak had been worked out in 1903. The surface of the ground slopes toward the head of Grass Gulch.

On the Summit claim the shaft is 106 feet deep, making the elevation of bed rock about 420 feet. From the foot of the shaft tunnels have been driven N. 82° E. and S. 82° W. along the pay streak, which lies on bed rock sloping slightly to the east. The whole deposit is frozen and the material thrown out in sinking the shaft contains a great deal of fine sand and silt, but only a few details regarding the section were obtained. In the gold-bearing gravel there are many bowlders and pebbles of well-rounded granite and gneiss, which must have come from a great distance and seem to indicate that the stream which deposited them was a long one. The pay streak of the claim is 6 or 7 feet thick and consists of gravels that contain many large, flat pieces of schist and schistose limestone. In places there is a layer of stratified clay above the pay streak, and at the west end of the mine some false bedding was observed and also small seams of graphitic material probably derived from the decomposition of graphitic schist.

The width of the pay streak is from 50 to 80 feet. On the south side a well-marked. rim in the bed rock has been noted, but on the north side the bed rock gives no indication of the limits of the pay streak. In general these gravels are more waterworn and fresher in appearance than those in the bench south of King Mountain; they also contain material transported from a distance. As a rule the gold is coarse and contains many well-rounded nuggets. The pay streak is reported to average about $1 to the bucket, which would probably be equivalent to $7 or $8 per cubic yard. Some spots in the gravel have been found to be very rich, and pans containing $150 are not uncommon. One nugget worth $138 is reported to have been found.

DRY-DEXTER CREEK DIVIDE,

The third notable high-bench deposit lies between Anvil and Newton peaks in the saddle which forms the divide between Dry Creek and Left Fork of Dexter Creek. The surface of this divide has an elevation of 570 feet above the sea. It is developed by a line of shafts extending nearly north and south for a distance of about 2,000 feet. The gravels were found in 1900 and some deep shafts were sunk nearly to bed rock, but the pay streak was not located until the following year. In 1901 some auriferous gravel lying west of the main pay streak was worked out. The richest ground was found during the winter of 1902-3 and considerable material was taken out, the richest of which was washed with rockers during the winter. In the following spring the tailings from these rockers and the waste material not rich enough to be rocked were sluiced with water obtained from melting snow and yielded fair returns.

The surface of the ground here is nearly level, but by careful measurement is found to be about 25 feet lower at the south end than at the north end. In the shafts near the north end of the deposit the depth to bed rock is reported to be from 70 to 75 feet, making the elevation of the bed rock nearly 500 feet above the sea; at the south end the shafts are only 35 feet deep, showing that the bed rock lies about 10 feet higher than at the north end. The prospect holes between these points show that the bed rock has a nearly uniform slope to the north. Prospect holes sunk several hundred feet to the east of this main line indicate that the bed rock there has a higher elevation, for most of the holes are not more than 30 feet deep. The material from these excavations is chiefly light-brown sand containing some well-rounded schist pebbles, as well as some pieces of vein quartz and limestone and a few granite bowlders. The following section is reported in a shaft near the middle of the bench:

Section in shaft on Dry-Dexter Creek divide.

	Feet.
Muck and slide rock	16
Well-washed gravel carrying some gold	12
Soft sandy soil	2
Soil, peat, and slide rock	22
Stream gravel	10
Decomposed schist bed rock.	

None of these layers are permanently frozen. Most of the bed rock is very much decomposed chloritic mica schist, but in the northernmost shaft of the group broken limestone bowlders were encountered and penetrated to a depth of 40 feet. More or less brown clay sediment containing some waterworn material was found throughout the crevices of this limestone, indicating that the bed rock consists of bowlders of limestone formed in place by decomposition along the walls of the crevices. The bed rock exposed in the hills surrounding this area consists of alternating beds of limestone and chloritic and feldspathic schists. The sandy part of the deposit resembles the decomposition products from these schists and limestones. It is often difficult to distinguish between the bed rock decomposed in place and the brown sandy part of the gravel which overlies it, but this difficulty is not encountered if even a few rounded waterworn pebbles are present. The granite bowlders occurring here and there in these deposits indicate that a part of the gravel has a distant source, and the rounded character of the gold also indicates transportation.

As a general rule, the bed rock rises more or less gradually east and west of the bottom of the shafts, indicating the presence of a rock channel. Toward the south end of the deposit the rim is wanting, and it is probable that if followed far enough to the west the bed rock will be found to dip down beneath the floor of Dry Creek, which heads only a few hundred feet away. In many places the eastern rim is well defined and in some of the workings it has been followed for 50 feet up a 30° slope. The pay streak lies immediately on the bed rock, ranges from 2 to 3½ feet in thickness, and consists of light-brown sandy material containing well-rounded pebbles of schist, with some vein quartz and limestone. In general appearance it does not differ materially from the gravels that overlie it, which are not rich enough to work under present conditions. The pay streak probably averages from $6 to $12 a cubic yard and occasionally pans yield from $3 to $4. The gold is rather coarse and occurs in well-rounded pieces. The gold-bearing gravel is not closely confined to the pay streak indicated by the line of mine workings, for a number of shafts to the east have also yielded profits, though these placers are not so deep nor so rich as those along the line of the shafts. Some deep shafts, now abandoned, about 200 feet west of the main line of holes, are said to have reached gravels that yielded fair returns.

There is an area of not less than 10 acres on this divide below which the bed rock seems to be approximately level, and through this area there is a bed-rock channel running north and south. The gravels carry gold on bed rock over the whole area, but the richest pay streak is along the incised channel. The unfrozen condition of the deposit is accounted for by the fact that the gravels and sands are porous and well drained throughout.

In working the deposit, shafts were sunk to bed rock at intervals of about 100 feet and were cribbed and timbered sufficiently to make them stand for a few months. From these shafts the gold-bearing gravel was mined out in all directions. Very little timber was used underground and the roof was supported by pillars of pay gravel, which were removed before the workings were abandoned. These claims were worked both in winter and summer, as owing to their unfrozen condition the gravels stood as well in one season as the other. Sluicing was done early in spring with water obtained from melting snow. In the underground workings no cars were used. The buckets were dragged to the foot of the shaft, where they were hoisted to the surface by a windlass. Usually two men worked underground and a third operated the windlass at the surface.

GENERAL CONCLUSIONS.

The occurrence of these high-bench gravels is difficult of explanation. The most plausible theory as to their general origin is that they are stream-channel deposits formed during an earlier epoch of erosion. When these channels were formed, the land probably stood considerably lower and their beds were not far above sea level. The overburden was contributed mainly by slides from the hills. After most of the overburden was deposited subsidence occurred and the granite and other foreign bowlders were floated over the surface by icebergs or ice floes.

CRIPPLE RIVER BASIN.

GENERAL DESCRIPTION.

Cripple River, which is about 25 miles long, flows nearly southward into Bering Sea 12 miles west of Nome. (See topographic map, Pl. VIII, in pocket.) It drains an area of about 90 square miles and has an average fall of 20 feet to the mile. Where it traverses the coastal plain, here 3 miles wide, the river takes a meandering course, with banks 30 feet or more high. About 4 miles from the coast the valley becomes very narrow for a mile and the hills rise abruptly on either side. The source of the stream lies in a broad basin, with gentle slopes and a wide flood plain. The upper end of the valley is separated by a low, flat divide from the flood plain of Stewart River, a tributary of the Sinuk. The upland surrounding the valley is characterized by flat ridges and hilltops having an average elevation of about 800 feet, and the higher hills rise above this level as dome-shaped buttes. In a few places broad gravel-covered gaps interrupt the continuity of the ridges. Though these gravels are attributed to old drainage channels, none of them as yet have been found rich enough in placer gold to justify mining. On some of the valley slopes there are broad gravel-covered benches, in the main less than 400 feet above the sea, some of which contain placers of economic value. Erratic granite bowlders strewn over the surface occur up to elevations of 800 feet. These have already been described (p. 95), and are attributed to the action of floating ice during a comparatively recent period of submergence. In several of the tributary stream beds along Cripple River these bowlders are so plentiful as to interfere seriously with the working of the placers.

The rocks of the region are mainly interbedded limestones and schists belonging to the Nome group. (See Pl. X, in pocket.) They include some massive crystalline limestones, which are differentiated on the geologic map and are provisionally correlated with the Port Clarence, though none of them have yielded fossils. Along the channel of Cripple River granite bowlders from the Kigluaik Mountains are mixed with gravels of local origin.

On Cripple River itself almost no successful mining has been done, though colors are found in many places, and some areas which will probably pay if worked by economical methods have been located in the bed and low terraces of the river. A ditch 4½ miles long was built in 1903 to bring water from Willow Creek for hydraulicking at a point just below the mouth of Elizabeth Creek. The capacity of this ditch is 3,000 miner's inches and the head for hydraulicking 157 feet, but its maintenance has been almost impossible, as in many places it runs over beds of ground ice, which melts away under the water. To avoid the ground ice, a new ditch is proposed to run at a level 100 feet higher. This will necessitate an extension of about 11 miles to the head of Cripple River, where, it is claimed, about 3,000 miner's inches of water can be obtained.

At the mouth of Stella Creek some sluicing, yielding not more than wages, has been done in gravels that contain many granite and gneiss bowlders. The deposit lies in an old channel cut in a bed rock of quartz-chlorite schist about 10 feet above and parallel to the present river bed. A small ditch to bring water from the head of Stella Creek has been constructed, but as it was built across ground ice the bottom soon gave way and no mining was done at this place during the season of 1903.

OREGON CREEK.

Oregon Creek, the most important tributary of Cripple River, is between 5 and 6 miles long and drains about 15 square miles of territory. It joins Cripple River 10 miles from the coast. The valley is broad, flat, and gravel floored for the lower 2 miles, above which it gradually narrows to its head.

Gold was discovered on Oregon Creek in the winter of 1898–99, but it was not worked systematically until 1900. In the latter year Collier made a hasty examination of the placers of this region,[a] and in 1903 it was visited by Hess. Eight creek claims and two or three bench claims were in operation on the creek and its tributaries during the season of 1903. The production up to the end of 1903 had probably not exceeded $100,000. Supplies are hauled from Nome by a wagon road along the beach to Penny River, thence up Penny River to Willow Creek, and across the hills to Oregon Creek.

Along the lower 2 miles of the creek's course bed rock is not exposed. One-half mile below Nugget Creek decomposed chlorite schist is reached at a depth of 2 to 5 feet. From Nugget Creek to Mountain Creek the bed rock is heavy limestone, dipping downsteam and jointed and broken into irregular fragments. Frequently at stages of low water the creek is nearly lost in underground channels through this bed rock. At Mountain Creek there is a narrow belt of chloritic schist, but above that point the creek runs through massive limestone which seems to form an anticline.

The placers of the creek are said to be "spotted" and on some claims small portions pay well, while the remainder will not yield expenses. This is probably due partly to the varying character of the bed rock, rich spots being found on that which affords good natural riffles, and partly to local origin of the gold. All the paying claims of the main creek lie in its middle part in a belt about 2 miles long. In the lower 2 miles the creek is spread over wide gravels and a number of claims that have been worked by primitive methods have failed to pay. Half a mile below the mouth of Nugget Creek there is some ground that has been fairly productive.

The gravel is from $1\frac{1}{2}$ to 6 feet thick and is composed of schist, limestone, and granite pebbles. Most of the pay lies upon a tough yellow clay immediately over the bed rock, but some gold is found in the crevices of the limestone to a considerable depth and is difficult to obtain. The gold is bright, coarse, and slightly worn; the

---

[a] Brooks, A. H., Richardson, G. B., and Collier, A. J., Reconnaissances in the Cape Nome and Norton Bay regions, Alaska, in 1900, a special publication of the U. S. Geol. Survey, 1901, p. 92.

largest nugget found was worth about $4.50. The concentrates contain a large amount of octahedral magnetite, some hematite, scheelite, bismuth, and garnet.

One-fourth mile below the mouth of Nugget Creek a strip of gravel about 20 feet wide along the creek bed was worked in 1900. Here the bed rock is hard, thin-bedded limestone, in the crevices of which the gold is found. About 2 feet of washed gravel, consisting principally of rounded limestone pebbles and a few granite bowlders, lies upon the bed rock. This gravel forms the pay streak, and, together with about a foot of the bed rock, was put through the sluice boxes. At the mouth of Nugget Creek the limestone bed rock is broken and irregular. Some gold has been taken from a " false bed rock " consisting of fragments of limestone embedded in reddish clay. Preparations were being made in 1903 for hydraulicking this claim. The following section was measured in a cut bank shown on the south side of the creek:

Section on Oregon Creek at mouth of Nugget Creek.

	Feet.
Moss and muck	1½
Yellowish red sandy clay	6
Gravel, consisting of broken fragments of limestone	2
False bed rock, consisting of clay with fragments of limestone embedded.	

Near the mouth of Mountain Creek the bed rock is chloritic schist dipping downstream, and the pay gravel is from 2 to 5 feet thick. It is reported that in 1900 two men, with rockers, obtained 2 pounds of gold in less than a day, getting as high as $27 in one pan. About one-fourth mile above Mountain Creek, on a claim worked in 1900, the pay streak was confined to the creek bed and was not more than 20 feet wide and less than a foot thick. Here the best pay occurred in the crevices to a depth of 18 or 20 inches. The gold is coarse, but most of the grains are well rounded and bright. One nugget, worth $130, and two smaller ones, worth $42.60 and $13.20, were obtained. Above this point, which is 3½ miles from the mouth of Oregon Creek, no gold has been mined and there is probably little, if any, to be found, for the bed rock consists of massive limestones in which there is little evidence of mineralization. The rounded form of the gold in the upper claim worked indicates that it may have been brought down from this part of the creek; but in swift flowing streams with shallow gravels nuggets may be quickly rounded by battering pebbles, without themselves traveling far.

On the east side of Oregon Creek a broad gravel-covered bench, 50 to 75 feet above the creek, extends from Short Gulch to Nugget Creek. This bench, which is about 400 feet above the sea, is probably an old valley floor of Oregon Creek and is partly covered with flood-plain gravels. Several claims on this bench have been productive, but only in a small way. One of these claims was in operation in September, 1903. The pay streak, which seems to follow a definite channel, is about 65 feet above the present bed of Oregon Creek and is 70 feet wide. It is about 6 inches thick and rests upon bed rock, but very little gold is found in the crevices. The overburden is gravel containing some bowlders of granite, gneiss, and greenstone and is from 3 to 10 feet thick. The gold is smooth, bright, and coarse, averaging probably from one to three pieces to the cent. One nugget weighing about 38 pennyweights has been found. In the concentrates there is much magnetite and hematite, with some bismuth, garnet, rutile, and scheelite.

On several of the claims worked by drifting during the winter the gravel is from 20 to 25 feet thick. One claim is said to have yielded $4,500 at an expense of $4,000. Mining here has been done mostly by pick and shovel, under heavy costs, and has probably barely paid expenses. The ground is well situated for hydraulicking, but water for this purpose can probably not be obtained without a long and expensive ditch.

NUGGET CREEK.

Nugget Creek, less than 2 miles in length, heads in a limestone mountain and flows northwestward, joining Oregon Creek 2½ miles above its mouth. The prevailing bed rock is limestone, but a dike of decomposed pyritiferous greenstone cuts the limestone near the mouth of the creek. A second greenstone dike is reported to occur a mile above. On a claim at the mouth the gravel is composed largely of angular limestone fragments, with some rounded schist and green-stone pebbles. The gravel lies from 6 to 20 feet below the surface, in a channel not more than 25 feet wide. The pay rests upon a stiff yellow clay above a limestone bed rock and does not extend into the crevices of the limestone. The gold is all coarse enough to be easily saved, but contains few nuggets as large as a pennyweight.

The concentrates are made up largely of octahedral magnetite in small grains. One specimen obtained has a small piece of gold attached. The concentrates also contain garnet, specular hematite, rutile, scheelite, bismuth, and scattered crystals of pyrite partially oxidized to hematite. One piece of bismuth the size of a pea was collected in which a fleck of gold is embedded. The claim was economically worked as follows: Ten men were worked in two shifts of five each. The gravel was piped to the sluice box by a stream from a 2-inch nozzle under a head of less than 20 feet, and water from a lower ditch was allowed to run over the upstream edge of the bank, thawing and washing down the gravel and supplying more water to the sluice boxes. One string of 18-inch sluice boxes, with pole riffles, was used. The tailings were removed by a scraper pulled back and forth by wire cables from drums run by a 4-horsepower gasoline engine, a man being required to guide the scraper.

One other claim, about a mile above the mouth of Nugget Creek, was worked during 1903. Several men were employed part of the season, but the pay streak "pinched out" before the end of the season and operations were suspended. · The gravel deposit is from 3 to 6 feet deep and from 30 to 50 feet wide. The average size of the gold pieces is about 1 cent, and the concentrates are similar to those at the mouth of the creek.

MOUNTAIN CREEK.

Mountain Creek heads in a limestone mountain and flows southward, emptying into Oregon Creek about $3\frac{1}{2}$ miles from its mouth. It carries less than a sluice head of water. It is reported that along the lower part of the creek there are extensive deposits of schist gravel carrying values in gold. No mining has been done along this creek, but it is said to prospect well and preparations had been made to work one claim during the season of 1904.

HUNGRY CREEK.

Hungry Creek, about 2 miles long, flows in a northwesterly direction and joins Oregon Creek 1 mile from Cripple River. Its valley is broad and gravel filled and the creek has trenched its floor to a depth of about 20 feet in gravel and bed rock. Gold was discovered on this creek in July, 1900, and three mines were in operation in September, 1903, although previously the creek had been worked unsystematically throughout nearly its whole length. The following section was exposed in one of the mines:

Section at mine on Hungry Creek.

	Feet.
Moss and muck, about	1
Gravel, consisting mainly of chloritic schist fragments, and containing bowlders; pay dirt of mine	4
Decomposed chloritic schist which can be cut with the shovel	1
Chloritic schist, containing no gold in crevices.	

The bowlders are largely of gneiss and granite from the Kigluiak Mountains and reach $2\frac{1}{2}$ feet in diameter. Associated with the gold there was a small amount of bismuth in well-rounded nuggets, some of which weighed as much as an ounce. The concentrates were about half magnetite, the remainder being mostly garnet, in large, well-rounded pieces and small crystals, with some limonite, pyrite, ilmenite, rutile, and bismuth. Gravel-covered benches occur at several places along Hungry Creek, but none of them have produced placer gold.

MAY GULCH.

May Gulch, the south fork of Hungry Creek, heads in a broad, low divide about 700 feet above the sea. Near its head prospect holes have been sunk to a depth of 20 feet through washed gravel without finding bed rock. Colors are reported all the way from the surface down.

TRILBY CREEK.

Trilby Creek, a small tributary of Hungry Creek from the northeast side, about a mile from Oregon Creek, is reported to have yielded $5,000. It does not cut down to the bed rock, and it seems probable that the gold has been reconcentrated from the bench gravels. During 1903 nothing more than assessment work had been done, on account of lack of water.

STREAMS TRIBUTARY TO BERING SEA BETWEEN CRIPPLE AND SINUK RIVERS.

Three small streams—Rodney, Sonora, and Quartz creeks—flow into Bering Sea at distances of 13, 15, and 17 miles, respectively, from Nome, or 1, 3, and 5 miles from the mouth of Cripple River.

The hills rise steeply from a narrow coastal plain, which is here from a quarter to half a mile broad. The bed rock exposed along the beach and in the face of the bluffs consists mainly of chloritic mica schist, but contains some beds of graphitic schist and limestone. Small quartz veins are common in the face of the bluff and some prospecting has been done, but nothing of economic importance has been discovered. A short distance west of Quartz Creek a large irregular vein several feet thick at the widest place occurs between beds of chlorite and graphite schist. The vein thins out rapidly. It contained a thin streak of highly mineralized quartz in which gray copper ore seemed to be predominant. A specimen obtained from one of the men interested in the development was assayed by E. E. Burlingame & Co., of Denver, Colo., who found only a trace of gold and silver. No placer mining is in progress on either Rodney or Sonora creeks, though colors of gold have been found and some development work was done on Sonora Creek in 1901. On Quartz Creek sluicing is reported about 1 mile from the coast, where two men were making a little more than wages in 1903, but the claim was not visited by members of the Survey party.

SINUK RIVER BASIN.

GENERAL DESCRIPTION.

Sinuk River rises in the Kigluaik Mountains, flows westward in a valley parallel with them for about 10 miles, and then, turning south, reaches Bering Sea 25 miles west of Nome. The lower course of the river lies within the area of chloritic schists and limestones of the Nome group that are the source of placer gold in other parts of the precinct.

Colors of gold have been reported from many localities in the Sinuk basin, and in 1900 some sluicing was reported to have been done on Charley Creek, a tributary from the south near its headwaters. No gold has been produced along the main channel of the river. Of its tributaries Washington, Rulby, Charley, Boulder, Independence, and Coal creeks are reported to have produced small amounts. Extensive gravel deposits carrying low-grade values, mantling the hills between Sinuk and Cripple rivers, are reported to occur from 12 to 18 miles from the mouth. These deposits have been staked for hydraulic mining, but most of them lie at such high elevations that it will be a difficult and expensive undertaking to bring water to them, as the only sources of supply are the streams of the Kigluaik Mountains, and pipes would be required in crossing the Sinuk and Stewart valleys. Washington and Rulby creeks cut across these high gravels and probably reconcentrate their gold from them.

WASHINGTON CREEK.

Washington Creek, about 6 miles long, enters Sinuk River from the east 12 miles from the mouth. The placers of this creek have not been visited by members of the United States Geological Survey, but they are reported to consist of high-bench gravels surrounding Irish Hill, a mountain about 1,300 feet high, in the Sinuk-Cripple River divide. Good prospects, including one $10 nugget, are reported to have been found in this bench several hundred feet above the level of the river.

RULBY CREEK.

Rulby Creek, which enters Sinuk River from the east 18 miles from the mouth, is also reported to cut across high gravel deposits. A small amount of gold has been produced here in prospecting, but the gravels lie at high elevations, are of low grade, and can probably be worked only by very economical methods.

CHARLEY CREEK.

Charley Creek, a southern tributary of Sinuk River, entering 28 miles from its mouth, flows with a steep gradient in a narrow valley cut in chloritic schists. When the creek was examined in 1903 no mines were found and the workings had evidently long ago been abandoned, but a few colors of gold were panned from the gravels. It is reported that when this creek was worked a great deal of native bismuth was obtained in the concentrates.

BOULDER CREEK.

Boulder Creek, which enters Sinuk River about 9 miles from the coast, is reported to have yielded a small amount of gold in 1902 and 1903, but it has not been examined by geologists.

INDEPENDENCE CREEK.

Independence Creek, another tributary of the Sinuk about 10 miles from the coast, is also reported to have produced some gold though it was not worked in 1903. The creek is cut for the greater part of its length in a massive limestone bed rock and along it there are many evidences of attempts at mining and prospecting, such as crosscut ditches and bed-rock drains.

COAL CREEK.

Coal Creek, which joins Sinuk River about 15 miles northeast from its mouth, rises in the plateau region and flows down the gentle slope of the valley, exposing the bed rock in only a few places. The bed rock for about half a mile consists of coal-bearing sediments younger than the Nome group, which are described on page 83. In exploiting the coal deposits gold has been found in the gravels of the creek, which are believed to be rich enough to justify hydraulicking, and a ditch for this purpose will probably be constructed from the Kigluaik Mountains. About 2,000 miner's inches of water can be obtained from Glacial Lake at an elevation of about 400 feet above the sea. If it is found that a higher elevation is required, smaller amounts can be obtained from other sources in these mountains. The placer on Coal Creek must be regarded as in part a reconcentration from sediments of Tertiary or Mesozoic age.

REGION WEST OF SINUK RIVER.

GENERAL DESCRIPTION.

West of the lower waters of Sinuk River lies a triangular area comprising about 100 square miles, bounded by the Sinuk on the southeast, Bering Sea on the southwest, and the Port Clarence precinct on the north, which has received some attention from prospectors and has produced a small amount of gold, probably not exceeding a few thousand dollars. Topographically this region is made up of (1) a coastal plain along Bering Sea about 4 miles wide, (2) a number of high ridges to the northeast, attaining an elevation of 1,600 feet and extending in a northwest-southeast direction approximately parallel to the coast, (3) between these ridges and the Kigluaik Mountains an undulating surface that has an average elevation of 1,000 feet and has been correlated with the Kougarok Plateau. The coastal plain is covered by gravel deposits which may be of considerable depth, but in some places are known to be only a thin layer not over 30 feet thick, resting upon a limestone bed rock. The ridges parallel to the coast are made up of limestone and schists. The more massive limestones, which are crystalline, but, as a rule, not so highly metamorphosed as those in the immediate vicinity of Nome, have been provisionally correlated with the Port Clarence. The schists are in places graphitic, but the larger masses are feldspathic, chloritic schists of a character similar to the larger bodies exposed in Anvil and Newton peaks near Nome.

The surface of the plateau region is, for the most part, mantled with moss and muck, beneath which there are gravels that contain, as a rule, a great many bowlders of granite derived from the Kigluaik Mountains and probably directly or indirectly of glacial origin. (See pp. 94–99.) The bed rock where it is exposed consists of limestones and schists, which are almost as highly metamorphosed as those at Nome. In the northwestern portion of the region under discussion they contain many intrusions of greenstone. At a point near Sinuk River a basin of younger sediments of Mesozoic or Tertiary age is infolded, and it is possible that rocks of this character have a considerable extent underneath the surficial deposits. Except for the fact that the bed rock in general is slightly less metamorphosed than in the Nome region there is no reason why workable placers should not exist here. Colors of gold have been found in a number of small streams, and mining for gold is reported at several places; though no active mines were seen by members of the survey party. Of the small streams which flow into Sinuk River from the plateau several are known to carry colors of gold, but none have yet been found to contain gold in paying quantities for small outfits working with sluice boxes or rockers. It is probable, however, that some of the gravels may be worked by more economical methods, if water be brought from the Kigluaik Mountains.

IGLOO CREEK.

Igloo Creek rises in a range of hills back of the coastal plain and flows westward to Bering Sea, which it enters just north of Cape Woolley. Although colors of gold have been reported from this creek it has not produced gold in commercial quantities, and was deserted by prospectors during the season of 1903.

FAIRVIEW CREEK.

Fairview Creek, a large stream having a length of about 12 miles, also heads in the range of hills parallel to the coast near the heads of Independence and Boulder creeks, and flows northwestward into the lagoon north of Cape Woolley. The bed rock of this creek consists of graphitic and feldspathic schists and some massive limestone. For about 4 miles the stream flows in a trench cut across the coastal plain, and it is not known whether it cuts the bed rock at any point within this belt. The bed of the creek contains many bowlders of granite, which came with the glacial drift from Kigluaik Mountains. Colors of gold have been reported in the gravels at various points, though no pay streak has been found.

TOMBOY CREEK.

About 6 miles from its mouth several small tributaries enter Fairview Creek from the north side. Some mining has been done on one of these, known as Tomboy Creek, which has a length of about 2 miles and flows in a westerly direction. About 1 mile above its mouth a deposit of several feet of gravel containing a great many bowlders and pebbles of granite rests upon soft chloritic mica schist bed rock dipping to the northeast at a high angle. Two shovelfuls of the loose decomposed bed rock were panned, yielding two colors of gold, each equal to about one-half cent. This test, so far as it goes, would seem to indicate a layer of decomposed bed rock, carrying about $1\frac{1}{2}$ cents to the pan, or $2.25 to the cubic yard. Evidently some mining was done here in 1902, but early in 1903 no one was seen on the ground and the claims appeared to be abandoned.

FEATHER RIVER.

Feather River, which also enters Bering Sea through the lagoon north of Cape Woolley, has a length of about 16 miles. Six miles from the coast it forks, the north fork being called Johnston Creek and the south fork Livingston Creek. Johnston Creek is the larger stream and derives its water from a number of tributaries which head in glacial valleys in the Kigluaik Mountains. Livingston Creek, the southern tributary, heads in a broad basin-shaped depression in the plateau already described. The gravels of both streams are made up largely of pebbles and bowlders of granite and other rocks derived by glacial agencies from the Kigluaik Mountains. It is reported that colors of gold have been found along these streams, but no mining has been done.

The gravels of the coastal-plain belt of this region are not known by the writers to have been prospected for gold. It is certain that no promising deposits of gold have been found in them. The beaches that border these coastal-plain deposits on the west would afford an excellent opportunity for determining whether or not the coastal-plain gravels carry gold values. So far as known, no gold has been found in paying quantities at any point along this beach and no mining has been done on the beach northwest of the mouth of Sinuk River, although a few colors of gold can be found at many places.

TISUK CREEK.

Tisuk Creek, which heads in the west end of the Kigluaik Mountains and flows southwestward to the lagoon north of Cape Woolley, carries a large volume of water, and its channel is trenched from 100 to 200 feet deep, across a low plateau between the mountains and the sea. The bed rock in the plateau portion of the creek's course consists of chloritic schists, limestones, and greenstone intrusives. These rocks belong to the Nome group, but are much less metamorphosed than those in the Nome region. No gold has been produced along the main stream and scarcely appreciable amounts in its drainage basin. However, some production is reported on two tributaries that were lying idle in 1903.

HUME CREEK.

Hume Creek is a small northern tributary of Wesley Creek, which enters Tisuk Creek from the south side, 4½ miles from the coast. Two men who worked here with rockers in 1901 are said to have made from $2 to $3 per day. The workings were abandoned when examined in 1903. A section of the deposit exposed in a cut bank of the creek and in a drain ditch is as follows:

Section on Hume Creek.

	Feet.
Moss and black muck	6
Angular gravel, composed of chloritic and graphitic schist, gneiss, and granite	5
Chloritic mica schist bed rock.	

On a small southern tributary of Tisuk Creek, name not known, 6 miles from the coast, there is said to be near the level of the river's flood plain 5 feet of gravel which carries from 3 to 10 cents a pan. It has been worked with rockers, yielding from $3 to $10 per day to the man. The ground was idle in 1903 and was not examined.

In addition to these, several other small streams tributary to Bering Sea north of Tisuk Creek are reported to yield colors of gold, but none of them have been worked. On one of these, Sourdough Creek, which enters the north end of the lagoon south of Cape Douglas, Collier panned in 1901 and raised one color of gold. The bed rock consists of dark slaty schist containing many intrusions of greenstone.

ELDORADO RIVER REGION.

GENERAL DESCRIPTION.

Eldorado River has its source within a few miles of Salmon Lake, from which it is separated by a low divide. (See topographic map, Pl. VIII, in pocket). It flows southward through a broad gravel-filled valley and empties into Port Safety Lagoon. The bed rock of the whole basin consists of the schists and limestones of the Nome group, but its headwaters are in limestones correlated with the Port Clarence.

Placer gold has been reported from a number of the tributaries of Eldorado River. Some mining is said to have been done on San Jose, Mulligan, and Fox Creeks, but in 1903 mining was confined to Venetia Creek, though large areas staked for placer-mining purposes are said to be held by assessment work in other parts of the drainage basin.

VENETIA CREEK.

Venetia Creek is an eastern tributary of Eldorado River, which it enters about 23 miles from the coast, and can be reached by a rough wagon road from Nome. Gold was discovered and some mining was done on this creek in 1900. Since then it has been worked every year and the total production has probably not been far from $6,000. Although nearly all the claims are owned by Italians, the name Venetia is not after the historic Italian city, but after the discoverer of gold on the creek.

In the lower 2 miles of its course Venetia Creek flows in a sharply cut, narrow gorge, whose walls rise almost perpendicularly for 100 feet to a series of comparatively narrow benches. Above this gorge the valley widens and flood plains 100 to 300 feet wide border the stream bed. In the productive portion of Venetia Creek the bed rock is limestone and calcareous chloritic schists, into which the limestones appear to grade. The strikes of these rocks are variable, but mainly northeast. The dip also is not constant, the bedding being in many places nearly horizontal. The rocks contain small veins of quartz, in general parallel to the foliation, but locally cutting across it. The creek gravels consist of fragments of schist and limestone, a few pebbles of greenstone, and some vein quartz. Little gold has been mined on the lower part of the creek, but for 3 miles above the canyon the gravels have been productive.

The placers are all of the creek type, and though the topography indicates that there are terraces along the valley slopes, it is not known whether they carry gold. The flood plain above the canyon is from 100 to 200 feet wide, but the pay streak is only 10 to 50 feet wide. A large part of the placer gold is found in the crevices of the bed rock. In the lower claims the gravel and overburden have a thickness of 5 or 6 feet and a considerable amount of gold is obtained from the gravel, but in the claims near the upper end of the productive portion practically all the gold is contained in the crevices of the bed rock to a depth of 3 or 4 feet.

The mining has all been done by the " shoveling in " method. Near the lower end of the productive portion the whole creek has been turned out of its bed and confined in a ditch along one side of the valley. A system of well-built ditches, which are practically sod flumes, is one of the interesting features of the workings. The gold obtained along Venetia Creek is mostly fine, with some nuggets worth several dollars. In general it is bright and is characterized by flat pieces shaped like pumpkin seeds. The assay value is said to be $19.40, the bright-yellow color being indicative of the high percentage of gold. Most of the nuggets are sold to jewelers in Nome for souvenir making.

SOLOMON RIVER REGION.

INTRODUCTION.

Solomon River heads close to the Casadepaga, from which it is separated by a low divide, and, flowing southward for 20 miles, empties into Port Safety Lagoon about 30 miles east of Nome. Its more important tributaries, Shovel Creek, Big Hurrah Creek, East Fork, and Coal Creek, enter at right angles and flow approximately east or west. The placer mines of the region are accessible from the settlements of Solomon and Dickson, at the mouth of the river. Dickson is the terminus of the Council City and Solomon River Railroad. The harbor facilities here are no better than at Nome, and at the present time landings are fully as difficult, but it has been planned to dredge out the inlet to Port Safety Lagoon, through which Solomon River discharges, in order to permit the entrance of lighters, and to construct docking along the land side of the lagoon. If the other openings into Port Safety Lagoon should be closed, compelling all the tributary waters to pass out by the inlet at the mouth of Solomon River, a fair channel might be maintained at small expense. It should be borne in mind, however, that with the development of hydraulic mining that is almost certain to occur along Solomon River, much sediment will be contributed to the stream, and this will make it difficult to keep a channel open.

For 5 miles above its mouth Solomon River flows across the coastal plain in a broad trench. Its gradient here probably does not exceed 6 or 8 feet to the mile. Above the coastal plain the river occupies a comparatively broad valley and the gradient increases to at least 50 feet to the mile. In the floor of this valley the river is entrenched, leaving a system of gravel terraces from 20 to 50 feet above the water. Gravel bars from 100 to 500 feet wide, only partially covered at ordinary stages of the water, fill the river bed.

The river was named by Pierce Thomas,[a] who staked Discovery claim in June, 1899. In the same season the river and its tributaries were prospected, and in 1900 probably $10,000 worth of gold was mined in this district.

The bed rocks along Solomon River are limestone and schist of the Nome group. The strikes generally have a north-south direction, but the dips are low and in some places the rocks lie nearly flat. On the eastern side of the valley a number of intrusive masses of green igneous rock have been observed.[b] Veins and stringers of more or less mineralized quartz are common in the schists, both parallel with and cutting across the schistosity. Many of them carry gold, and at Big Hurrah Creek a quartz mine is in successful operation. The following assays made for the Survey from specimens collected in 1900 and 1903, none of which are to be regarded as commercial samples, are useful as indicating the sources of the placer gold.

Three samples were obtained from float quartz found on the surface near the cropping of a large, partially developed quartz vein in graphitic schist on the hill north of the mouth of Big Hurrah Creek. Of these, one carried a trace of gold and no silver, another carried 16 ounces of gold and 1 ounce of silver, and the third contained 1.6 ounces of gold and 0.24 ounce of silver to the ton.[c] The material is bluish-white quartz, with black bands, and some of the specimens show free gold in fine grains. A specimen taken from a large vein in the Big Hurrah quartz mine yielded an assay return of 2.56 ounces of gold and 0.24 ounce of silver, and a specimen of the graphitic-schist wall of this vein contained 0.16 ounce of gold and 0.08 ounce of silver.[c] Most of the gold is free, and picked specimens showing large amounts of free gold have been obtained both in the mine and in the wash of Big Hurrah Creek below the mine. A specimen from a stringer of quartz and pyrite nearly a foot thick exposed in the bed rock of the placers about a mile below the Big Hurrah mine yielded an assay return of 0.02 ounce of gold and a trace of silver, and some ore rich in antimony said to have come from a vein on Last Chance Creek, a tributary of Big Hurrah Creek, was found to contain only traces of gold and silver.[c] Quartz obtained from a small

[a] Brooks, A. H., Richardson, G. B., and Collier, A. J., Reconnaissances in the Cape Nome and Norton Bay regions, Alaska, in 1900, a special publication of the U. S. Geol. Survey, 1901, p. 100.

[b] Brooks, A. H., et al., op. cit., pp. 29–31.

[c] Assay by E. E. Burlingame & Co., Denver, Colo.

stringer in the schist near the mouth of Nugget Creek, a tributary of
Solomon River, contained no trace of either gold or silver.[a]

Although gold has been found along Solomon River through its
whole length, the placers operated in 1903 and 1904 are all located
within 12 miles of the coast. The workings are in benches along the
river, where ordinary mining methods are employed, and in the
gravel bars of the river bed itself, where several dredges have been
used. The mines along the lower part of Solomon River, where it
crosses the coastal plain, have not been examined in detail by the
writers, but data regarding them were furnished by Messrs. Purington and Paige, who visited them in 1904. These placers are of the
river-bar and gravel-plain type.

A system of ditches to bring water from the head of the river and
its tributaries to the placers located near the mouth is in construction,
and it is probable that about 1,000 miner's inches of water under a
head of from 250 to 400 feet will be available.

SOLOMON RIVER VALLEY.

The excavations indicate that most of the gravels of the coastal
plain are comparatively shallow and that the bed rock under the river
bars is not deeply covered. A river-bar deposit about 3 miles from
the coast has been worked for two seasons by a dipper dredge. The
pay streak here has an average width of about 200 feet and a thickness of $9\frac{1}{2}$ feet, most of the gold being concentrated in the lower
part. It rests upon a soft mica-schist bed rock which is mined by the
shovel to a depth of about 1 foot. Overlying the bed rock is a layer
of clay several inches thick that carried good values. The pay streak
is said to be spotted, but has been estimated to contain from 12 to 30
cents to the cubic yard.

Near this point mining is in progress in the river's flood plain,
which is about 400 feet wide. The deposit, which resembles that just
described, rests upon a bed-rock floor about 3 feet higher than that of
the river bed, and the gold is mostly contained in a soft clay that overlies the bed rock. Two ditches, one 3 miles, the other 9 miles long,
supply about 1,000 inches of water. Mining is done by the shoveling-
in process, but it is proposed to use the water for hydraulicking in
the future.

In 1903 mining was done on a bench on the left side of the river
about 6 miles from the coast. The section exposed shows 4 or 5 feet
of muck and sand above 4 feet of pay gravel resting upon bed rock,
about 8 feet above the water of the river. The placer gold has penetrated the crevices of the limestone bed rock to a depth of about 4
feet. The gravels carry values for about 200 feet away from the

---

[a] Assay by R. H. Officer & Co., Salt Lake City, Utah.

river. The gold is described as consisting of coarse nuggets mixed with fine gold, all of which is worn smooth.[a]

A deposit worked by a dredge near the mouth of Big Hurrah Creek, about 8 miles from the coast, was examined in September, 1903, after work had closed for the season. The river bed here is about 150 feet wide and consists of gravel bars, most of which are above water except during floods. The gravel is from 2 to 6 feet thick. Several pans tested by Mr. Collier yielded 3 or 4 cents in comparatively coarse gold. These can not be regarded as average samples, however, as they were taken from parts of the gravel known to be rich. One cut about 40 feet wide and several hundred feet long, in the middle of the channel, has been dredged out. In dredging below the river bed frozen gravel, which the dredge was unable to handle, was found in some places, and on this account the dredge was moved near the end of the season, but the gravels at the new location were not found to contain sufficient gold to pay for mining. The machine was equipped with 18 or 20 buckets, having a capacity of 1 cubic foot each, on an endless chain, and gravel from the buckets was dumped first into a revolving trommel from which the finer gravel and sand fell into two lines of sluice boxes, each about 60 feet long, supplied with ordinary iron riffles. This plant was said to have a capacity of 400 to 500 yards a day of twenty-four hours.

Near the mouth of East Fork, about 15 miles from the coast, the river bed and flood plain have a width of 300 to 500 feet, and the gravels, which are from 5 to 8 feet thick, rest upon schist and limestone bed rock. The gold is fine, flat, and smooth and is said to be uniformly distributed through the gravel, which is reported to carry upward of $1 to the cubic yard. A ditch taking water from Solomon River at the mouth of Coal Creek has been constructed, and mining with a hydraulic nozzle and elevator began late in the season of 1904.

Above Coal Creek, near the point where the trail leaves Solomon River to cross the divide to Ruby Creek and the Casadepaga, several claims were worked in 1900. In the bed of the river about 4 feet of gravels overlie a calcareous mica schist. The gold is fairly coarse. Several $3 and $4 nuggets were found, but few of the larger pieces exceeded 25 cents in value.

JEROME AND MANILA CREEKS.

Jerome and Manila creeks are two short western tributaries of Solomon River within a few miles of the coast, heading within the coastal plain. Some development work on these creeks was reported as early as 1900.[b] The localities were not again visited in 1903.

[a] This information was obtained by Joseph Edge, packer for the Survey party.
[b] Brooks, A. H., Richardson, G. B., and Collier, A. J., Reconnaissances in the Cape Nome and Norton Bay regions, Alaska, in 1900, a special publication of the U. S. Geol. Survey, 1901, p. 100.

15604—Bull. 328—08——15

These occurrences of gold, even though they were not worked profitably, are of interest, because they go to prove that there is gold in the coastal plain. They suggest that this tundra belt may contain placers of the gravel-plain type.

SHOVEL CREEK.

Shovel Creek flows into Solomon River from the west about 4 miles from the coast. It heads in a divide about 2 miles from the Casadepaga and flows nearly southward to its junction with Solomon River. Although not a very long stream, it has a large drainage area and carries during ordinary seasons about 500 inches of water. The topographic map shows that the creek has a gradient of about 100 feet to the mile. Through the greater part of its course it spreads over wide gravel bars, none of which have produced placer gold up to the present time. The bed rock of the Shovel Creek basin consists for the most part of limestones and calcareous mica schists. Near the head of the creek there are some intrusions of greenstone. No pebbles that could not be traced to these bed rocks have been found in the creek gravels. Mining operations in the Shovel Creek basin are confined to three small tributaries—Mystery, West, and Kasson creeks.

MYSTERY CREEK.

Mystery Creek flows into Shovel Creek about 2 miles from Solomon River. In 1900 mining in this basin was limited to Problem Gulch, a small tributary 2 miles above Shovel Creek, and the placer was confined to the bed of the gulch, in which 2 to 3 feet of gravel rested upon a decomposed schist bed rock containing placer gold to a depth of about 3 feet. The gold was rather coarse and of a bright yellow color. Crosscuts in an ill-defined bench on both sides of the gulch for about 20 feet were said to pay, but not so well as the creek bed. Mystery Creek was not visited by Collier in 1903, but it was reported that three or four outfits of about six men each were sluicing along its course.

WEST CREEK.

West Creek, a tributary to Shovel Creek, 4 miles from the Solomon Valley, flows across the strike of the schist and limestone bed rock, here dipping to the northwest at an angle of 45°. In the creek bed from 3 to 4 feet of gravel overlies bed rock. The gold near the mouth of the creek is rather fine, but increases in coarseness toward the head. Garnet, magnetite, and some pyrite, chalcopyrite, and arsenopyrite are associated with the gold in the sluice boxes. It is reported that in 1903 work was in progress on nine claims, with probably only three or four men employed in shoveling in on each claim.

KASSON CREEK.

Kasson Creek, which enters Shovel Creek from the east, flows over a limestone bed in which the water sinks so that throughout much of its length it is dry, even after heavy rains. In 1903 mining was being carried on in the creek bed for about 1 mile up from its mouth. The bed rock here is massive limestone fissured and broken up into frag-ments. The gold-bearing sediments lie upon this limestone and pene-trate the crevices to an unknown depth. During the mining opera-tions excavations have in some places penetrated the bed rock to a depth of 20 or 30 feet, giving the workings the appearance of quarries rather than of placer mines. The pay streak, which is in the main confined to the creek bed, has a width of 16 to 100 feet. For about a mile along the creek the claims are under one control and are worked as a unit. Water is supplied from Shovel Creek by means of a ditch about 2 miles in length, and the bed rock, in which the gold occurs in crevices and fissures, is quarried out in irregular blocks and washed in sluice boxes, together with the sediments from the fissures.

PENNY CREEK.

Penny Creek is a small tributary to Solomon River from the west side, about 5 miles from the coast. Its valley floor is about 200 feet wide, and the creek has cut its channel in this floor to a depth of 20 feet, leaving a series of benches. The volume of the stream during wet weather is equal to about a sluice head, but in dry seasons the water sinks beneath the bed rock.

The creek deposits are limestone and mica-schist pebbles cemented with clay. The gravel in some places rests directly upon broken lime-stone bed rock that has been penetrated by the alluvial gold, but more commonly a bed of clay intervenes between the limestone and the pay streak. In a placer 2 miles above the mouth of the creek 2 feet of gravel was observed resting upon a mixture of broken limestone and clay that had been excavated to a depth of 4 feet. Although the creek has produced considerable gold since 1900, the scarcity of water makes it difficult to work. Nearly all the gold has been obtained by the use of rockers, and sluicing can be done only for a few weeks each year after periods of heavy rain.

BIG HURRAH CREEK.

General description.—Big Hurrah Creek, which has produced more gold than any other tributary of Solomon River, is a confluent from the east about 7 miles from the coast. At its mouth it carries over 500 inches of water. Through the lower 5 miles of its course the gradient of the stream does not exceed 50 feet to the mile. The creek flows in a comparatively straight, narrow canyon incised to a depth of 100 feet in the floor of a broad valley that represents an older period of erosion. The gravel deposits of the older valley are left as bench deposits resting upon a bed-rock floor 10 to 30 feet above the creek.

The schists and limestones composing the bed rock have low dips, and though the strike is not uniform it is generally parallel with the course of the valley. Quartz veins are of common occurrence, both cutting across the bedding and parallel with it. Some of these veins which are of economic importance are described on page —. The creek bed ranges in width from 100 to 500 feet, and is filled with gravel deposits derived in part from the immediate bed rock and in part from the older gravels of the benches. Gold was discovered and mining operations begun on Big Hurrah Creek in 1900, but the production of that year was comparatively small. In 1903 mining was confined to the wide gravel bars and flood plain for 3 miles above the mouth, but some prospecting was done farther upstream.

The methods of mining along Big Hurrah Creek have been extremely primitive, most of the mining having been done by lay-men, who leased small areas by the season. The waters of the creek are usually turned to one side of the valley while the shallow gravels are worked either with rockers or by shoveling into sluices. As a general rule the laymen who have worked for 75 per cent of the returns have taken out hardly more than $5 per day. The creek placers appear to be well adapted for exploitation either with steam shovel or with horses and scrapers, but hydraulicking is not to be recommended, because of the low gradient of the stream and the thinness of the deposit.

Some mining was done in 1903 in the creek bed and on a low bench about 10 feet above the creek bed at the mouth of Lion Creek. The bed rock is a soft schist which dips about 20° NE. The pay streak in the creek bed was from 10 to 20 feet wide and consisted of 2 or 3 feet of gravel overlying the decomposed bed rock, in which most of the gold was found. The pay streak probably does not average more than $1 to the cubic yard. On the bench claim the bed rock was overlain by 2 or 3 feet of black gravel. Mining on the creek claim was done by shoveling in, but an attempt had been made to hydraulic the bench with water brought in a ditch from the south fork of Big Hurrah Creek with a head of about 200 feet. In September, 1903, operations had been suspended.

Quartz veins.—Since mining began on Big Hurrah Creek a great many specimens of vein quartz containing free gold have been found in the placer gravels. Some of these have been traced to their source, as in the Big Hurrah mine, which has been a producer for several years. This mine is located on one of the southern benches of Big

Hurrah Creek, near the mouth of Little Hurrah Creek. This bench, which is from a quarter to half a mile wide, is not covered with gravel and rises very gently toward the hills to the southeast. The country rock at the mine is hard, siliceous, graphitic schist, lying nearly horizontal, and the quartz veins seem to be confined to bed rock of this type, for they have not been traced into the schists and limestones that overlie them about a quarter of a mile to the east. In the graphitic schist the quartz veins fill fissures that cut across the bedding, are more nearly vertical than horizontal, and strike northwest and southeast. In 1903 three veins were being developed

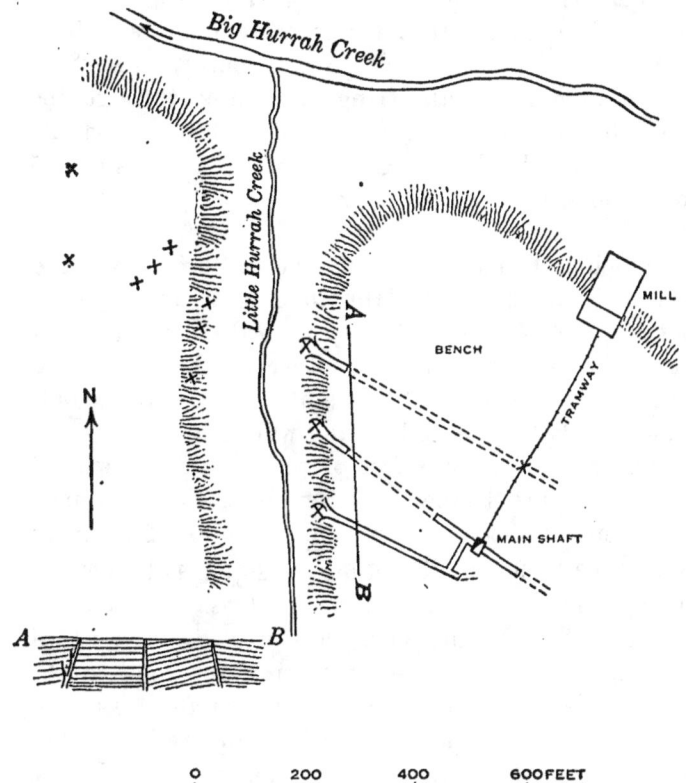

Fig. 13.—Sketch map and section of Big Hurrah mine.

on this property by short tunnels. (See fig. 13.) The southernmost vein strikes N. 60° W., the middle vein strikes N. 55° W., and the north vein strikes N. 60° W. These directions were determined, however, from short exposures of not exceeding 100 feet, and they may be more nearly parallel than the notes indicate. The distance between the south vein and the middle vein is about 100 feet, and between the middle vein and the north vein about 125 feet.

The main shaft of the mine is located on the bench east of Little Hurrah Creek, 60 feet above the level of the creek bed, close to the middle vein, and has a depth of 135 feet (1904). On the 60-foot level the main drift follows the vein for about 70 feet on each side of the shaft. Apparently there has been some movement along this fissure since the quartz was deposited, for there is slickensiding on the hanging wall and a gouge of crushed schist and quartz. The lode itself, which is from 4 to 5 feet wide, consists of quartz arranged in ribbons and bands, some of them more than 1 foot wide. Here and there are horses of what appears to be black schist in the vein and these are deformed parallel to the banding of the quartz. Both the hanging and foot walls show some irregularity. Along the foot wall of the vein there is no gouge like that of the hanging wall, but the quartz is in close contact with the apparently unaltered schist.

A crosscut near the main shaft leads to the south vein, which was found to consist of many small stringers running through the bed rock. A drift follows this vein for about 250 feet from its outcrop on the bank of Little Hurrah Creek. Here it shows 4 or 5 feet of ribbon ore in one body and dips to the south parallel with the dip of the middle vein.

The third vein, which is the most northerly one, has been developed by a tunnel from the bank of Little Hurrah Creek. It ranges in width from about 4 to 7 or 8 feet. The dip is to the northeast, but the vein seems to be more irregular than either of the others and the hanging wall is cut by many small seams of quartz. At the outcrop a large stringer leads off on the hanging-wall side and is separated from the main vein by 2 or 3 feet of schist. At the end of the tunnel the vein splits into smaller veins on either side of a mass of black schist. Where exposed in a crosscut on the surface about 150 feet southeast of the mouth of the tunnel this vein is broken up into many small stringers, but here also carries values.

A number of prospect holes and open cuts were made west of Little Hurrah Creek for the purpose of tracing the lode to the northwest, and it is believed to have been recognized for 1,000 feet along the strike. On the west bank of Little Hurrah Creek, in what appears to be the outcrop of the northern portion of the lode, the quartz occurs in small irregular veins, having a maximum thickness of 2 feet and a dip to the northeast parallel to that of the same vein on the west side of the creek. Excavations on the line of extension of the middle vein had caved in, so that it was impossible to see the rock in place, though there was considerable ore on the dump. For about 400 feet west of Little Hurrah Creek excavations have been made on the surface of the bench, exposing many small masses of vein quartz from 2 to 6 inches thick, which seem to represent the western extension of the lode, but the whole surface is covered with a débris of decomposed schist, and it is possible that the quartz veins that are so broken up on the surface may be found to be united into larger masses in depth.

No statement of the assay value of this ore has been given out by the operators of the mine and no attempt was made to take commercial samples. A large part of the gold is free and before the mine was developed the outcrops yielded many specimens in which free gold could be seen. A specimen that did not show any visible gold was assayed and found to contain 2.56 ounces of gold and 0.24 ounce of silver to the ton.[a] The ore does not appear to contain pyrite or other metalliferous impurities, but is banded with layers of graphite, which seem to have been derived from the graphite schists forming the walls. It is reported that this graphite interfered with milling operations, so that it was difficult to save all the gold. A sample of the tailings from the stamp mill was found by assay to contain 0.92 ounce of gold and 0.16 ounce of silver to the ton,[a] indicating that the milling process had not extracted all of the gold.

The ore from the main shaft of this mine, as well as that from the various prospecting tunnels and holes, is crushed in a 20-stamp mill. Water power for the mine is supplied from a ditch leading from the south fork of Big Hurrah Creek to a penstock 100 feet above the creek, from which the water is conducted in a sheet-iron pipe to a Pelton wheel at the level of the creek. The free gold is amalgamated in the battery and on copper plates, after which the coarser and heavier tailings are caught on iron-wire riffles and the finer slimes are held in a large pool made on level ground near Big Hurrah Creek by a dam.

On the summit of the hill between Big Hurrah Creek and Solomon River, about three-fourths of a mile northwest from the Big Hurrah mine, similar veins outcrop and have been developed by a prospecting shaft, about 7 feet square, wholly within the ore body. In September, 1903, the only exposure of the quartz in place was in this shaft, and attempts to trace the course of the lode had not been very successful. From this opening the Big Hurrah mine bears S. 61° E., and it is believed that the course of the vein or lode is toward this mine. Some shallow trenches that have been cut across this course, several hundred feet to the southeast of the main opening, fail to show any defined vein, though a great deal of ore has been thrown out. The prospect holes have not penetrated the deep cover of partially decomposed schist débris, and they offer little evidence with regard to the course of the quartz ledge. Much of the quartz here is coated with iron rust, but shows a banded or "ribbon" structure similar to that at the Big Hurrah mine, and many specimens showing spangles of free gold were obtained.

[a] Assay by E. E. Burlingame & Co., Denver, Colo.

Three assays made from material collected at this place, none of which can be regarded as representative, as the small amount of development made it impossible to obtain an average sample, are as follows:

Assays of gold ore from prospect holes on hill between Big Hurrah Creek and Solomon River.

		1.	2.	3.
Gold_____ounces per ton__		1.6	trace	16
Silver_____do____		.24	none	1

There appeared to be sufficient gold-bearing quartz here exposed to justify further search.

The discoveries of gold ore already made have encouraged prospecting for quartz and many claims have been staked. One of these, which cuts across Big Hurrah Creek about a mile above its mouth, is on a vein about a foot wide, parallel with the bedding of the schist, which here strikes nearly north and south. The ore consists of quartz strongly mineralized with pyrite, and when assayed was found to contain an ounce of gold to the ton and a trace of silver. A specimen rich in antimony, reported to have come from a vein on Last Chance Creek, a tributary of Big Hurrah Creek on the north side, yielded a trace of both gold and silver. Another specimen, which consisted largely of some mineral in which antimony was predominant, contained 0.24 ounce of gold and 1.10 ounces of silver to the ton.

LION CREEK.

Mining was going on in 1903 in the basin of Lion Creek, a tributary to Big Hurrah Creek, 5 miles from its mouth. Throughout the workings the gravel averaged about 4 feet in thickness and carried gold more or less irregularly distributed from the surface down. Though generally richer on the bed rock, the gold was not found in the crevices to any considerable depth. The gravels at a point half a mile from the mouth of the creek are reported to have yielded about $1.70 in gold to the cubic yard, and 2 miles above the tenor was about $2 a cubic yard. A prospect tunnel 100 feet long, driven into the bench deposit on the north side of the creek 2 miles from the mouth, follows the surface of the bed rock, which here lies about 20 feet above high water. This tunnel shows that the bed-rock floor upon which the bench gravel rests is nearly horizontal. Though colors of gold were found in these gravels up to 1905, no pay streak yielding as much as 1 cent to the pan had been located.

PINE AND CACHE CREEKS.

Pine and Cache creeks flow into lagoons connected with Bering Sea about 5 and 10 miles, respectively, east of Solomon. They are both relatively small streams which head in the highland and take tortuous courses across the coastal plain. These streams have not been examined, but their gravels are said to be auriferous, and some prospecting has been done along their valleys.

NOTES ON RECENT DEVELOPMENT.

In 1906 the Solomon River region was hastily visited by Philip S. Smith, of the Geological Survey, just before the freeze-up, when mining was almost at a standstill for the season. The following notes furnished by him give the latest facts regarding the work in progress on Solomon River below East Fork:

At the mouth of Quartz Creek two small outfits had been working the gravels lying from 10 to 30 feet above the river level. The size of the dumps indicated that not much ground had been turned over in the last season, but data were not available as to values, etc., as all the camps had closed down at the time of my visit. During the season a drill had been used in prospecting the gravels near Quartz Creek, but was not in operation when I passed.

A short distance above the mouth of Penny Creek a small amount of work was in progress on gravels which occur near the contact of chloritic schist and limestone. Downstream, between Penny and Shovel creeks, an outfit was hydraulicking gravels. Two pipes were in operation and four two-horse scrapers. It is reported that a steam shovel has been in operation on Solomon River this season, but at the time of my visit it had been removed to Shovel Creek.

The most active work in the district was at the dredge on Solomon River near the mouth of Rock Creek (Pl. IV). Mining had been in progress all summer and was being carried on, owing to the late date of the freeze-up, even as late as the first of October. The lode mine on Big Hurrah Creek was not visited, but, although continuously in operation, no new developments from that place have been reported during the last season.

COUNCIL PRECINCT.

INTRODUCTION.

The Council precinct includes all the drainage tributary to Golofnin and Norton bays as far east as the divide between Kwiniuk and Tubutulik rivers. This is an area of extensive lowlands and flat-topped uplands from 800 to 1,000 feet high, together with some mountains from 2,000 to 3,500 feet high. (See map, Pl. VIII, in pocket.) A broad, gravel-covered plain extends up Fish and Niukluk rivers for 25 miles from tide water and another depressed area, almost surrounded by hills, lies on upper Fish River. The Bendeleben Mountains form the northern boundary of the district and the Darby Mountains, a lower group, lie to the southeast. The deposits of these lowland areas have been described under the heading " Gravel-plain deposits " (pp. 86–92), in the section on general geology, and the origin of the basins themselves has been briefly discussed. The flat-topped hills form the dominant topographic type in the district and are the remnants of a partially base-leveled area that was uplifted and later dissected into a system of benches that break the valley slopes and mark erosional surfaces intermediate between the plateau and lowland levels. Most of the precinct drains into Fish River, which heads in the Bendeleben Mountains and empties into Golofnin Sound. Its principal tributary, the Niukluk, in whose drainage basin lie nearly all the placer mines now being operated, joins Fish River from the west about 20 miles from Golofnin Sound.

GEOLOGY.

Two larger subdivisions of the bed rock have been recognized in this region, the Kigluaik group being the older and the Nome group the younger. (See geologic map, Pl. X, in pocket.) The Nome group contains a massive limestone of Silurian age, called the Port Clarence, which has, however, only in part been differentiated on the map from the rest of the terrane. Intrusives are plentiful in the older series and greenstones are found everywhere in association with the rocks of the Nome group. The Kigluaik group occupies a broad east-west belt in the northern part of the precinct. It consists of dark-colored biotite schists, quartzites, and very highly crystalline limestones interbedded with sills of acidic igneous rocks which have been altered to gneiss. The whole series is again cut by dikes and sills of unaltered granite. Although these rocks also contain some quartz veins the region underlain by them has not been found to include workable placers.

A broad belt of granite touches the southeastern part of the precinct and has been described by Mendenhall [a] as an intrusive in the metamorphic rocks. This granite is not known to be gold bearing and hence lies outside of the province of the present discussion.

The bed rock of the greater part of the placer-bearing region consists of metamorphic schists and limestones of the Nome group lying south of the Kigluaik Mountains. The schists are of various kinds, including siliceous mica schist, calcareous mica schist, feldspathic chloritic schists, and graphitic schists. The limestones vary in their degree of metamorphism and purity, ranging from moderately altered fossiliferous rocks to marbles and calcareous schists. The

---

[a] Mendenhall, W. C., Reconnaissances in the Cape Nome and Norton Bay regions, Alaska, in 1900, a special publication of the U. S. Geol. Survey, 1901, pp. 204-205.

schists include sills and dikes of basic rocks, whose igneous origin is evident. In many places these appear to shade off into chloritic schists, but this has not been certainly determined.

Within the region here under discussion are several areas of limestone correlated with the Port Clarence (see geologic map, Pl. X, in pocket) of irregular outline. These rocks vary in character from blue fossiliferous limestone to highly crystalline marbles. While they are not generally believed to be gold bearing, and the unaltered masses certainly are not, zones of mineralization are often found at their contact with the schists.

The metamorphic rocks of the Nome group in this region contain numerous small veins and stringers of mineralized quartz and calcite, many of which carry gold, some of it visible but more commonly inclosed in sulphide minerals. In a few places the schists contiguous to the quartz veins also carry free gold. In these veins and mineralized zones the placer gold has its source. Copper ores have also been found in some of the veins and veinlets. No veins of either gold or copper large enough for mining have yet been developed, but in the northeastern part of the precinct, in crystalline limestone, there is a deposit of silver-bearing galena which was discovered and to a certain extent developed before the placer deposits were known. During the epochs of erosion that have been referred to, great quantities of rock were broken down and carried away, and their gold content, together with other heavy minerals, was left as concentrates in the valley, and bench gravels form the present placer deposits. Most of the placers are of the creek and bench types, though river-bar and gravel-plain deposits have received some attention. No high benches assignable to abandoned drainage systems have yet been discovered.

DEVELOPMENTS.

Although colors of gold were found in the gravels of Niukluk River by members of the Western Union Telegraph survey as early as 1865, the first discoveries of gold in paying quantities were made in March, 1898, on Melsing and Ophir creeks, tributary to the Niukluk. In 1899 probably $40,000 in gold was taken from the Ophir Creek placers, but the season of 1900 practically marked the beginning of active mining in the district.[a]

Council, the principal town and recording office, is situated on the north bank of Niukluk River 15 miles from Fish River. Both these streams are navigable for small power boats as far as Council, but the usual method of freighting supplies is by steamer to White Mountain, at the head of tide water, and thence by flatboat. A narrow-

---

[a] Brooks, A. H., Richardson, G. B., and Collier, A. J., Reconnaissances in the Cape Nome and Norton Bay regions, Alaska, in 1900, a special publication of the U. S. Geol. Survey, 1901, p. 112.

gage railroad leads from Council to the junction of Dutch and Ophir creeks, the latter of which is the most important gold-bearing stream in the precinct.

The placer mines of this region were examined and reported on by the Brooks party in 1900, but by 1903 the operations had been greatly extended and many additional facts bearing on the occurrence of placer gold were gathered. Although auriferous gravels occur on a majority of the streams of the district, the producing placers are confined to Fox River and to Niukluk River and its tributaries, Mystery, Melsing, Ophir, Goldbottom, and Elkhorn creeks and Casadepaga River. The total production of gold up to 1903 was between $5,000,000 and $6,000,000, but the many undeveloped placers in this field give promise of a greater production in the future. Many large ditches are under construction, and mining machinery is being rapidly introduced, which will greatly increase the area of profitable mining. When completed the railroad from the mouth of Solomon River to Council will lower the cost of transportation and thereby increase the placer-mining operations.

DESCRIPTION OF PLACERS.

FISH RIVER AND ITS EASTERN DRAINAGE.

Fine colors of gold may be found everywhere on the bars of Fish River, both above and below the mouth of the Niukluk, but up to the present time no successful attempts at mining have been made along the main stream. The eastern tributaries of Fish River have also been so far nonproductive of placer gold, though the first attempts at mining on the peninsula were at the Omalik silver mine, which is located on an eastern tributary near the head of the river.

FOX RIVER.

In the region south of Niukluk River there are a number of small creeks and rivers tributary to Fish River and Golofnin Sound, which have received some attention from prospectors, but have as yet produced very little placer gold. This region is marked by a rolling plateau surface, about 800 feet above the sea, in which the river valleys are incised. It is traversed by a well-beaten trail from Council to Solomon River.

Fox River is the only stream in the region south of Council that has been examined in detail by Geological Survey parties. It heads about 20 miles north of Topkok Head, and after flowing northeastward to the Niukluk lowland turns southeastward and skirts the edge of the highland to its junction with Fish River. In the lower 10 miles of its course it has a very sinuous channel filled with sand

and gravel bars, making the water shallow and navigation for small boats difficult. Above this portion the valley narrows down and the river channel has nearly a straight course. The gradient of the valley from the head of the river to the mouth is very slight. Where the stream is confined by valley walls its fall does not exceed 20 feet to the mile, and below, where it crosses the lowland, the grade is probably less than 10 feet to the mile. The valley of Fox River is broad, with the bed trenched from 50 to 100 feet below the valley floor, leaving a system of benches. The river bed itself is a broad expanse of gravel and sand bars, not entirely covered by water except at times of freshet.

The bed rock consists of a series of light-gray chloritic micaceous schists, interbedded with which are limestone and graphitic schists. Sills and dikes of rather massive greenstone are very generally distributed along the creek. The alluvium includes pebbles derived from bed rock, light micaceous sands, and in some places quicksands. Prospecting has not been carried on systematically, and there is little evidence of such work.

Some prospecting has been done near the mouth of I X L Gulch in the margin of one of the many benches common along the Fox River valley. In this bench, which is 50 feet above the river, a hole was sunk 17 feet through the gravel to bed rock, consisting of broken limestone and clay. Colors of gold were found on the bed rock. A pit about 50 feet square in the edge of the bench near this place yielded $150 in gold. The section exposed in this pit shows $3\frac{1}{2}$ feet of muck and soil, and 6 feet of gravel resting upon decomposed schist bed rock, which was pierced through to broken limestone about 11 feet below it. On the surface of the bench, about a quarter of a mile west of the river, there is what appears to be a well-marked old channel about 100 feet above the present river bed. This channel, which can be traced for at least a quarter of a mile, is probably worthy of investigation. Preparations were being made in 1903 for working claims here on a more economical scale with water brought in a ditch from another tributary of Fox River, known as Blue Rock Creek. The plan was to use this water for ground sluicing the muck from the surface, after which the pay gravel could be shoveled into sluice boxes. No successful mining has been done in the creek gravels of the main river bed, though the whole stream has been staked and is held largely by one company.

I X L Gulch is a small tributary which heads about a mile southwest of Fox River. The bed rock is mica schist underlain by limestone, probably a thin lenticular mass interbedded with the schist. Above the bed rock the gravels are from 2 to 6 feet deep. In the bed of the gulch near its mouth 2 ounces of gold were taken in 1903 from about 4 cubic yards of pay dirt. A layer of muck and soil about 2 feet thick above the pay dirt had been ground sluiced off.

No definite information of the occurrence of placer gold in other creeks of the neighborhood could be obtained.

NIUKLUK RIVER BASIN.

As has been noted, nearly all the productive placers of the Council precinct are found within the drainage basin of Niukluk River. Deposits of the river-bar, gravel-plain, creek, and bench types are represented, but nearly all the gold obtained comes from the latter two. The lower course of this river is entrenched in the broad gravel plain of the Niukluk lowland. (See " General geology," pp. 60–69.) Above this lowland portion the stream flows for some distance in a narrow valley along the sides of which there is a system of benches or terraces. The bed rock consists of the limestones and schists of the Nome and the Kigluaik groups, and the gravels therefore include pebbles of many varieties of rocks—of limestone and schist as well as of greenstone, granite, and vein quartz.

NIUKLUK RIVER PROPER.

River bars.—Although the first discoveries of gold in Seward Peninsula were probably made on the gravel bars of Niukluk River and small grubstakes were obtained with rockers from some of the richer bars, no systematic attempt to work them was made until the summer of 1903, when the bed of the river for a number of miles was controlled by a company which operated a dredge a short distance below Council.

The gold is well distributed through the bar gravel, but it was found to be somewhat richer near bed rock. There are no data at hand for estimating the gold content of the gravels in any particular place, but the operators near Council report that they carried from 50 cents to $1 to the cubic yard. The gold is, as a rule, bright in color and of a character approaching flour gold. A sample gave 96 colors to the cent. Associated with the gold are many fine particles of pyrite, magnetite, and garnet. Although some of the gold may have been derived from the local bed rock, most of it was probably contributed by the great number of streams which flow into Niukluk River, the light flaky gold being easily carried by running water.

The dredge operating near Council was reported to have a capacity of about 3,000 cubic yards in twenty-four hours. Gravel was elevated by buckets on an endless chain, and dumped into a sluice box $3\frac{1}{2}$ feet wide and 20 feet long, riffled with railroad iron. From this box it passed to a grizzly 20 feet long, which was constantly shaken. The fine material thus separated was distributed to a number of sluice boxes having an area of 10 by 24 feet. These sloped off on each side from the center of the boat and emptied into two 20-foot sluices running parallel with the boat and discharging into the river at the stern. The coarse material, sorted out by the grizzly, passed onto an endless belt and was stacked about 40 feet away. Water for the sluice boxes was raised by a centrifugal pump. Except for the railroad-iron riffles used where the coarse unsorted material was dumped, Australian riffles were used in the sluice boxes, as it was found that the sand packed less in these than in the Hungarian riffles with which the machine was first equipped. Although this machine was not able to penetrate the bed rock to any great depth, loose pieces were often brought up in the buckets. A large amount of black sand was associated with the gold in the riffles and sluice boxes at the clean-up, which showed about 50 per cent of magnetite, 20 per cent of garnet, 10 per cent of pyrite, and 10 per cent of limonite from which the gold had to be separated. No scheelite or stream tin is reported. Only one dredge has been used on Niukluk River, and though this machine has demonstrated the possibility of working low-grade bar gravels, it is probable that only by the most careful management can profits be obtained from gravels of as low grade as 50 cents a yard.

During the summer of 1904 this machine was moved from the bars near Council up Niukluk River to the mouth of Ophir Creek, where somewhat richer bar gravels have been reported, but the result of the operations there is not known.

Gravel plains.—An attempt was made in 1903 to work a placer in the gravel plain south of Niukluk River near the point where the dredge was located. The surface of the plain is about 15 feet above high water in the river and the excavations show several feet of sandy gravel overlain by several feet of sandy silt, muck, and vegetation. A steam shovel was used to dig the gravel and dump it into a large sluice box that was supplied with water by a ditch from Bear Creek. The shovel had a capacity of about 3 tons of gravel and sand a minute, but the gravel was dumped in large masses into the sluice and was probably not thoroughly washed before it passed out of the boxes. The work could be more economically done by first ground sluicing the deposits nearly to the level of the river. A more thorough washing of the gravel could be easily obtained by the use of a hopper, from which the gravel could be fed into the sluice more gradually. Although the enterprise was not successful and work was suspended before the end of the season, it is evident that the gravels here contain a considerable amount of fine gold.

MYSTERY CREEK.

Mystery Creek, 5 or 6 miles in length, empties into Niukluk River from the north about 5 miles below Council. It carries from 50 to 200 miner's inches of water and flows in a southerly direction through a wooded valley cut in calcareous mica schist with local thin beds of graphitic siliceous schist. Both of these rocks contain small stringers of quartz. The prevailing dip is to the southeast and in places is as high as 34°. The gravels of Mystery Creek are made up of calcareous mica schist with some vein quartz. The placers are all of the creek type, and are of no great extent.

One mile above the mouth of the creek the gravel is about 7 feet thick, but most of the pay is in the lower 3 feet. Nuggets of values up to $6 and $8 have been obtained, but the greater part of the gold is much finer. Part of it is bright and part is rusty, but all is rough and angular. About 2 miles above the mouth of the creek the gold is nearly all coarse and bright.

Most of the mining on Mystery Creek is done by shoveling into sluices, though horses and scrapers have been used on one claim. The gold was evidently derived from the calcareous mica schist bed rock and its uniformly angular character shows that it has not been transported far from its source. In the schist it probably occurs mainly in the small quartz or calcite veins and blebs. No veins containing gold are known to have been discovered.

MUD CREEK.

On Mud Creek, a small rivulet flowing into Mystery Creek from the west side, about 2 miles from the mouth, one claim was worked in 1903. Here about $2\frac{1}{2}$ feet of muck overlies 3 or 4 feet of angular schist gravel. The bed rock is decomposed calcareous mica schist. Gold is found both in the gravel and to a depth of 3 or 4 feet in crevices of the bed rock. It is very rough, spongy, and somewhat rusty, and is coarse and easily saved.

MELSING CREEK.

Melsing Creek flows into Niukluk River at Council. It is about 7 miles long and flows in a southwesterly course. It occupies a broad asymmetrical valley whose western side slopes gently from the creek up to the flat-topped hills, the eastern side rising more abruptly. This creek carries possibly from 100 to 300 miner's inches of water. The gradient for the first 4 miles from the creek's mouth is probably not over 50 feet to the mile; farther up it increases to more than 100 feet to the mile. The placers are confined to the creek bed and to low terraces little higher than the flood plain.

Gold was discovered on Melsing Creek in 1898 by Libby, Melsing, Mordaunt, and Blake, the four pioneer prospectors of the region, and since that time some mining has been done every season. In the summer of 1900 40 men were employed on the creek near the end of the season.[a] Although no bonanzas have so far been developed on this creek, it has probably yielded grubstakes and wages to many prospectors and miners. It is impossible to estimate closely the amount of gold produced, but it has probably not exceeded $50,000 all told.

The valley of the creek, from the water's edge to the tops of the hills on both sides, is well wooded with spruce. Some of these trees grow to 14 or 16 inches in diameter and 30 to 40 feet in height, and although not large enough to make first-class lumber, they were whip-sawed to make the first sluice boxes of the region, and are used as saw logs in a mill at Council.

The country rock on Melsing Creek is largely a gray calcareous mica schist, showing in places very little calcite and in other places passing into limestone. In a low terrace on the north side of the creek, half a mile above the mouth, there are gravels which have yielded some gold. The workings show the following section:

Section on Melsing Creek one-half mile above the mouth.

	Ft.	in.
Sandy muck	4	
Gravel	3–4	
Clay seam (bed rock)	4	
Gravel.		

The gold is found in the gravel within a few inches above the clay seam, below which it is reported there are no values. The smaller pieces of gold are nearly all smooth and bright, but the larger pieces are round and iron stained. In a placer 1 mile above the mouth the gravel ranges in thickness from 18 inches to 5 feet, and the sandy muck above it also varies in thickness. Here also the gold lies on a clay seam, and it resembles that just described, except that it is less iron stained. About 1¼ miles from the mouth of the creek the workings show the following section:

Section on Melsing Creek 1¼ miles above the mouth.

	Feet.
Moss and muck	1
Gravel	6–8
Calcareous mica schist.	

---

[a] Brooks, A. H., Richardson, G. B., and Collier, A. J., Reconnaissances in the Cape Nome and Norton Bay regions, Alaska, in 1900, a special publication of the U S. Geol. Survey, 1901, pp. 112–114.

The pay is confined to the lower part of the gravel and extends about 1½ feet into the crevices of the bed rock.

At the mouth of Basin Creek, which joins Melsing Creek 1½ miles from the mouth, the placer is confined to the creek bed and flood plain, and the workings show gravel 3 to 4 feet thick resting upon calcareous mica schist bed rock. Gold is distributed through the gravel, but is more abundant near bed rock and has penetrated the crevices to a depth of 12 to 18 inches. The average yield in gold to the man was reported in 1903 to be about $50 a day. If we estimate 4 cubic yards per day to the man, the pay gravel on this claim must carry not far from $12 to the cubic yard. The gold is mostly coarse, nuggets as large as a dollar being common. The largest nugget found was worth about $10. The nuggets are as a rule well rounded and iron stained. One of the nuggets examined showed a small square hole filled with hydrous iron oxide, probably left by the weathering out of an iron-pyrite crystal, and this indicates the probable occurrence of the gold with pyrite, which is widely disseminated through the schists and limestones of the valley.

Above the mouth of Basin Creek no gold in paying quantities has yet been found on either Melsing or Basin creeks and no mining has been done. The waterworn condition of the gold at the mouth of Basin Creek indicates that it may have been carried for some distance, and suggests the possibility of there being other placers farther up the creek, although the wearing may be due entirely to re-sorting or reconcentration of the gravels or exposure to the pounding of rolling pebbles. No auriferous quartz veins or stringers are reported in this basin.

OPHIR CREEK.

Ophir Creek is a northern tributary of Niukluk River, about 2 miles in a direct line above Council. For about a mile above its mouth Ophir Creek flows parallel to and within 100 yards of the Niukluk. A channel has recently been cut through this neck, called "the portage," turning the creek water directly into the river. This point is probably not more than 100 feet above sea level. Above this point the general course of Ophir Creek is north and south, and for a distance of about 8 miles it flows in a broad valley with gently sloping sides. Above this stretch the valley contracts to a narrow canyon, 3 miles long, extending to a point 12 miles above the portage. (See Pl. V, A.) The slopes of the valley are for the most part mantled with talus and gravel deposits to an elevation of about 600 feet above the sea. Within the broad part of the valley the creek winds back and forth with long, sweeping meanders, which, especially in the lower reaches, are incised below the general valley level, leaving projecting spur benches that rise gently from the convex bends of the meanders to the valley walls. (See fig. 14, p. 247.) The gold placers of Ophir Creek are of the creek and bench types.

Although gold was discovered on Ophir Creek during the summer of 1898, yet in 1900, when the region was first visited by a party from the United States Geological Survey, placer mining was still in its initial stages.[a] Since 1900 developments have been rapid and the deposits of gold-bearing gravel have been found to be of large extent, making the stream second in importance in Seward Peninsula only to Anvil Creek, and giving it a rank among the first gold producers. In 1900 only a few claims were worked, but in 1903 the entire creek from the mouth to the head of the canyon had been prospected. Though the immediate creek bed has been mined out, and in some places yielded only moderate returns, large deposits of auriferous gravel have been found in benches, and the indications are that they will yield a considerable output of gold.

In 1903 more than 1,000 men were employed on Ophir Creek and its tributaries. The mines are accessible from Council by two wagon roads—one up Niukluk River to the mouth of the creek, and one over the hills to the claims above Dutch Creek. A narrow-gage railroad, 7 miles in length, runs from Council to claim No. 15 Ophir, at the mouth of Dutch Creek.

The bed rock on Ophir Creek, from the mouth to the head of the canyon, belongs to the Nome group, and consists of limestone which in places has graphitic phases and shades off gradually into graphitic mica schist. Though the strikes and dips of the rocks are variable, the prevailing strikes are approximately parallel to the general course of the creek valley and the dips in almost all places where they have been observed are to the east at angles ranging from 10° to 90°, the average being not far from 45°. The hills along the river rise to 600 or 800 feet and mark a surface of erosion above which rise dome-shaped buttes. Most of these buttes consist of limestone in which it is at many places difficult to determine the structure, since the rock is so jointed and fissured as to obscure the bedding. The limestone of the buttes is apparently similar to that interbedded with the schists along Ophir Creek itself. This rock may be thick lenticular masses in the more thinly bedded schists, or perhaps closely folded portions of comparatively thin limestone beds. There was no opportunity to settle this question.

In the schists and limestones exposed in the valley many quartz veins, in general occurring as stringers parallel with the bedding or schistosity, have been observed, and similar quartz veins have been found in the limestone buttes just described. Assays of the stringers

[a] Brooks, A. H., Richardson, G. B., and Collier, A. J., Reconnaissances in the Cape Nome and Norton Bay regions, Alaska, in 1900, a special publication of the U. S. Geol. Survey, 1901, pp. 111–113.

show that as a rule they carry small amounts of gold and silver. A stringer several inches thick exposed in the bed rock of a claim at the mouth of Crooked Creek consists of quartz heavily mineralized with pyrite. An assayed sample of this ore yielded 0.06 ounce of gold and a trace of silver to the ton.[a] Samples were taken near the mouth of Ophir Creek from several small stringers of quartz that showed no mineralization and from the schist contiguous to them. When assayed the quartz showed no trace either of gold or silver, but the schist was found to contain a trace of gold and no silver.[b] Samples from this locality crushed in a hand mortar and panned were found to carry free gold. The prospectors who have searched for quartz veins in the Ophir Creek basin have not infrequently found quartz carrying free gold. Although no deposits of auriferous quartz large enough for mining have yet been found, these small veins and stringers are sufficient to account for the origin of the placer gold.

In the schists and limestones of the lower part of the Ophir Creek basin no rocks of proved intrusive origin have been found in place. The nearest occurrence is a greenstone outcrop on the divide between Warm and Sweetcake creeks.

For 2 miles above the canyon of Ophir Creek no bed rock is exposed, the prevailing formation consisting of Pleistocene gravels. The bed rock at the head of the creek consists of the schists, limestones, and gneisses of the Kigluaik group, with characteristic sills and dikes of granite. The Pleistocene gravels of the upper basin consist largely of materials derived from the Kigluaik group, and some pebbles and bowlders of this kind have been carried down Ophir Creek and are mixed with its gravels all the way to the mouth. In the lower portion of the creek such transported pebbles occur in the bench as well as in the creek gravels and have been found at an elevation of at least 100 feet above the creek bed.

The gravels proved to carry gold values occur from the mouth of Ophir Creek to the head of the canyon, a distance of about 12 miles in an air line, or following the bends of the creek probably not less than 20 miles. It is probably true that throughout this length there is no part of the creek-bed gravel which was entirely barren of placer gold. Above the canyon the bed rock lies at a greater depth and no gold deposits that have warranted development have yet been found, although considerable prospecting has been done. Of the tributaries, Sweetcake, Dutch, Crooked, and Oxide creeks have been producers of gold. As in many places along Ophir Creek the stream-bed and bench deposits merge, they will be described together, beginning at the mouth.

---

[a]Assay by E. E. Burlingame & Co., Denver, Colo.
[b]Assay by R. H. Officer & Co., Salt Lake City, Utah.

Near the mouth the bed rock lies at considerable depth, and until the waters were diverted it was not practicable to mine the gravels in the creek bed. Colors of gold are plentiful in these gravels, and it has been estimated that in some places they carry as much as $3 or $4 in gold to the cubic yard. The gold is finer than that found farther upstream. In the summer of 1903 no mining was in progress, but small grubstakes had previously been taken out with rockers. It is believed that the whole area between the old bed of Ophir Creek and the south bank of Niukluk River will be found rich enough to pay for dredging

The Ophir Creek valley for one-fourth mile above the portage is from 300 to 350 feet wide. On the west side, 30 to 50 feet above the creek bed, there is a bench, which rises gently toward the west and continues up the creek to the mouth of Sweetcake Creek, a distance of about half a mile. About one-fourth mile from the portage the creek has been confined to a ditch along the west valley wall, and here excavations in the old creek bed showed 4 feet of sand and muck underlain by 6 feet of auriferous gravel lying upon bed rock. In an upper pit this bed rock consists of blue clay, probably derived from decomposed graphitic limestone or calcareous schist, in a lower pit limestone was exposed. Although the horizontal extension of the pay gravel had not been determined, it is believed that it covers practically the whole of the valley floor and has a width of 300 to 400 feet. The bed rock where exposed has a very irregular surface, with some deep holes that have to be bailed out before the gold can be obtained. Where the bed rock is made up of schist it is soft and can readily be shoveled into the sluice boxes. The limestone is fissured, and the placer gold has penetrated along the crevices, but it has not been found profitable to mine it to a depth of more than 1 or 2 feet. Though the basis for estimate is indefinite, it is probable that the gold tenor in these gravels will average about $5 to the cubic yard.

The mine excavations at this place are about 10 feet below the creek bed, and the seepage water collects in a sump, from which it is pumped by means of a gasoline engine. Three lines of 12-inch sluice boxes were used in each pit. Usually two men shoveled the pay dirt from the bottom of the pit to a table about 5 feet above the pit floor, and a third man, standing on this table, shoveled the dirt into the sluice boxes.

About a quarter of a mile above this place, near the mouth of Sweetcake Creek, the gravel appears to be less extensive, for limestone crops out on the left bank. No mining was in progress here in 1903.

The gravels on the bench 300 feet west of Ophir Creek and south of the mouth of Sweetcake Creek have been mined with water brought in a ditch from the latter stream. Here the following section was measured:

Section of bench gravels on Ophir Creek near mouth of Sweetcake Creek.

	Feet.
Moss and muck	1½
Brown sandy soil	3
Gravel	2
Decomposed mica schist.	

The gravel is made up of schist fragments mixed with rounded pebbles of granite, vein quartz, quartzite, and greenstone similar to the wash of Ophir Creek.

At a point about a mile above the mouth of Sweetcake Creek the valley of Ophir Creek widens and the stream flows in a succession of deeply incised meanders. (See fig. 14.) The placers of the creek bed, which appear to have been mined out, are shallow and not more than 100 to 200 feet in width. Several large forces of men were at work mining gravels that occur in a spur bench covering an area of about 30 acres, almost surrounded by the meanders of Ophir Creek. This bench slopes gently down from the west side of the valley to the apex of the meanders, where it stands about 20 feet above the creek. The section is as follows:

Section of bench gravels on Ophir Creek 1 mile above Sweetcake Creek.

	Ft. in.
Black loam	3
Brown soil with some sand and a few pebbles	2
Sand and gravel	10
Broken and fissured limestone bed rock.	

The gravel is generally not well sorted, but contains angular pieces mixed with well-rounded material ranging from fine sand to bowlders one-half foot in diameter. The pebbles, many of which are much weathered, consist of schist, vein quartz, limestone, and some pieces of coarse-grained greenstone. The gravel layer is reported to yield from 10 to 15 cents to the pan, which would be approximately the equivalent of $15 to $20 a cubic yard. At the time the claim was visited nine men were shoveling into a 10-inch sluice box and the tailings were being removed by a team of horses and a scraper.

The gold is bright in color. One nugget worth $3.75 had been found, but nearly all the gold is fine and flaky. The limestone bed rock is broken and fissured and along the fissures has weathered out to such an extent that its upper part consists of large, more or less angular residual bowlders. The pay dirt penetrates between the bowlders and in some places has been worked to a depth of 2 or 3 feet.

At another locality on this bench a pit showed 1 foot of moss and soil resting upon about 8 feet of gravel and sand. Here the pay gravel was being shoveled into sluice boxes and the tailings dumped over the edge of the bench into the stream bed. Near by about 1½ acres of the bench had been ground sluiced, leaving in the pit about 2 feet of concentrated gravel. A section here showed 1½ feet of brown sandy clay and moss resting upon about 6 feet of gravel. In some of the bench workings a considerable number of granite and other pebbles of distant origin were mixed with the gravels; in others only pebbles of local origin were seen. This is a fair sample of a number of the bench deposits that are being mined along Ophir Creek. For 1½ miles above these bench workings the only mining carried on at the time of visit was on a small scale, and was confined to the creek bed proper. The bench gravels had not been opened up and it is not known whether they are auriferous.

About 4 miles from its mouth Ophir Creek winds back and forth

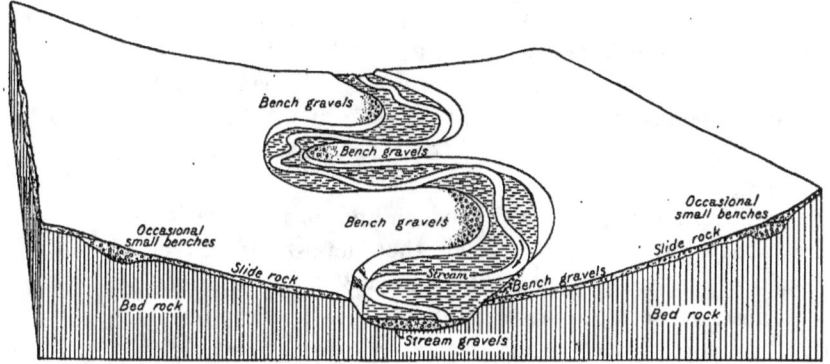

Fig. 14.—Diagram of part of Ophir Creek valley, showing position of bench-placer deposits.

across its valley to such an extent that a ditch crosses the stream four times within a mile. Between these meanders are bench deposits of gravel similar to that described above. An account of one of these will suffice. A meander of the creek channel includes a bench of about 20 acres lying 15 or 20 feet above the water on the left bank. Bed rock lies at the surface near the valley slope, but at the other end of the meander the bed-rock surface is only slightly above the creek bed. The nature of the deposit and its relation to the meanders are shown diagrammatically in fig. 14. A section was measured as follows:

Section of bench gravels on Ophir Creek 4 miles above its mouth.

	Feet.
Tundra, muck, and soil	5
Pay gravel	5
Bed rock, partly limestone and partly schist.	

The gold has penetrated the crevices in the bed rock to an unknown depth, the limit to which it is mined being determined by the cost of extraction. The pit was drained by a gasoline pump that raised the water from the sump at one side of the workings. Two strings of sluice boxes were used, and the gravel was shoveled first to tables and then reshoveled into the boxes.

About one-half mile above this point, or 4½ miles above the portage, is a bench of somewhat different contour, not being outlined by the meanders of the stream, which here flows nearly straight. This bench is on the east side and probably about 30 feet above the creek bed. An area about 250 feet square had been ground sluiced in 1903. The section exposed shows a thin layer of vegetable matter resting upon about 6 feet of silt or light-brown soil above the pay gravel, the full thickness of which was not disclosed, nor was the bed rock exposed. This pay streak consists of sand and gravel, containing an abundance of rounded granite pebbles from 1 to 3 inches in diameter and many rounded pieces of schist, all much decomposed. The mining method employed was as follows: A strong stream of water flowed through a sluice box, about 2 feet wide, which was set in the ground on a level with the surface of the pay streak. The pay dirt was carried to the sluice boxes by a number of scrapers, which completed their circuit by passing the lower end of the sluice box and scraping away the tailings.

The hill rises rather steeply above this bench and at an elevation of about 120 feet above the creek there is another deposit of gravel of the pocket-bench type. Of this deposit an area 300 feet square had been ground sluiced off to an average depth of 6 feet. At one place 7 feet of muck and soil were observed overlying 4 feet of gravel that rested upon 2 feet of decomposed gray calcareous mica schist containing some gold. The gold is fine, but round, and may be described as "shot gold." It is easily caught in the sluice boxes.

About 6 miles above the portage the creek bed is entrenched to a depth of only a few feet, the creek-bed gravels merging with those of the terraces. For a mile or more above there was much mining in progress, in which probably more than 500 men were employed. At the lower end of this strip a pit 300 feet long, 100 feet wide, and 17 feet deep was examined. Here 2 to 6 feet of sandy muck and soil rested upon 12 to 16 feet of well-stratified and some cross-bedded gravel. The gravel contained many well-rounded pebbles and bowlders of granite up to 8 inches in diameter. The bed rock was limestone with some thin beds of schist. The limestone was weathered and fissured and the gold-bearing sediments penetrated it to an undetermined depth. In schist bed rock, however, the gold has not penetrated more than a foot at most.

A quarter of a mile above this point a section was measured show-ing 2 feet of muck and soil underlain by 12 feet of gravel containing well-rounded pebbles of schist, limestone, and granite, mixed with yellow or brown sand. The limestone bed rock is mined to a depth of 4 to 6 feet, and some holes have been dug to an even greater depth. The deepest reached open crevices filled with water, in which there was a slow current. It is reported that the sinking of the water into underground channels on Ophir Creek has decreased since its bed has been filled with silt from the mining operations, and that the stream now carries more water than it formerly did.

About 6½ miles above the portage, excavations show from 5 to 14 feet of gold-bearing gravel resting upon a broken limestone bed rock. Three-fourths of a mile to the north 6 feet of sand and muck rest upon about 12 feet of gravel, of which the upper 2 or 3 feet carry very little gold. A section 1¼ miles further upstream shows about 1 foot of muck and soil over 6 feet of gravel that lies on a graphitic calcareous schist. This schistose bed rock on decomposition yields a clay called by the miners "hard pan," into which gold does not penetrate.

These deposits extend along the creek for nearly 2 miles and include a pay streak ranging from 4 to 20 feet in thickness and probably from 200 to 1,200 feet in width. It seems probable that this pay streak underlies an area of at least 160 acres. It has been impossible to obtain the facts necessary to determine even approxi-mately the average gold content of this gravel, but in some places it is known to have been very rich. It seems conservative to estimate the average at between $3 and $5 to the cubic yard.

In this part of Ophir Creek nearly all the mining is done by mechanical devices of various kinds, including several track and incline systems (Pl. V, B), derrick systems, and hydraulic elevators. Such devices have been described in detail by Purington.[a]

Above this large area of rich gravel Ophir Creek flows over a com-paratively broad bed bordered by flood plains, making the total width of the valley floor over 300 feet. In a few places along the margins of the flood plains there are benches, some of which are rock cut and others covered with gravel. In absolute elevation they correspond with the high pocket benches along the lower part of the valley, but are only 30 feet above the creek bed. On the east side one of these benches has been opened up, exposing bed rock about 10 feet above the creek. This bed rock forms the rim of a channel parallel with Ophir Creek. It is reported that the rich gold-bearing gravel lay on top of the rim and not in the old channel. The section shows 3 or 4 feet of muck and clay, below which is 10 feet of gravel containing pebbles of granite, resting upon a pay streak of gravel 4 or 5 feet

[a] Purington, C. W., Methods and costs of gravel and placer mining in Alaska: Bull. U. S. Geol. Survey No. 263, 1905.

thick, chiefly made up of pebbles of local origin. Along its margin some of the pay streak was strongly cemented.

In 1903 no work was being done for about a mile above this point, and the next working mines were found near the mouth of Crooked Creek. Here terrace gravels on the left bank 10 to 15 feet above the creek were being exploited. The bed rock of the deposit is probably little above the present creek. A section showed 2 or 3 feet of muck overlying 5 or 6 feet of gravel, which rested upon calcareous schist. The gravel was being washed down by a hydraulic giant to a hydraulic elevator. After the gravel was stripped the bed rock was cleaned by hand to a depth of 1 foot. Mining was also in progress at the mouth of Crooked Creek and is described in the account of that stream (p. 252).

For about 4 miles above Crooked Creek there was no mining in 1903, though some of the gravels doubtless carry gold. In the canyon, 3 miles long, whose lower end is a mile above Crooked Creek, the creek bed has an average width of 100 feet and contains shallow deposits of gravel, in which very little mining has been done. A characteristic view of this canyon is given in Pl. V, A.

On the left bank of the creek, near the upper end of the canyon, a small body of gravel is exposed on a bench possibly 20 feet above the water. Here a cut shows 5 feet of muck and soil, resting upon 2 to 3 feet of pay gravel. The bed rock is calcareous mica schist, some of it graphitic. Mining was done by stripping the muck by hand and then shoveling the pay gravel into sluice boxes. No estimate of the gold content of the pay streak was obtained, though it must have yielded at least wages to the seven or eight men employed. About one-half mile above this point the valley of Ophir Creek merges with the gravel plain of the upper Fish River lowland. So far as known no mining has been done above this point, but there has been some prospecting, and the claims all the way to the head of the creek are held by annual assessment work. The gravel deposits of the lowland consist mostly of pebbles derived from the mountains to the north, a region believed not to be auriferous, and it does not seem likely that workable placers will be found in them.

SWEETCAKE CREEK.

Sweetcake Creek, a tributary to Ophir Creek from the west a mile above the portage, is about 3 miles long. Near its head the creek flows through a narrow valley incised in the upland, but it widens out below, and near its mouth a flat bench already described (p. 245) separates it from Niukluk River.

Gold was discovered on Sweetcake Creek on August 12, 1898, by S. E. King, N. S. Vestal, Charles Phillips, Tom Baer, and Nicholas

Credel, and what was probably the first successful placer mining in the precinct was begun there. About $36,000 is said to have been taken in that year from one claim. The pay streak was found to. be confined to a narrow strip along the creek bed. The following section was measured near the mouth of the creek:

Section on Sweetcake Creek near mouth.

	Feet.
Fine micaceous sand	5
Gravel and sand	2
Angular schist gravel (pay streak)	1
Bed rock exposed	1

The gravels contain much. mineralized quartz and calcite, and the gold was formed in a layer made up of angular pebbles from the immediate bed rock. Average pans taken on bed rock ran from 25 to 50 cents and a $9 nugget was the largest found. By 1903 the richest gravels on this creek had been exhausted, and little, if any, mining was in progress. It is reported, however, that there is still considerable gold left along the creek, which will be reclaimed on the completion of the ditch now in construction from the head of the Ophir Creek canyon.

DUTCH AND SNOWBALL CREEKS.

Dutch Creek joins Ophir Creek from the northeast about 5 miles from the portage, in the center of the most extensive mining operations on the main creek. It forks about 2 miles up, the west branch being called Snowball Creek. Both forks rise on the surface of a plateau about 800 feet above the sea, on the south side of the Fish River lowland. They have not been studied, but it is reported that the heads of their valleys are incised in gravel deposits resembling those of the Niukluk lowland. Not much gold has been produced on Dutch Creek, though there has been some mining, and prospectors report that colors of gold are to be found both in the creek and on benches. A bench resembling one on the headwaters of Crooked Creek, described below, has been located near the source of Snowball Creek.

CROOKED CREEK.

In 1903 there was more mining on Crooked Creek than on any other tributary of Ophir Creek. Crooked Creek has a length of about 3 miles and a general southeasterly course, joining the main stream 8 miles from the portage. Two small tributaries, known as Balm of Gilhead and Albion Gulch, enter Crooked Creek from the southwest about 1 and 2 miles, respectively, from its mouth. The valley walls, which are covered with talus and gravel, rise gently from the creek floor to altitudes of 200 or 300 feet, where limestone crops out and the ascent steepens. The placers are generally confined to the creek bed and the immediate flood plain, but some occur in benches near the head of the creek.

The flood-plain deposit at the mouth of Crooked Creek widens out and merges with that of Ophir Creek. The gravels from the two sources are readily distinguished, as those of Ophir Creek contain granite pebbles, whereas those from Crooked Creek are made up entirely of limestone and schist. The gravels of Ophir Creek extend about 400 feet up the valley of Crooked Creek, and the deposit here is wider than that which belongs wholly to Crooked Creek. It is said that the limits of the Ophir Creek gravels are marked by a distinct rise in the bed rock which has been channeled by Crooked Creek. The bed rock is graphitic calcareous schist, much of it mineralized, and stringers of vein quartz carrying graphite were observed at a number of places. One of these is a mineralized belt 12 feet wide, which strikes northwest. In this impregnated zone vein quartz is associated with pyrite. It is reported to assay as high as $8 to the ton, but a specimen obtained by Mr. Collier showed only traces of gold and silver.[a] At the lower end of this claim the pay streak has a width of 250 feet, but one-fourth mile above it narrows down to about 6 feet. It has been worked out and no section could be measured, but its average thickness is reported to have been about 6 feet. The gold tenor of the gravel mined is estimated at $4.50 to the cubic yard.

The gold is comparatively coarse and the pieces well rounded. Some are bright and others iron stained. In the sluice boxes are found heavy concentrates consisting principally of garnet and magnetite, but including some topaz. It is impossible by ordinary means to separate all of the gold from this concentrate, for after all of the apparent free gold has been removed it still contains from 40 to 60 cents to the pound. An assay of some of these concentrates shows 89.4 ounces of gold and 11 ounces of silver per ton.[a] In these concentrates the gold is probably contained in limonite, an alteration product of pyrite. The pay streak along Crooked Creek from the upper end of this claim is in the main not more than 24 feet wide, and the excavation along it shows it to be nearly straight. About three-fourths of a mile up the creek the pay streak is 20 feet wide, and the section shows 6 feet of black mucky clay, 2½ feet of yellow sandy clay, 2½ feet of black clay mixed with broken limestone, resting upon schistose limestone. The gold is found in the lower black clay and yellow sandy clay.

About 1½ miles above the mouth of the creek the pay streak is from 150 to 200 feet wide and the bed rock is calcareous mica schist. Immediately above the bed rock there are many limestone bowlders

---

[a] Assay by E. E. Burlingame & Co., Denver, Colo.

that appear to have reached their present position by a landslide from the side of the valley. These limestone bowlders are of large size and a derrick is required to move them. (See Pl. VI, *A*.) Here the pay streak rests either on a yellow clay above the schist or directly on and in schist. Very little work was in progress in 1903 above this place and the creek was not examined.

Above Albion Gulch the valley of Crooked Creek broadens out into a comparatively wide basin and the slopes on the side of the valley become very gentle. Between Albion Gulch and the main fork of Crooked Creek there is a broad bench, nearly level in some places but generally sloping gently to the southeast, on the surface of which gravels and sand have been found at a great many places. Prospecting had not in 1903 gone far enough to determine definitely the limits of this gravel deposit, though it apparently extends from the mouth of Albion Gulch and covers the whole hillside to the northwest for nearly half a mile; its upper limit is marked by a sort of swale that extends from a point near the head of Albion Gulch northeastward to Crooked Creek.

Prospect holes have been sunk in many places and in all of them colors of gold have been found, while in a few the gravels contain sufficient gold to warrant mining. One open cut showed 7 feet of yellow and blue sandy clay, containing pebbles of quartz and schist, resting upon a nearly flat floor of limestone. In another prospect the bed rock slopes to the northwest and appears to be the rim of an old channel. Concentrated gravel taken from a space that had been ground sluiced yielded about 2 cents in bright angular gold to the pan. On the basis of 150 pans to the cubic yard, this is equivalent to about $3 in gold a cubic yard. Between the point where this gold was obtained and Crooked Creek a hole has been sunk to a depth of 82 feet without reaching bed rock. The shaft passed through sand and gravel beds, some of which were deemed to be low-grade pay streaks.

Prospecting has not gone far enough on this bench to determine the origin of the deposits. The fact that the pebbles and the gold are angular points toward a very local source for all the material. In some places, however, the deposit appeared to be distinctly water-laid. It seems probable that the greater part of the material was derived from the decomposition of the bed rock almost in place, but these decomposition products have been in part sorted by water. In a few claims the surface deposit of the bench has been ground sluiced and the gravels underneath shoveled into sluice boxes. Water is brought in a ditch from the head of Crooked Creek and also collected from the seepage of the hillside, but the supply is insufficient for mining purposes.

BALM OF GILEAD GULCH.

Balm of Gilead Gulch, which enters Crooked Creek from the west about 1 mile from its mouth, is only one-half mile long. Some work done in 1903 exposed a section of 5 feet of brown soil resting upon 2 to 4 feet of broken limestone that gradually became more solid in depth. Gold occurs here from the surface down, but is richest in the crevices of the limestone. The gold is rough and angular, with sharp corners. Three or four men were employed and the gravel was washed in sluice boxes, which were supplied for a few hours each day with a small sluice head of water collected by a dam across the gulch above.

ALBION GULCH.

Albion Gulch, which enters Crooked Creek from the west side about 1½ miles from its mouth, is about three-fourths of a mile long and flows across the strike of a series of schists and limestones. The gravels of this gulch are auriferous throughout its course. Near the mouth an exposed section revealed 3 feet of soil overlying 3 feet of gravel, resting upon limestone; the gravel consists of pebbles derived from the immediate bed rock. The claims along this gulch were worked by stripping the soil and muck from the top, either by horses and scrapers or by hand, and then shoveling the gravel into sluice boxes.

GOLDBOTTOM CREEK.

Goldbottom Creek joins Niukluk River from the north about 9 miles above Council. It is about 10 miles long and drains an area of about 50 square miles. Its floor is comparatively flat and from 500 to 1,000 feet wide, and the valley slopes rise more gently on the western than on the eastern side. The stream has a grade of about 100 feet to the mile and at its mouth carries about 1,000 inches of water. Warm Creek, which joins Goldbottom Creek 1½ miles from its mouth, is the largest tributary and at present the only one on which mining is in progress.

Goldbottom Creek is reached from Council by trail along the Niukluk, but most of the supplies are towed up the Niukluk in flat-boats to the mouth of the creek, whence they are hauled to the claims in wagons. Though mining was begun on this creek as early as 1900, the total output is small.

The bed rock is green and gray schist with interbedded, crushed, and crumpled limestones, which in general strike a little west of north and dip to the east; below the mouth of Warm Creek the strike is east of north. Many coarsely granular pebbles and bowlders of greenstone occur in the gravels of the creek and seem to have been brought in from the west. Fine-textured greenstones outcrop at the mouth of Warm Creek. The schists contain numerous veins and lenses of quartz, both parallel to the schistosity and cutting across it at various angles. Many of these are mineralized with pyrites and other sulphides, but no samples have been assayed by the Survey.

The placers are confined to the creek bed and flood plain, and only two claims were in operation in 1903. One of these, near the mouth of Warm Creek, had been worked since 1900. A section here shows 5 to 6 feet of fine gravel resting upon decomposed chlorite schist. The values are contained in the lower 2 feet of gravel and about 1 foot of the schist bed rock, which are passed through the sluices. The gold is coarse, well rounded, and mostly bright, though some of the pieces are stained with iron oxide. There are not many nuggets, and the largest was valued at $6. The concentrates consist of garnet, hematite, and ilmenite, with some scheelite in small rounded white grains. In mining a sluice box 2 feet wide is used, into which the gravel is dumped by horse scrapers, as shown in Pl. VI, B.

On a claim near the head of the creek about 2 feet of soil and muck were removed by hand and about 3 feet of gray mica-schist gravel was shoveled into the sluice box. The gold is fine and angular. Although no other claims were in operation, a large amount of prospecting has been done in the neighborhood and colors of gold have been obtained at many places.

WARM CREEK.

Warm Creek, the principal tributary of Goldbottom Creek, joins it from the east side to about 1½ miles above the Niukluk. It is about 6 miles long and occupies a V-shaped valley that gradually widens from its head to its mouth. It carries possibly 60 inches of water.

Supplies for the placer mines which lie in the upper part of the valley are hauled from Council, a distance of about 8 miles, by a trail crossing the hills from the mouth of Ophir Creek. All the placer deposits yet discovered are of the creek type.

The bed rock, which is limestone and schist, with some greenstones, has in most places an easterly dip. Mineralized quartz veins occur at a number of places along the creek. Near the mouth of the creek there are two quartz veins, one about 3 feet wide and the other about 1 foot wide, striking N. 30° E. and standing nearly vertical. These veins cut a gray siliceous mica schist near its contact with a greenstone mass. They show little mineralization and no sample was taken for assay. The creek gravel has all been derived from rocks exposed within the drainage basin.

The workings show a thin layer of muck and moss overlying about 7 feet of barren gravel, below which some gold is found in the crevices of the weathered bed rock. In places the lower foot of the gravel also carries gold.

Mining was begun on this creek in 1900, and though no bonanzas have been found and the distribution of the gold seems rather irregular, yet many claims have yielded fair profits, and the entire output is probably about $100,000.

Most of the gold is rough and iron stained, and some of it is almost black. One nugget, worth $45.10, at $16 per ounce, was found in 1902; and one worth $12.33 in 1903. The concentrates contain ilmenite, scheelite, magnetite, garnet, and some hematite and rutile. A sample of these concentrates, from which the free gold had been extracted by panning, was assayed for the United States Geological Survey, yielding 22.40 ounces of gold and 4.20 ounces of silver to the ton.[a] This result is suggestive, as Warm Creek heads near Crooked Creek, from which similar rich concentrates are obtained.

It is reported that a number of claims near the mouth of the creek have been bought by a company that purposes to work them with a dredge. The gravel deposit is several hundred feet wide and is said to be from 6 to 9 feet thick; the bed rock is mainly a soft mica schist, but there is also some limestone. The weathered surface of the schist could be easily cut by a dredge, but to obtain the gold in the limestone would probably be more difficult.

CAMP CREEK.

Camp Creek flows into the Niukluk from the south about a mile below Goldbottom Creek. Several claims were worked on this creek in 1904. The auriferous gravel is from 50 to 100 feet wide and about 3 feet thick, with an overburden of about 3 feet, and is said to carry from .75 cents to $1 a cubic yard. Most of the mining was done by the shoveling-in process, but one claim was hydraulicked with a canvas hose carrying water under a 70-foot head.

ELKHORN CREEK.

Elkhorn Creek, a small stream which joins the Niukluk from the south about 5 miles above Ophir Creek, is incised in a broad terrace bordering the river. Its trench ranges from 150 to 500 feet in width. Mining here began in 1900 and the placers were then examined by Richardson, who reported as follows:[a]

Near the mouth of the creek 2½ feet of gravel overlie 6 inches of clay and disintegrated bed rock. It is reported by miners that the pay streak is in patches and that the average yield of pans is 5 cents. The bench near the mouth gives colors but has not been developed. The bed rock is mica schist, interbedded with limestone, and the strike is at right angles to the course of the stream, with almost vertical dip, giving favorable conditions for concentration of gold.

---

[a] Assay by E. E. Burlingame & Co., Denver, Colo.

[a] Brooks, A. H., Richardson, G. B., and Collier, A. J., Reconnaissances in the Cape Nome and Norton Bay regions, Alaska, in 1900, a special publication of the U. S. Geol. Survey, 1901, p. 110.

The gold is medium coarse and bright yellow in color. Some very coarse gold has been found stained with iron. The average assay shows its value to be $19.12 per ounce. Quartz is often found attached to the placer gold, and one nugget was attached to a piece of mica schist. This goes to show that it is of local origin. One nugget worth $55 has been found and several worth from $12 to $16.

Since 1900 the placers for about half a mile have been entirely exhausted, but farther up the creek work is still in progress both in the creek bed and on the benches. It is estimated that the total production of the creek up to date has been from $110,000 to $120,000.

CASADEPAGA RIVER.

Casadepaga River,[a] the largest southern branch of the Niukluk, has a length of about 30 miles and a general northeasterly course. The writers have had no opportunity to examine this basin and the following notes are compiled from the published report of A. H. Brooks [b] and from additional data kindly furnished by C. W. Purington and Sidney Paige, who made a hasty trip through the Casadepaga Valley in 1904.

The gradient of this river does not exceed 13 or 14 feet to the mile and from its mouth to a point within 2 miles of the head of its longest tributary the whole fall is not more than 400 feet. The drainage area is large and the stream is navigable for small boats and canoes for a distance of about 15 miles. The upland in which the Casadepaga Valley is incised ranges in elevation from 800 to 1,800 feet. The bed rock of the whole basin consists of the limestones and schists of the Nome group, including many veins and stringers of quartz, some of which are known to be auriferous. Though none of these yet discovered are large enough to mine, they are ample to account for the origin of the placer gold in the basin.

The river occupies a broad and deeply gravel-filled valley, in the floor of which the present stream bed is trenched to a depth of 30 to 150 feet, leaving well-marked gravel terraces and benches through nearly the whole length. In the lower part of the valley the stream has cut to bed rock through the gravel deposits in only a few places, but in the headwaters region the gravels overlie broad rock-cut benches on both banks.

Although the gravels of the river bed undoubtedly contain considerable gold, it has probably not been successfully extracted, both because of the volume of water and the slight grade of the river bottom. The placers that have been operated up to the present time are in the small tributary streams at points where they cut the gravel

---

[a] By a recent decision of the United States Geographic Board, the correct name of this river has been declared to be Casadepaga instead of Koksuktapaga.
[b] Op. cit., pp. 107–110.

terraces and benches.   These are creek placers that have derived most
of their metallic contents by reconcentration from the benches and
terraces. The gravels of the benches are auriferous, but their tenor
does not appear to be high enough to mine under the present cost of
labor and supplies.   There is a well-marked terrace 30 feet above
the river bed and there are traces of others up to altitudes of 600
feet.   The following is quoted from Brooks's report already cited:

The Koksuktapaga [Casadepaga] was first prospected in 1898 by Mordaunt,
Libby, Nelson, and Blake. These pioneers are known as the, Big Four, from
whom the largest tributary of the Koksuktapaga receives its name. Claims as
far up as Goose and Quartz creeks were located in 1898, but most of the
staking on the Koksuktapaga and its tributaries was done in 1899. Quartz,
Boulder, Dixon, and Spruce creeks are reported to have yielded a few thousand
dollars in 1899. This was taken out in the course of prospecting rather than
in systematic mining.

In 1900 there was a general delay in getting into the country, and later on
the low water, consequent to the dry season, delayed transportation of supplies.
In the fall, but a short time after sluicing had begun, floods washed away many
dams, ditches, and sluice boxes. The season of 1900 must be regarded chiefly
as a period of further prospecting. It is expected that the actual possibilities
of the country will be more clearly shown in 1901. The output of the Koksuktapaga region for 1900 is estimated to be $15,000. This has been taken mostly
from the tributary streams, as little mining has been done on the main
river. * * *

Dawson Gulch joins the river nearly opposite Big Four Creek. The 30-foot
bench of the river valley across which the gulch flows carries colors. On the
mica-schist bed rock, which was reached by test pit, fine gold is found associated with garnet.

Several gulches on the east side of the creek between Big Four and Dixon
creeks were reported to give good prospects. Of these Thorpe Gulch, opposite
Dixon Creek, was being worked. Four men had dug a ditch about 40 feet long
by 4 feet deep in the bed of the gulch, which is a channel across the terrace of
the Koksuktapaga. Bed rock was not reached. Mica schist, graphitic quartz
schist, and limestone are exposed at the head of the gulch. The gold is fine
and is associated with quartz and magnetite. The claim was reported to be
paying small wages.

Dixon Creek was dry in July and no work was being done on it. A claim
near its mouth is reported to have yielded a few thousand dollars in 1899. The
workings were at the base of the terrace in the creek channel. Massive gray
crystalline limestone caps Mount Dixon, north of the creek. Below the limestone is mica schist carrying small quartz veins.

Dry Gulch is a small channel incised in the terrace of the main valley. No
measure was obtained of the thickness of the terrace gravels, but it may reach
50 feet. The workings in the terrace have reached a depth of 8 feet, and colors
have been found from the grass roots down. The gold is usually fine; one
$2 nugget was obtained. Pans are said to average 3 cents.

An important consideration relative to the gold resources of the Koksuktapaga
is the fact that the gold of Dry Gulch and similar streams is not found on bed
rock, but occurs in the gravels of the terrace and is usually concentrated on clay
seams. These bench gravels are said in many places to carry 3 cents to the
pan. If this proves to be the case, the ample water furnished by the main
river would offer good possibilities for hydraulic mining.

At the time of Mr. Richardson's examination the lower valley of Spruce Creek was buried in a remnant of ice of the previous winter and no work was being done. It was reported that 40 or 50 ounces of gold were taken from this creek in 1899.

Some development was made on Penelope Creek during 1900, but in the lower part of the creek bed rock had not been reached. Pans are said to average 5 cents. The adjacent country rock is quartz-muscovite schist and graphitic quartz schist. This creek seems to have ample water for sluicing at all times.

Goose Creek, which joins the Koksuktapaga opposite the mouth of Penelope Creek, was dry in July. Mining operations were going on, however, to be ready for the fall rains. Gravel is from 4 to 8 feet thick. The lower 2 to 3 feet usually carry values, and pans are said to run from 2 to 5 cents. Quartz Creek is a tributary of Goose Creek. The gravel is about 3 feet thick and is said to average about 8 cents to the pan. Mr. Richardson saw one pan taken from bed rock which yielded 80 cents. The gold is flat and coarser than Goose Creek gold. Mint receipts give its value at $18.60 an ounce. The concentrates from sluices carry much garnet and magnetite. Bed rock is quartz-muscovite schist with thin-bedded limestone.

A mass of ice averaging 10 feet high filled the mouth of Canyon Creek and extended up its bed for a mile. The lower course of the creek lies in the river terrace, through which it has cut its valley into the quartz-mica schists. About 2 miles from its mouth some work has been done, exposing 3 feet of gravel on bed rock.

Boulder Creek, a northerly tributary of Canyon, is said to have yielded 50 ounces in 1899 as a result of a few days' work. In July work on one claim showed about 2 feet of gravel on bed rock, of which about 18 inches are said to carry values.

Banner Creek had some water in it in July, but not enough for sluicing. One claim in 1899 is reported to have yielded $400 to four men who sluiced two days and a half. When Mr. Richardson visited this creek two men were at work near the mouth, where the gravel is 8 to 10 feet thick. The gold is rather coarse, though flattened, and assays $19.20 an ounce. Farther up the creek more or less work had been done and the prospects were reported good. Limestone is bed rock in the mouth of the creek and outcrops on the divide to the north. Graphitic quartz schist caps the hill at the head of the creek, and mica schist forms the divide between Banner and Ruby creeks.

Left Fork [a] of the Koksuktapaga is about 5 miles long. Like Canyon Creek, in its lower course it cuts through the wide terrace of the main river, while above the valley it is confined by adjacent hills. The rocks in the lower course of the Left Fork are mica schists dipping S. 65° W. at an angle of 30° to 60°, and jointed N. 70° E. Between Wilson and Willow creeks is a belt of gray limestone about 2,500 feet thick interbedded in the schistose series. Small quartz veins are frequent throughout this series. Pebbles of garnetiferous greenstone are common in the creek, but no outcrop of this rock was seen. No developments have been made on this stream.

Willow Creek is a small branch at the head of the Left Fork, from which good reports came toward the end of the season. A low, narrow bench, about 6 feet high, extends along the creek. Gravel is shallow in the creek bed. Preliminary development, it was claimed, showed that a man could average $10 a day with a rocker. One $8 nugget and another worth $4.35 have been found. The gold is coarse and dark. In the lower part of the creek, bed rock is limestone, while farther up it is mica schist.

---

[a] Now known as Lower Willow Creek.

Ruby Creek, so named for the numerous small garnets found in it, is about 3 miles long and flows in a comparatively broad valley. Bed rock is mica schist dipping S. 65° W. at an angle of 45°, and jointed east and west. Numerous small quartz veins parallel both to the schistosity and to the joints occur in the schists. Pebbles of garnetiferous greenstone are common in the creek, but the rock was not seen in place. At the head of the creek gray limestone and graphitic quartz schist are exposed. In July there were about twenty men on the creek. Almost every claim had at least one representative on it, though prospecting and preliminary work rather than actual mining was being done. Several ditches have been dug in the creek bottom, and crosscuts have been made into the adjacent bench. In the lower part of the creek there is a depth of 2 or 3 feet of gravel through which the gold is pretty well distributed. Many garnets and a little black sand are associated with the gold. The gold is coarse, rather dark colored, and is said to assay $19.35 an ounce. The largest nugget found was worth $5.50.

NOTES ON RECENT DEVELOPMENTS IN THE COUNCIL REGION.

By PHILIP S. SMITH.

The following notes were gathered by the writer during a week's journey in the Council district in 1906. They are intended to supplement the descriptions given in the foregoing pages.

MELSING CREEK.

In 1906 but little work had been done on Melsing Creek, owing to the exceedingly dry season. Toward the last of September active operations were in progress only near the mouth of Basin Creek, although two outfits of three to six men each had been at work earlier in the season with scrapers below this point.

A force of six to ten men had been employed immediately below Basin Creek, but had ceased work, owing to frost, just before the visit of the writer on September 28. The course of the pay streak, which lies only a few feet above the present stream level, is very sinuous and suggests that these gravels were deposited by a stream of relatively small size meandering widely on a flat slope. A feature of some interest is the occurrence of large granite and quartzite bowlders, many of them 18 inches in diameter, in a layer of mud and decayed vegetable matter lying above the gravels. The granite is but slightly decomposed, and the bowlders are rather angular, suggesting a transporting agency other than running water. Associated with the auriferous gravels in many places are thin strata of cemented gravels, in which the cementing material is mainly calcite. The cemented character prevents the separation of gold, so that if much of this sort of gravel should be encountered recourse to some method of crushing would be necessary.

At this claim a method of preparing the sluice boxes not in use in other parts of the district was noted. This consisted of nailing a strip of canvas or cocoa matting on a plank slightly narrower than the bottom of the sluice box. On top of the canvas a strip of galvanized-iron screen, with about a one-fourth inch mesh and the same width as the plank, was fastened. In use the plank was placed in the bottom of the sluice box and the riffles laid on top, thus holding it in position. To clean, the plank was taken out of the sluice box, turned upside down and pounded with a hammer or mallet. Although no comparative figures were available to prove the added efficiency of the sluice boxes thus equipped, the operator was completely satisfied with the results, as he was convinced that the additional saving of gold was very great. It is not necessary that every box in a string should be equipped with such a false bottom. Individual practice and study will determine the most effective number for different kinds of gold.

Just at the mouth of Basin Creek work was in progress during the summer of 1906 on gravels that lie in the creek and on a bench a few feet above the stream. The pay-streak seems to be a direct continuation of the one on the claim next downstream, already described. In addition, at points where these slightly higher benches are cut by the creek the present stream gravels contain values that repay working. The ground is developed by the use of horse scrapers and sluices. Operators all along the creek were much hampered owing to the lack of water in the early part of the summer. Another difficulty encountered in the district is due to the low grade of the creek.

OPHIR CREEK.

Since the date covered by the report on the Seward Peninsula (1903) the development of the Ophir Creek placers has continually called for additional water and additional head. To meet this demand high-level ditches have been constructed and maintained. It was recognized, however, that the basin of Ophir Creek could not, even under the most favorable circumstances, afford sufficient water to meet the growing demands. Consequently it has been necessary to lead water from other drainage areas into that of Ophir Creek. The largest operation of this kind projected has been successfully carried out, and undoubtedly permitted mining which the dry weather of the summer of 1906 would have otherwise prohibited. The ditch takes water from Parantulik River at Helen Creek, a small tributary, about 2 miles north of the summit of Chauik Mountain. It is 11 miles long, and in many places where the slopes are excessive a flume has been constructed. The water is led around the east flank of Chauik Mountain and then across the divide into the Ophir Creek basin. In order to obviate additional ditch construction, the water is discharged into Ophir Creek, and taken up again farther downstream by one of the existing ditches. It was estimated that about 8,000 miner's inches were available from Parantulik River, but during the dry period of 1906 only about 500 inches were delivered by the ditch.

Another project for leading water from Parantulik River to Ophir Creek is under way, but as yet actual ditch construction has made little progress. Up to the present time the work of this company has been mainly devoted to surveying the course of the ditch, making preliminary observations, and acquiring rights of way. The proposed ditch will take water from a point considerably below the completed one, and for that reason should have more water available.

A ditch of less size than either of the foregoing has been constructed to collect water from the northern tributaries of the Niukluk about 2 miles west of Ophir Creek. This ditch is at a low level, is of small cross section, has only a small available drainage area, and is to be used only by the properties near the mouth of Ophir Creek, below Sweetcake Creek.

In addition to the ditch building, another feature of interest in the district has been the location of a gold-bearing lode on the divide between Goldbottom and Ophir creeks, near the head of Crooked Creek. The lode occurs near the contact between limestone and schist, and the specimens show considerable free gold. Values of nearly $40 to the ton are reported, but it is not known that they were from commercial samples. This discovery seems to be significant in connection with the fact that the gold at many parts of Crooked Creek is very sharp and angular and much of it has quartz fragments attached. A specimen of gold seen near the mouth of Crooked Creek was of such fragile shape and crystalline form that it seemed impossible for it to have been carried more than a few feet from its source.

Work on Ophir Creek during 1906 was carried on less by individuals and more by large companies than in previous years. The most active work has been done by the dredge at the portage, by elevators near Sweetcake and Dutch creeks, by derricks a little above Dutch Creek, and by shoveling in near the mouth of Crooked Creek. No work was done on Ophir Creek above Crooked Creek during the summer of 1906. Some work was in progress on a few of the tributaries of Ophir Creek. A little mining has been done on Sweetcake Creek, but the values do not seem to run much more than a mile above its mouth. On Dutch Creek also a little mining has been done. The small stream joining Ophir Creek near claim "19 above" has been prospected, but does not seem to carry values above its mouth. Along Crooked Creek for a distance of 2 miles the creek was worked almost continuously all summer by parties of two to fourteen men. The fact that the side streams almost all carry gold has led to an enrichment of the main-stream gravels. Practically every one of the bonanzas of Ophir Creek has occurred at the junction of a side stream. The recognition of this feature, which prevails also on many other streams, should be of some assistance in prospecting undeveloped regions.

NIUKLUK RIVER.

Above Ophir Creek on the Niukluk very little work has been in progress in 1906. The placers on Richter Creek, the first tributary from the west, seem to be nearly exhausted. On the broad bench below the junction of Camp Creek and the Niukluk a company has been engaged in working gravels by means of a shaft 40 feet deep. The section of this shaft is as follows:

Section on bench below junction of Camp Creek and Niukluk River.

	Ft.	In.
Tundra	2	
Pure ice	10	
Sand and ice	15	
Rock fragments, etc.; much mica		2
Sand and ice	1	
Frozen gravel	12	

Owing to the ice and frozen ground, much difficulty has been experienced in maintaining the shaft. Water is not easily available at the mine. This property has evidently been developed to meet the increasing demand for winter work in the vicinity of Council, and is to be regarded in its present state as little more than a method of prospecting the gravels of the benches along the Niukluk.

On Elkhorn Creek a little desultory work has been in progress during the summer, but at the end of September none of the properties were in operation. The largest amount of work has been done near the mouth, where the section was as follows:

Section near mouth of Elkhorn Creek.

	Feet.
Tundra	2
Clay or muck	4
Sands and gravels	4

The lowest member showed considerable cross-bedding in the sands associated with the gravels, thus indicating the variable character of the water by which they were deposited. Numerous pieces of wood in a more or less decomposed condition were found in the gravels. The surface form and internal structure of the deposit at this point suggest that it is an alluvial fan of Elkhorn Creek rather than part of the flood plain of Niukluk River. Owing to the absence of miners from the creek, no estimate of the production or tenor of the gravels could be obtained.

One-half mile southeast of Post Creek, a tributary of the Niukluk from the north, a quartz vein carrying gold has recently been located. This vein occurs at the contact of schist and limestone and is about 8 feet in width. According to reports, it shows considerable free gold, and the values obtained by crushing and panning are from 25 to 75 cents a pan. If these figures are reliable, the vein should run nearly $35 to the ton. This vein serves as one more indication of the origin of the gold in mineralized zones near the contact of the limestone and schists.

CASADEPAGA RIVER.

The most important development in the Casadepaga basin was the extension of the Council City and Solomon River Railway to the mouth of Penelope Creek. This puts the district into close communication with the ocean transportation routes.

Work along the Casadepaga in 1906 seemed to be carried on more by prospectors than by active settled companies, so that the production would be small. In the lower course of the river, as far as Bonanza Creek, no mining has been in progress. Near Bonanza Creek two camps have been established to work low-bench gravels of the Casadepaga, but as these camps employed only from two to five men not much work has been completed A little work has also been done on Bonanza Creek, but it was not visited.

From Bonanza Creek to Penelope Creek the river gravels have been extensively prospected during the summer by a drill with a crew of six men, with a view to determine whether a dredge could be installed. No statement as to the results of this work can yet be made. One peculiar feature noted in drilling below the mouth of Penelope Creek was that on certain of the river bars gold occurs on the surface and not on bed rock. There is no false bed rock of clay at these places, and the surface concentration is due to the washing away of the gravels of the bars during periods of flood, the particles of gold previously contained in the gravels being left behind. On Big Four Creek, a tributary of the Casadepaga from the south between Bonanza and Penelope creeks, the summer of 1906 saw only assessment work. On Birch Creek, which flows into Big Four Creek about 5 miles above the Casadepaga, two camps have been engaged in working creek gravels below Shea Creek. At Dixon Creek, 2 miles above Big Four Creek on the Casadepaga, there has been some development work. The bed rock at this place is schist and limestone, the creek appearing to follow the contact more or less closely. As this contact is in many other places the seat of mineralization, it would seem desirable to investigate further the gravels of this creek and of the Casadepaga near its junction.

Penelope Creek is now the terminus of the Council City and Solomon River Railroad, and by this line is about 32 miles from Solomon. There have been only two camps on this creek, one near the mouth and one about a mile above. The upper camp has been the most active this summer. Four men have been employed, and the gravel has been handled by horse scrapers. A short ditch has been constructed at a low level, but, as in other parts of the peninsula, considerable difficulty has been experienced from lack of water.

On Goose Creek only two men have been mining this summer, and, according to local reports, not much more than wages has been produced. Three-fourths of a mile above Goose Creek a broad bench of gravels trenched by the Casadepaga shows good values. Mining in this flat, however, has been inactive pending the completion of a ditch from Moonlight Creek. (See p. 266.)

No mining except assessment work was done in 1906 on Canyon Creek. On Banner Creek also work was practically at a standstill. It is reported that all the gravels on the latter creek have been turned over and that the only values left are those that have been lost by the primitive methods in vogue when the creek was first worked. Certain claims, however, are held by annual assessment work, though they seldom yield more than wages.

Willow Creek, which is noted on certain of the Survey maps as Left Fork, is now known as Lower Willow, in order to distinguish it from Upper Willow Creek, also a tributary of the Casadepaga. Upper Willow Creek enters the river from the south about a mile west of Johnson Creek; Lower Willow Creek has its mouth nearly opposite Ruby Creek. At the mouth of Lower Willow Creek two men have been mining all summer. A mile above the mouth two men have been at work, but have been much hampered by lack of water. A mile above Wilson Creek two men have been doing some work, but operations were suspended late in the season to allow the installation of a California grizzly. About $1\frac{1}{2}$ miles above Wilson Creek two men had been employed all summer. They stated, however, that the claim had been previously worked out and that their operations this summer consisted merely in saving some of the values that had been lost in the earlier mining. A short ditch at a low level takes water from the upper part of the creek and carries it to the discovery claims, a distance of about 2 miles. The geology of the region at the mouth of Wilson Creek is complex, the bed rock consisting of limestone and chloritic and graphitic schists. The gold of this part of the stream has probably been derived from a near-by source. Mineralization is evident in at least two places at the schist and limestone contacts on the south side of Lower Willow Creek. At one point sulphides were recognized in a quartz vein, and numerous copper stains on weathered vein stuff were found on the summit of the divide between Lower Willow Creek and the Casadepaga near the head of Ptarmigan Creek.

On Ruby Creek two parties have been at work during the summer, but the creek is now exhausted. It is said that the values have been more completely exhausted from the gravels of this creek than from any other stream in the Casadepaga drainage, so that reworking these gravels in the future will not be remunerative.

On Moonlight Creek the main activity during the last two years has consisted in ditch building. This creek heads in a series of bare limestone hills with steep slopes, so that the run-off is high. The ditch has an intake at an elevation of about 500 feet. It is proposed to carry the ditch across Canyon Creek to the broad bench of Casadepaga River about three-fourths of a mile southwest of Goose Creek. The supply from Moonlight Creek will be augmented from Upper Willow Creek by a ditch line that delivers water to the ditch at Moonlight Creek at an elevation of 500 feet. It is estimated that the ditch will have an average delivery of 1,500 to 2,000 inches.

An eighth of a mile below the junction of Moonlight Creek and Casadepaga River there has been some slight exploration of the bench gravels that occur a few feet above the level of the river. The gravels seem to be typical river gravels, but the floor upon which they rest is rather uneven. Old channels in this district have been reported, but the rumors could not be investigated. There have been no mining operations during the last season on the Casadepaga above Moonlight Creek.

KRUZGAMEPA RIVER BASIN.

GENERAL DESCRIPTION.

Kruzgamepa River [a] has its source in Salmon Lake, which lies in a broad valley at the southern base of the Kigluaik Mountains. It flows to the northeast and, sweeping around the east end of the range, reaches the head of Imuruk Basin by a northwesterly course.

The geology (compare the geologic map, Pl. X, in pocket) in this basin affords considerable variety. The core of the mountains is made up of massive granite, associated with crystalline limestone and schists. Above, a formation correlated with the quartzites of the Kuzitrin consists of schists and graywacke carrying much graphite. This formation is overlain by the Nome group, including a great thickness of flaggy limestones and calcareous schists correlated with the Port Clarence limestone. The rocks resemble those which in other places produce placer gold, and in many localities

---

[a] The drainage basin of the upper Kruzgamepa was first organized as the Golden Gate mining district, and is now included in the Kougarok mining district. This river is sometimes called the Pilgrim. The placers of the Kougarok basin are described on pp. 306–328 of this report.

show evidence of mineralization. Large quartz veins are abundant, and some of them carry gold, though probably none have been proved to contain commercial values. Two specimens taken from large veins near the east end of the Kigluaik Mountains were found on assay to contain traces of gold and silver, but a specimen from a large mass of quartz near the mouth of Slate Creek contained neither. A large ledge said to outcrop near the mouth of Iron Creek is heavily miner- alized with iron and copper pyrites and specimens are reported to assay well in gold and silver; the deposit, however, has not been examined by members of the United States Geological Survey.

Kruzgamepa River flows in a broad gravel-filled valley. Above the mouth of Iron Creek broad gravel terraces occur on both sides of the stream about 50 feet above the water. Much of the gravel, as has been explained elsewhere (p. 94), was contributed by glaciers that flowed from the Kigluaik Mountains and probably does not contain placers of value. What little prospecting has been done has not led to encouraging results, though colors of gold are widely distributed throughout the Kruzgamepa basin, and it is quite possible that with a reduction in cost of transportation many of the low-grade aurifer- ous gravels may be mined at a profit. Iron Creek, the largest eastern tributary of the Kruzgamepa and the most important from the stand- point of mining, is described on pages 329–336 by P. S. Smith.

WILLOW AND SLATE CREEKS.

Willow Creek, which is about 3 miles long, joins the Kruzgamepa from the south about 8 miles above Iron Creek. Near its mouth this stream flows through a small rock canyon about 50 feet deep, above which is a bench on either side covered with gravel. This bench rep- resents an old valley floor into which the stream has cut its present valley. The bed rock consists of limestones with interbedded quartz schists of sedimentary origin and with intruded greenstones. The strikes are almost directly across the course of the stream and the dips upstream. Gold has been found on this creek, and pans from bed rock have yielded as high as 25 cents. The gravel deposits on the bench are of the same character as those in the stream, and there- fore probably carry gold. During the dry months the creek would probably not afford a sluice head of water.

Slate Creek, which is about 2 miles long, joins the Kruzgamepa from the south 2 miles above Willow Creek. The character of the bed rock is the same as that of Willow Creek. Twenty-five cents to the pan has been obtained. The gold from both streams is bright colored and flat.

Both of these creeks have been worked to some extent every season since 1901, and some gold has been produced, though probably not more than a few thousand dollars in all. A high gravel bench on the east side of the Kruzgamepa, between Slate and Willow creeks, was prospected in 1904 by a company proposing to bring water for hydraulicking from Salmon Lake. Colors of gold are reported to have been found, but the results of the prospecting operations are not known.

PORT CLARENCE PRECINCT.

INTRODUCTION.

The Port Clarence precinct embraces the west end of Seward Peninsula. It is bounded on the east (see fig. 2, p. 42) by a sinuous line, running first northward, then around the basin of Agiapuk River along the divide that separates that basin from streams flowing into Goodhope Bay, and then turning northward and reaching the Arctic Ocean at the west end of Shishmaref Inlet. On the north and west the precinct is bounded by the Arctic Ocean, and on the south by the summits of the Kigluaik Mountains. Recording offices have been established both at York, near the west end of the precinct, and at Teller, on Port Clarence, but the former office is now discontinued.

Within this area, which embraces about 2,000 square miles, are extensive gravel-covered lowlands and flat-topped uplands. (See topographic map, Pl. IX, pocket.) The uplands are clearly erosional surfaces that have been uplifted and dissected. They do not all represent the same epoch of erosion, but indicate a succession of cycles of degradation that have been separated by periods of uplift.[a] The intervals of erosion are marked in the present topography by plateaus, benches, and plains standing at different altitudes. The gold-bearing gravels have been deposited since the youngest of these features were formed.

The entire stratigraphic succession of Seward Peninsula, already presented in the account of the geology (p. 60), is found in the Port Clarence precinct. (See geologic map, Pl. XI, pocket.) To the south the Kigluaik uplift has brought to the surface the limestones, crystalline schists, and intruded granites, here called the Kigluaik group. North of these and forming the bed rock of the peninsula, between Imuruk Basin and Port Clarence, are the schists and limestones of the Nome group. These rocks stretch for about 10 miles north of Grantley Harbor to the region where they pass underneath the heavy limestones, here termed Port Clarence. To the west this limestone is succeeded by Devonian (?) slates, and these in turn at the apex of the peninsula by a Carboniferous (?) limestone. Intru-

---

[a] The physiography of this part of the peninsula has been discussed by A. J. Collier in A reconnaissance of the northwestern portion of Seward Peninsula : Prof. Paper U. S. Geol. Survey No. 2, 1902, pp. 34–43.

sives are common throughout the district, with two dominant types—a granite, which occurs in large stocks, and a green basic rock, here termed greenstone, which occurs both as stocks and dikes. The Quaternary is represented by gravel deposits mantling the extensive lowlands that fringe the Arctic coast and form inland basins, and also by gravels of the present stream beds. Some of the terraces and plateau remnants are covered with thin layers of gravels, but high-bench deposits are relatively rare in this part of the peninsula. A few small areas of recent lavas occur north of Port Clarence and form outliers of the more extensive extrusives found to the east. Though few details are known, the bed-rock structures in general trend north-eastward, and the folding is open. Variation from this type occurs along the margins of the intruded stocks.

Placer gold here, as in other parts of the peninsula, appears to be confined to the areas of mica schists and limestones belonging to the Nome group, which are more metamorphosed than any of the younger terranes, but less so than their equivalents to the east and south. The slates carry some gold, but the placers derived from them appear to be of no great commercial importance. The Port Clarence limestone is not a gold-bearing formation and the areas in which it forms the country rock should not attract the placer miner. Along the margins of some of the granite masses cutting this lime-stone mineralization has taken place in the form of veins carrying cassiterite, galena, etc. An account of these occurrences has been given by Collier.[a]

It was probably in 1898 that the first prospectors visited this region, but they confined their attention to the Agiapuk basin, where they found no workable placers. In the following year [b] auriferous gravels were discovered in Buhner Creek, a tributary of Anikovik River, in what was then organized as the York recording district. A little gold has been taken out in the York region, but it has now been practically abandoned as a gold placer district, though it is growing in importance as a tin producer. Practically all the gold-producing creeks are confined to the Bluestone and Agiapuk basins, except a few tributary to Grantley Harbor. The district has not been a large producer, the entire output up to 1903 being estimated at $200,000. With the introduction of better mining methods, no doubt many mining enterprises now abandoned will be revived and can be made to yield an adequate return.

Teller, located on a spit between Grantley Harbor and Port Clarence, is the recording office, the seat of a commissioner, and a general

[a] Collier, A. J., Tin deposits of the York region, Alaska: Bull. U. S. Geol. Survey No. 229, 1904.

[b] Brooks, A. H., Richardson, G. B., and Collier, A. J., Reconnaissances in the Cape Nome and Norton Bay regions, Alaska, in 1900, a special publication of the U. S. Geol. Survey, 1901, p. 137.

supply point. It was founded in 1900, when the Bluestone placers, 15 miles to the south, were discovered. A town named Bering, on the shore of Port Clarence, 5 miles south of Teller, was the first settlement, but the newer town of Teller drew its people from it. The latter town, which, during the boom of 1900, had over 1,000 inhabitants, now has not over 100.

Port Clarence, the best harbor along the entire coast line of the peninsula, affords shelter to deep-sea vessels, but storms coming from different directions require changes of anchorage. Large vessels can anchor within about a mile of Teller. Grantley Harbor, almost landlocked, can be entered by vessels drawing not more than 12 feet of water.

CREEKS TRIBUTARY TO GRANTLEY HARBOR FROM THE NORTH.

Several small streams rising in the plateau on which Mukacharni Mountain stands and flowing southward into Grantley Harbor are all within a few miles of Teller and easily accessible. Prospects were found here in 1900, and mining has been done each year since, but no very rich placers have yet been developed and the amount of gold produced has been small. In 1903 very little mining was done on any of these streams.

In general, the bed rock can be described as a succession of chloritic mica schists and interbedded limestones containing some intrusive greenstone and small veins and stringers of quartz.

All these creeks have the same general features. They head in the upland, which they traverse through rather narrow valleys running nearly north and south. Below, they cross the coastal plain, made up of silts and gravels, in which they have trenched shallow channels. From west to east these streams are Bay, Sunset, Igloo, Dewey, McKinley, and Offield creeks. On all of them except Bay Creek, which has not been very carefully prospected, the stream gravels are known to be auriferous, and Sunset, Igloo, and Offield creeks have yielded a small gold output.

Some mining has been done on Sunset Creek since 1901, and the workings have been briefly described in a previous report.[a] The creek is about 5 miles in length. For several miles from its mouth the bed is about 700 feet wide, but the pay is in general confined to narrower channels. At one place three such channels are reported. The workings about 3 miles from the coast show from 2 to 7 feet of rather fine gravel overlying the bed rock. The gold is found both in the gravel and to a depth of about 1 foot in the crevices of the bed rock.

---

[a] Collier, A. J., Reconnaissance of the northwestern portion of Seward Peninsula, Alaska: Prof. Paper U. S. Geol. Survey No. 2, 1902, pp. 45, 46.

The gold, most of which is more or less iron stained, is coarse and rough in the upper claims and finer and smoother toward the mouth of the creek. The largest nuggets are worth about $4. A few well-formed crystals of spinel have been found in the concentrates.

No work was done in 1903 on Igloo, Dewey, and McKinley creeks. A small force of men worked during part of that summer on Offield Creek, 9 miles from Teller, but no estimate has been obtained as to the amount of gold produced. It is known that in two of these creeks there are deposits of gold-bearing gravels of commercial value, though no exact estimate of their gold content is available. It is probably true, as reported, that with economical methods of handling the gravels all these creeks could be made to pay. A large ditch 27 miles long to bring water from Agiapuk and California rivers was under construction in 1904. These rivers carry an abundance of water, which can probably be brought to Sunset Creek at an elevation of 250 to 300 feet above the sea. It is proposed to work the gravels of Sunset Creek by hydraulicking and elevating the tailings.

AGIAPUK RIVER BASIN.

INTRODUCTION.

Agiapuk River, which drains an area of 700 to 800 square miles, empties into the north side of Imuruk Basin 24 miles from Teller. Its headwaters are about 35 miles from tidewater. About 20 miles north of its mouth the river forks; the western branch, which is the smaller of the two, retains the name Agiapuk, and the northern branch is called American River.

Being easy of access and supporting a large Eskimo population, the Agiapuk was one of the first streams of Seward Peninsula to receive attention from prospectors, but up to the present time no rich or extensive deposits of auriferous gravels have been discovered in its basin. Colors of gold have been found in many of its tributaries and nearly all of them have been staked and prospected. Small amounts of gold have been produced on Allene Creek, a southern tributary of the western branch of the river and on Lawson Creek, a tributary of American River.

In general the rocks of the Agiapuk basin are less metamorphosed than the gold-bearing rocks in other parts of the peninsula. They consist mainly of Silurian limestones that are regarded as equivalent to the Port Clarence member of the Nome group and generally have not been so productive of gold as the lower members of that group. United States Geological Survey parties traversed part of the Agiapuk basin in 1901, and again visited the region in 1903, but very few prospectors were seen on either trip, though prospect holes, claim stakes, and other evidences of white men were everywhere abundant.

A somewhat detailed description of many of the tributaries of the Agiapuk has been published in a previous report[a] and need not be repeated. The western branch of the river rises in the limestone hills about 25 miles north of Grantley Harbor, and flows southward to a point within 10 miles of that harbor, thence eastward to its junction with American River. Through the east-west part of its course it drains a broad, gravel-filled lowland, but about the headwaters the valleys are narrower and the gravel deposits are of small extent.

Several tributaries which enter the Agiapuk from the south within 10 miles of Grantley Harbor have yielded colors of gold, but only one of them has been worked.

ALLENE CREEK.

Allene Creek heads within 4 miles of Grantley Harbor and flows northward into North Creek, tributary to the Agiapuk. Its comparatively broad valley is carved out of the plateau on which Mukacharni Mountain stands.

When visited in September, 1901, only one claim was in operation, although some mining had been done on several others. The pay dirt was confined to the creek bed and rested upon a clay foundation, below which the true bed rock had not been reached.

A prospect hole about halfway up the creek is said to have been sunk 65 feet without reaching bed rock. This shaft showed gravel at the top, below which there was blue clay containing bark and driftwood. The gold in Allene Creek is bright, flaky, and not much waterworn. Although it is generally fine, it includes a few small nuggets. It is reported that some sluicing was done in 1902 and 1903, but the production of the entire creek was probably not more than $2,000.[b]

AMERICAN RIVER.

American River rises about 40 miles north of Grantley Harbor and flows in a westerly direction for 15 miles, then turns and flows nearly southward for 30 miles to its junction with Agiapuk River. Through the last 30 miles of its course the stream flows in a comparatively broad gravel-filled valley, in which it meanders tortuously. It is navigable for small boats and canoes for about 30 miles.

The only tributary reported to have produced gold in 1903 was Lawson Creek, which joins American River about 35 miles from tide water. In 1902, $1,000 to $1,500 in gold was obtained from its placers. The gold is of high grade, being reported to be worth $19 an ounce. A large quartz ledge from which samples assayed $2 a

---

[a] Collier, A. J., Reconnaissance of the northwestern portion of Seward Peninsula, Alaska: Prof. Paper U. S. Geol. Survey No. 2, 1902, pp. 56–59.
[b] Collier, A. J., op. cit., pp. 45–46.

ton in gold is reported to have been discovered on Collin Creek, a tributary of Portage Creek, which enters American River from the north. Neither of these localities have been visited by members of the Geological Survey and the information in regard to them consequently is very scanty.

BLUESTONE RIVER BASIN.

INTRODUCTION.

Bluestone River flows into Tuksuk Channel, which connects Grantley Harbor with Imuruk basin, about 13 miles from Teller. It drains the greater part of a region lying between Grantley Harbor and Bering Sea, usually referred to as the Bluestone region. This area is characterized by flat-topped hills on which two erosion levels can be recognized. The upper plateau encircles the west end of the Kigluaik Mountains, which rise rather abruptly from its surface and attain elevations of 3,000 to 4,000 feet. The streams have cut their valleys nearly to sea level. The very irregular drainage is due to causes which have not been studied in detail. For about 10 miles above its mouth Bluestone River traverses a rolling plateau that has an elevation of 400 to 600 feet. Through this part of its course the river valley is broad, and flood plains are developed to a width, in some places, of half a mile or more. Above this portion the valley contracts and for about 5 miles is a steep-walled canyon with sides rising from the river to flat-topped mountains 1,200 feet in height. Above the canyon the valley again broadens somewhat, but does not expand into a large flood plain. Here the river forks, the eastern branch being known as Gold Run and the western as Right Fork. Both branches are characterized by very crooked courses. For 6 miles above its mouth the valley of Gold Run has a nearly northwest course, then turns and has an east-west course for 6 miles to its head. Right Fork heads about 10 miles farther west and flows nearly due east to its junction with Gold Run; it has several large tributaries which enter it nearly at right angles. Below the forks Bluestone River carries approximately 3,000 miner's inches of water, derived about equally from the two streams. The average gradient of the river is about 30 feet to the mile.

Gold was discovered on Bluestone River early in the summer of 1900, and mining operations have been continuous since then. Gold has been produced on the main river below the forks, on Gold Run and its tributary, Alder Creek, and on Bering Creek, a small tributary of Right Fork. Both stream and bench placers have been worked. The region was visited by United States Geological Survey parties in 1900 [a] and in 1903.

The Bluestone basin contains mica and chlorite schists, with beds of limestone, some having graphitic phases and all belonging to the undifferentiated part of the Nome group. These schists contain numerous intrusive masses of greenstone which appear more commonly in the form of sills than as dikes. Generally these greenstones are only slightly altered, but in a few places they are schistose and are not always distinguishable in the field from the schists of sedimentary origin. Small quartz veins, both cutting across the foliation and running parallel with it, are common in the schists. Some of these veins are reported to carry values, but as yet none of them have been systematically developed. The results obtained from three assays of vein quartz collected by the Geological Survey seem to indicate that these veins are not so commonly auriferous as those of the Nome region. Two specimens, one from a quartz bowlder and another from a small stringer vein, carried no trace of either gold or silver. A specimen taken from the dump of a prospect hole sunk in the bed rock of Alder Creek yielded an assay return of 0.06 ounce of gold and a trace of silver. The material was soft and talcose and contained considerable nearly white pyrite.

BAR AND BENCH PLACERS.

Mining has been attempted at only one locality on Bluestone River, although there are evidences of prospecting in many places. Extensive preparations for mining were made in 1903 at a point about 4 miles below the forks of the river, to which the name Gilroy was given. In the canyon gravel deposits are of small extent, but below the canyon the valley widens and alluvium appears not only in the creek bed, but on low benches which at no very remote date were probably flood plains of the river.

Four miles below the forks the river bed is more than 150 feet wide and the pay gravel ranges in thickness from 2 to 20 feet or more. On the left bank at this place there is a flat about 20 feet above the river bed and nearly half a mile wide. The section along the river bank shows 10 feet of black schist bed rock underlying 6 to 7 feet of gravel. A series of samples taken across the river bed are said to have yielded from 2 cents to $1 to the pan. This gravel can possibly be worked by an extensive system of dams, which should turn the water from the river bed, but probably never will be on account of the volume of water and the occasional floods. It might also be handled by some form of dredge.

---

[a] Brooks, A. H., Richardson, G. B., Collier, A. J., and Mendenhall, W. C., Reconnaissances in the Cape Nome and Norton Bay regions, Alaska, in 1900, a special publication of the U. S. Geol. Survey, 1901, pp. 127–132.

In 1903 mining operations were confined to the gravel deposit on a bench that has an area of about 60 acres. The surface is nearly level and deeply covered with moss and muck. A section of the gravel on this bench is as follows:

Section of bench deposits on Bluestone River.

	Ft.	in.
Black moss and muck		6
Blue muck		3
Yellow clay	1	6
Bluish gravel	4	

Bed rock: Graphitic schist very much weathered at the surface.

The gravel is composed of pebbles of black schist and white and reddish vein quartz; next to the bed rock it carries a considerable amount of gold, which also extends into the fissures and crevices to a depth of about 1 foot. There is a smaller amount of gold in the upper part of the section, and the whole thickness of moss, muck, and gravel is said to carry sufficient to pay for shoveling into the sluice boxes. Although most of the gold is rather fine, it is heavy and easily saved. The largest nugget found was worth $12. It is proposed to bring water from a small western tributary of Bluestone River, called Ruby Creek, and to sluice the gravels near the river bank, dumping the tailings into the river. A ditch was built for this purpose about 1 mile in length, but near the middle of the season, when mining operations had well begun, the water from the ditch penetrated to a layer of ground ice beneath, which melted, undermining the ground for a distance of several hundred feet. This melting continued progressively and the ditch became wider as the season advanced, until finally it found an outlet to Bluestone River a quarter of a mile above the point where the miners had expected to use it. As the season was nearly spent and neither lumber for sluice boxes nor a supply of canvas hose for carrying water across the break could be obtained, work was necessarily suspended.

From the showing made here it seems probable that successful mining can be done over a part at least, if not all, of this bench, either by ground sluicing or hydraulicking. Ruby Creek is a small stream, but it is possible that sufficient water for a small hydraulic plant could be collected from several such tributaries on the west side of Bluestone River. The whole amount of gold produced during the summer of 1903 was probably not more than $300, and previous to 1903 the yield has been even less.

GOLD RUN.

Gold Run, the large eastern branch of the Bluestone, has a length of about 20 miles, but mining activities are confined to the lower 5 miles of its course. The principal settlement of the region is located at the mouth of Alder Creek, and is known as Sullivan. It is reached by a wagon road from Teller, the principal shipping point of Port Clarence, about 18 miles distant. The road crosses the upland plateau surface, and is very soft and marshy in many places. Another road crosses the low divide from Gold Run to Tisuk Creek, then follows the gravel bars in the bed of that stream to the coast at the old Shea road house, where Tisuk post-office is now located. During the summer months stages run daily from Sullivan to both these points.

The gradient of Gold Run probably does not exceed 20 feet to the mile. Along the valley slopes there is a series of benches, which toward the headwaters of the creek approach the level of the present bed, while near its mouth they are about 100 feet above it. Though the valley is nearly straight, the immediate creek bed, which lies in a narrow trench cut in an old valley floor, has a meandering course. The bed rock of Gold Run consists of chlorite and mica schists, in many places of a light-green color, with calcareous phases that grade into limestone and graphitic phases that result in graphite schists and black slates. Sills of greenstone intruded in the schists are common. The gravels of the creek bed contain many bowlders of greenstone and granite, which have been derived from the Kigluaik Mountains in some previous drainage epoch.

The placers of Gold Run include both creek and bench deposits. The gravels of the creek bed had an average width of about 60 feet and were mined for about 3 miles. They were richest in gold near the mouth of Alder Creek, where some small spots yielded more than $50 to the cubic yard; but the average was much less than that. At the mouth of Gold Run gold was found in both the run and Bluestone River.

In 1900 and 1901 the creek bed was mined for several hundred feet above its mouth and about $2,500 was produced, but the operations were difficult on account of the trouble in controlling the waters of the stream.

About 1 mile above the mouth the gravel in the creek bed is about 100 feet wide and 2½ feet thick. It carries values through its whole width, but most of the pay is found within 6 inches of bed rock and in the bed rock to a depth of 6 inches to 2 feet. What is called bed rock at this place consists of a sticky yellow clay, which forms a layer 2 to 3 feet thick above the schist. The method of working was as follows: The creek was turned to its right bank by a wing dam, and by this means the greater part of the gravels of the creek bed were exposed and freed from water. Sluice boxes were set with a grade of 18 inches to the rod, and the gravel was shoveled in. It is reported that a cut 18 feet wide and 75 feet long was worked out two years ago, producing $1,200. If we assume a depth of gravel of 3 feet, this would give about $8 to the cubic yard.

The stream bed at the mouth of Alder Creek has all been worked out. This ground was the scene of many legal controversies, and before it was exhausted it had changed hands a number of times in obedience to the orders of the courts, passing from one claimant to another. It is impossible to tell the amount of gold produced, though it was probably not far from $50,000, which was obtained in the years 1900 to 1902.

Above the mouth of Alder Creek the amount of gold produced on Gold Run has probably not been proportionate to the development work that has been done, although at one time claims were held for fabulous prices, and men who are now working for wages along Gold Run could in 1900 have sold their holdings at prices that would exceed the total production of the whole region. Probably not more than $10,000 has been obtained from all the workings on Gold Run above Alder Creek. About a quarter of a mile above Alder Creek a well-defined pay streak, 40 feet wide, was found extending nearly 1,500 feet. At the upper end, however, this pay streak apparently passed under the left bank of the creek and was lost. The gold was found on a layer of clay above the true bed rock.

In the course of the prospecting along Gold Run above Alder Creek a number of rich spots in the gravel were found which produced considerable gold, but were quickly mined out. One of these, which was abandoned in 1903, was at the mouth of Quartz Gulch, about 4 miles above Sullivan. The workings covered an area about 30 by 100 feet and were 5 feet deep. No reliable reports of the amount of gold produced were obtained. Although the bed of Gold Run has been prospected all the way to its head, no extensive deposits of gravel carrying sufficient values to justify mining under existing economic conditions have been discovered. The gravels generally carry some gold, but in general not enough to pay for working with rockers; moreover, the necessity of extensive wing dams to drain the creek bed and the liability of the stream to sudden freshets that destroy all dams make the expenses of mining prohibitive.

At the present time the attention of prospectors is directed to the benches, many of which are conveniently located for economic mining. On the left bank of the Bluestone, opposite the mouth of Gold Run, gold has been found in two benches, one about 10 feet and the other 50 feet above the present river bed. Ground sluicing in progress on the lower bench showed a section from the surface down as follows: Muck and red-brown clay, 2 feet; gravel, about 5 feet, resting upon massive greenstone bed rock. The bed rock on the upper bench is also massive greenstone and the excavations show gravel from a few inches to 7 or 8 feet thick. It is reported that on both these benches some of the gold is found in fissures in the bed rock.

On the right bank of Gold Run, near its mouth, there are also two benches in which placer gold has been found. One of these is only a few feet above the present creek bed, and was probably formed as a flood plain in a period not very remote. The other bench is about 100 feet above the creek bed and belongs to a well-defined system of benches that can be traced for several miles along the valley walls. Several prospect holes have been sunk on both these benches, showing 10 to 15 feet of well-washed gravel. The extent of the deposits had not been determined in 1903. From pannings seen by the writer it seems probable that these gravels may be made to yield $6 to $7 to the cubic yard. Attempts have been made to work them by use of a small stream of water collected from melting snow on the hillside above and also by carrying gravels down to the creek in wheelbarrows, but the claims are not at present in operation and can not be worked at a profit by either of these methods.

A bench 20 feet above the level of the creek bed, above the mouth of Alder Creek, probably also belongs to this system of benches. The bed rock, however, slopes gradually from the upper level of the gravels about 50 feet above the creek down to an escarpment only a few feet above the bed, producing a hillside placer. In 1903 this deposit was being worked by means of a ditch about 2 miles long, which brought water from the upper part of Gold Run to the hillside above, giving it an elevation above the Bluestone of about 50 feet. By the use of this water the surface materials were first ground sluiced off, after which the gravels were shoveled into the sluice boxes. At the upper end of the cut made by ground sluicing the section exposed was about as follows, from the surface down: Moss and muck, 1 foot; clear ice, from 6 to 10 feet; gravel of undetermined depth. At the lower edge of the bench, a few feet above the creek bed, several cuts had been mined out and sluiced. In these workings the gravels have a thickness of 2 to 6 feet and rest upon a bed rock consisting of graphitic schists with small sills of intruded greenstone. The gold-bearing gravels contain a large amount of sticky yellow clay sediment and the gold ranges in size from fine grains to comparatively large nuggets. Several nuggets worth more than $100 have been obtained. The gold is generally stained and coated with iron, and has a dark color. In one of the nuggets the gold was associated with green rock, probably decomposed greenstone. In the concentrates from this claim a great deal of limonite derived from the alteration of pyrite was found, but as the claim is located on the original camp site of the district a great many metallic things, such as nails and cartridges, have been found with the concentrates, and one loaded revolver is said to have been taken from the sluice boxes. A sample of these concentrates containing no free gold and consisting mainly of limonite pebbles was assayed, giving 0.12 ounce of gold and 0.16 ounce of silver to the ton. The product of this bench claim during the season of 1903 is unknown, but a clean-up aggregating $5,000 was reported after a three days' run with ten men shoveling into the sluice boxes.

About 2 miles above Alder Creek a bench deposit on the east side of Gold Run, 200 feet above the creek bed, has been prospected by a shaft 120 feet deep. Bed rock was reached at a depth of 115 feet, and the material on the dump consists of well-rounded gravel, most of it rather fine, but including many bowlders of greenstone and vein quartz, the largest being 2 feet in diameter. Coarse colors of gold were found on bed rock, but were not abundant enough to pay. This deposit lies in a low divide between the valleys of Gold Run and McAdam Creek. It is probably part of an abandoned drainage channel.

Benches at various levels are common features of the Gold Run valley, and it is evident from the developments already made that they are worthy of the prospector's attention.

ALDER CREEK.

Alder Creek, a tributary to Gold Run from the east side 2 miles from the Bluestone, has a length of about 2 miles. It heads near the general level of the plateau, at a point about 2 miles east of the forks of Bluestone River, and flows in a southwesterly direction. The gradient of the creek is comparatively steep. Its valley has gently sloping sides which do not give any superficial evidence of benches. The whole surface is deeply covered with muck and moss, probably in large part underlain by ground ice. The creek in ordinary seasons probably does not carry over 50 miner's inches of water at its mouth. In fact, near its mouth the water is just sufficient for a fair sluice head. Where not disturbed by mining operations the creek bed has a width of only 7 to 8 feet, and a section of peaty soil and muck is exposed in the banks.

Some mining has been done along this creek for about 1 mile above its mouth. Gold was discovered in 1900, and the creek for a quarter of a mile was nearly worked out at that time. Since then the workings have extended more slowly upstream, but it is reported that 1 mile above the mouth of the creek there is not sufficient gold in the gravels to pay for mining. In a claim at the mouth the gravels have all been washed, and the bed rock has been stripped for several hundred feet. The bed rock consists of chloritic mica schists containing some stringers and small veins of quartz, as well as some small intrusions of greenstone. Although the claim was practically exhausted in 1900–1901, some pay was found in 1902 extending under the left bank near the middle of the claim to a distance about 30 feet from the original creek bed. For this distance the bed-rock floor was found to be nearly on a level with the bed of the creek, but beyond this the bed rock gradually rose and the pay gravel thinned out on the edge. At the edge of the cut the section exposed consists of several feet of frozen moss and muck. In the bed rock of this claim there is a lode about 16 feet wide, which has been prospected to a depth of 60 feet. The ledge is made up of a gray talcose material containing some quartz in fine stringers and blebs. All of the ledge matter carries pyrite in streaks, and it is reported that fine colors of gold have been panned out of the ore taken from the bottom of the shaft. The hanging wall of this deposit is in a somewhat schistose limestone containing albite crystals; the foot wall is a graphitic schist. A sample of the ore assayed by E. E. Burlingame & Co., of Denver, Colo., yielded 0.06 ounce of gold to the ton and a trace of silver.

About half a mile from the mouth of the creek the pay streak is 25 feet wide and located under the left bank. The section shows 4 feet of muck above 4 feet of pay gravel resting upon a broken limestone bed rock. The gold on the lower half of the claim is coated with iron oxide; that on the upper half is bright. It is reported that a great many nuggets worth from 25 cents to $13.60 have been found, and that there is practically no fine gold. Although the concentrates in this placer contain cubes of iron pyrite that are not much waterworn, the gold is nearly all well rounded. Several pans of pay dirt washed by A. J. Collier yielded about 25 cents each.

During the season of 1903 a deposit of very rich pay dirt was found on one of the claims not far from the mouth of the creek, but probably $10,000 would be a fair estimate of the total produced on Alder Creek during the season.

RIGHT FORK.

Right Fork has been prospected from its mouth up, but though colors of gold have been found both in the creek bed and in a stream of benches like those on Gold Run, no placers that will yield profits by primitive mining methods have been discovered, except on one tributary, which was not examined.

BERING CREEK.

Bering Creek, a tributary of Igloo Creek, which enters the Bluestone from the south about 5 miles above Gold Run, is a short stream on which seven claims have been recorded. It is about 12 miles by wagon road from Teller. Gold in paying quantities was discovered here in 1901. It is reported that in 1902 the creek produced a few

thousand dollars and in 1903 probably somewhat more, but no definite estimate of the total has been obtained.   The gold is coarse, practically all in the form of nuggets, and it is reported that from $40
to $60 a day was often picked out of the gravel by the miners without
sluicing.

ANIKOVIK RIVER.

Anikovik River, which enters Bering Sea at the town of York,
about 12 miles east of Cape Prince of Wales, is 15 miles long.   It
heads in the York Mountains, and through the greater part of its
length flows across the York Plateau, in which it has cut a comparatively broad valley.   For several miles above its mouth the valley
and river bed contain gravels several feet deep and 200 to 300 feet
wide.   In 1900 the whole of this river was regarded as gold-placer
ground, but at the present time nearly all the workings have been
abandoned.   The conditions as they existed in 1900 were described
in Brooks's report, already cited,[a] and further notes were published
in the report of 1901.[b]

In 1903 only one party of prospectors attempted to mine the gravels of this river and the whole product for a month's work did not
exceed $600.   At a point about half a mile from the coast a ditch had
been dug along the side of the valley, and the river was turned
through this ditch by means of a dam, leaving the river bed exposed.
The gravel was about $2\frac{1}{2}$ feet thick and rested upon slate bed rock
dipping at an angle of about 70°, which contained placer gold to a
depth of 6 to 18 inches.   In size the gold ranged from very fine to
nuggets worth $10.   The fine gold was generally bright, but the nug-
gets were iron stained.   Cassiterite and magnetite were found with
the gold in the concentrates.   These gravels evidently carry a small
amount of gold, and if they could be worked economically by a company holding several miles of the river bed it is possible that adequate returns could be obtained.   Up to the present time, however,
the efforts to mine gold on Anikovik River have not been successful.
Small amounts of gold are also obtained from the tin placers of
Buck Creek, about 15 miles north of the Anikovik, which have been
described in a previous report.[c]

GOODHOPE PRECINCT.

The northern part of the peninsula, an area draining into the
Arctic Ocean between the west side of Shishmaref Inlet and the
head of Goodhope Bay is called the Goodhope precinct.   More

---

[a] Brooks, A. H., Richardson, G. B., and Collier, A. J., Reconnaissances in the Cape Nome
and Norton Bay Regions, Alaska, in 1900, a special publication of the U. S. Geol. Survey,
1901, p. 134.

[b] Collier, A. J., Reconnaissance of the northwestern portion of Seward Peninsula,
Alaska: Prof. Paper U. S. Geol. Survey No. 2, 1902, p. 48.

[c] Tin deposits of the York region, Alaska: Bull. U. S. Geol. Survey No. 229, 1904, pp.
29-35.

than half of this region is covered with the typical tundra, which skirts the Arctic Ocean. This coastal-plain area is built up of gravels and silts, but they are not known to be auriferous. Most of the streams are sluggish and flow in meandering courses over the plain. South of the coastal plain occur gravels, schists, and limestones of the Nome group, with one large area of granite, probably a stock. (See geologic map, Pl. XI, in pocket.) Devil Mountain, an isolated peak in the northern part of the precinct, is a basaltic lava mass rising to an elevation of nearly 900 feet above the coastal plain, whose gravels mantle its base on all sides. Placer gold has been found in a small area in the southern part of the precinct north of the divide at the head of Kougarok River. This region may be approached by a long overland trip up the valley of the Kougarok or by small boats up the rivers which flow into Shishmaref Inlet. It is one of the most difficult places of access in the peninsula and the cost of bringing in supplies is prohibitive to mining operations except on fabulously rich deposits. There has been considerable prospecting and a little mining here, and a small amount of gold has been taken out of Dick Creek, a tributary of Bryan Creek, which flows into Serpentine River. This locality has not been visited by Geological Survey parties, but from descriptions furnished by miners it seems probable that there is a considerable amount of rather low-grade auriferous gravel along Dick Creek and some of the neighboring streams. In general these deposits appear to be similar in character to those near the head of Kougarok River, described on page 326. A preliminary report on the placers of the Goodhope precinct has already been published.[a]

---

[a] Collier, A. J., Reconnaissance in the northwestern portion of Seward Peninsula, Alaska: Prof. Paper U. S. Geol. Survey No. 2, 1902, pp. 53–56.

THE BLUFF REGION.

By Alfred H. Brooks.

INTRODUCTION.

There is a deep bight in the southern coast line of Seward Peninsula, between Cape Nome on the west and Rocky Point on the east, and near the northernmost point of this reentrant, at the mouth of Daniels Creek, lies the small settlement called Bluff. The irregular group of buildings that makes up the settlement is already in part threatened with engulfment in the pit of the hydraulic mine close at hand (Pl. VII). Bluff is 50 miles east of Nome, with which it has a nominal weekly mail service and from which it can be conveniently reached by coasting vessels or by wagons along the beach. There is no harbor at Bluff, but it is a little more sheltered from westerly and easterly storms than many of the other settlements along the coast. The town, whose population is less than half a hundred people (1904), owes its existence to the rich gold placers of Daniels Creek, which are being systematically exploited by the Topkok Ditch Company. The total production of this region up to 1904 exceeds $1,000,000, practically all of which has been taken from Daniels Creek and the beach adjacent to its delta. Gold has, however, been found also on Koyana Creek to the east and on Eldorado, Ryan, and Little Anvil (sometimes known as Silver Bow) creeks to the west of Daniels Creek.

It is proposed to describe briefly the salient features of the gold placers of these creeks. This sketch is based largely on the writer's own observations, made during a few days' visit, but these have been supplemented by the results of G. B. Richardson's [a] examination, made in 1900.

---

[a] Brooks, A. H., Richardson, G. B., and Collier, A. J., Reconnaissances in the Cape Nome and Norton Bay regions, Alaska, in 1900, a special publication of the U. S. Geol. Survey, 1901, pp. 102–106.

TOPOGRAPHY.

The Bluff region, which will here be defined as the area including streams tributary to Bering Sea from Topkok Head to Koyana Creek, is of the same topographic type as the rest of the peninsula. Flat-topped highlands with gentle slopes are broken by broad gaps and shallow valleys, many of the latter widening out into broad amphitheaters.

A moss-covered plain stretches westward from the head of Golofnin Sound, through which Klokerblok River and many smaller streams meander in sluggish courses (Pl. VIII, in pocket). This lowland is bounded on the west by highlands that constitute a northwesterly extension of a range of hills trending along the seaward margin of the peninsula west of Golofnin Bay. A similar highland area stretches northwestward from Topkok Head, and the two are connected by several flat-topped, serpentine ridges, which separate the coastal from the Klokerblok drainage. The summits of all these uplands are flat and of strikingly uniform altitude, ranging only between 800 and 1,000 feet above the sea. This feature is very suggestive of a surface uplifted and dissected by erosion. The valley of Koyana Creek, the most easterly stream of the area here considered, is a wide, shallow basin, presenting a broad delta front to the sea. Steep, rocky bluffs, broken only by the small valley of Swede Gulch, face the sea for 3 miles to the west. At Daniels Creek the highlands recede and, with a roughly crescentic sweep, encircle the valleys of Daniels, Eldorado, and Ryan creeks, touching the sea again about 3 miles to the west. This crescentic lowland area has a gently rolling surface, caused by the incision of the three above-mentioned stream valleys and their tributaries. So gentle are the stream slopes and so shallow are the valleys that, when seen from a salient, the direction of the drainage is difficult to determine. The headwaters of these streams are separated by broad, low gaps from the drainage of the Klokerblok, and one of these gaps is traversed by the Topkok ditch, which brings water to Daniels Creek.

West of Ryan Creek the highlands again retreat to encircle the small valley drained by Little Anvil Creek, and approach the sea once more at the rocky bluff that marks the eastern side of Topkok Valley. Topkok River, the largest stream of the district, has a well-defined, steep-sided valley, which is incised to a depth of 400 to 500 feet below the highlands. It has a gravel floor and presents a broad delta to the sea. No commercial placers have been found within the basin of Topkok River, and it lies rather outside of the field of the present discussion.

GEOLOGY.

Investigations in this field have been confined to the gold placers, almost to the exclusion of the bed-rock geology, and even the auriferous gravels have not yet been adequately mapped. It is possible, therefore, to express the stratigraphic and structural geology in only the most general terms. Richardson, who through his acquaintance with adjacent areas had a broader knowledge of the Bluff region than the writer, divided the bed-rock terranes into three groups:[a] (1) A massive gray crystalline limestone; (2) a mica schist, with some interbedded graphitic limestone; and (3) a formation of massive limestone and mica schist. The writer's observations hardly bear out the correctness of this succession of beds, for it appears to him that there is only one massive white limestone, which is succeeded by a mica schist and graphitic limestone formation. Some facts also are presented below which would indicate that the mica schists are, in part at least, altered intrusives and hence do not mark any definite stratigraphic position. In the preceding pages

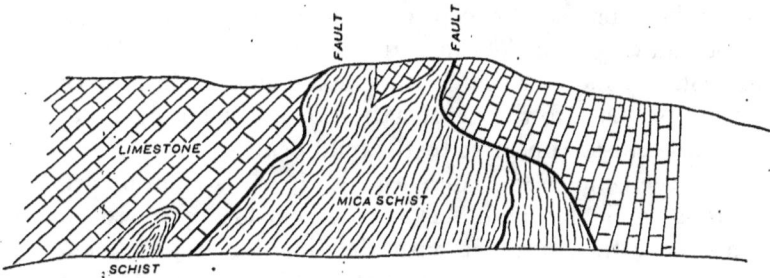

FIG. 15.—Diagram of cliff exposure near mouth of Daniels Creek.

Mr. Collier, in presenting the general geology of Seward Peninsula, includes this area as a part of the Nome group, and it is probable that the massive limestone belongs to the Port Clarence horizon. Mr. Collier's lower schist series is not exposed in the Topkok region, but his upper schists are probably represented by a belt of rocks which stretches northwestward from Rocky Point, and appears to overlie the limestone outcropping along the coast east of the mouth of Daniels Creek.

Mica schists occur as irregular masses within the limestone belts, and although they do not differ lithologically in any very essential way from the schists believed to be of sedimentary origin, their mode of occurrence strongly suggests that they are altered intrusives. The most striking example of this is seen in the cliff exposures just east of the mouth of Daniels Creek (fig. 15). Here an irregular mass of mica schist is inclosed in limestone walls. Lines of faulting have obscured the original relation of the two rocks; but the outline of the schist mass is very suggestive of an intrusion. The schist, what-

---

[a] Op. cit., p. 103.

ever its original character, has been intensely deformed and at this particular locality has been intruded by many quartz veins. It is now composed essentially of quartz and mica, with some chlorite, and its mineral composition might be that of either an altered sediment or an altered igneous rock. Further evidence of the intrusive character of some of these schists is found in the fact that at various localities the limestone walls near the contact with the schists are more or less metamorphosed. These facts, together with the irregular distribution of the schist, indicate an igneous origin, though it must be confessed that the evidence is by no means conclusive.

The larger structures of the Bluff region appear to be simple, but there are many minor complications. The heavy limestone has been uplifted into a low dome, whose longer axis stretches approximately N. 70° E. About 3 miles northeast of Bluff this structure carries the limestone underneath the younger mica schists, which probably mantle the limestones to the west also. The major folds are all low, the dips being from 10° to 30°, but locally there has been intense deformation. In this region, owing to the great scarcity of outcrops, few of the local features can be determined except along the sea-cliff exposures, and there minor faulting and folding can be seen in many places. These exposures plainly indicate that if the mica schists are intrusives they were injected previous to the deformation of the inclosing limestones. These mica schists appear to be the gold-bearing rocks of the district, and hence their mode of occurrence and their distribution are of economic interest. So far as observed, the schists appear to be mineralized only near their contact with the limestones. At these places quartz veins cutting the foliation of the schist are not uncommon. The individual veins appear to be of small extent, but at some localities a stockwork forms a considerable mass of low-grade ore. No assays have been made, and it is not known that these bodies carry commercial values, but the mineralization is of interest as showing the source of the placer gold. The ores appear to be chiefly iron pyrite, with some chalcopyrite and arsenopyrite. Three miles east of Daniels Creek attempts have been made to open up one of these ore bodies. The deposits in both structural and mineralogical character are closely similar to those of Anvil Creek. Though we have no direct evidence that there are lode deposits in the Bluff region which could be profitably mined, the occurrence of the phenomenally rich placers along Daniels Creek, the gold of which, as will be shown, is unquestionably of local origin, makes it not improbable that the stripping of the gravels may eventually reveal a workable ore body.

The distribution of the alluvium is of more immediate interest to the miner than the geology of the bed rock. The upper limit of the unconsolidated sands and gravels is less than 200 feet above sea level, but it is difficult to determine the exact line of demarcation, because along the valley slopes the waterworn material and the talus have intermingled. As the sands and gravels were, for the most part, laid down by streams, the upper limit is not a horizontal line but varies with the gradient of the streams by which they were deposited. For the same reason the alluvium, in many places, has a very perceptible dip, as on Daniels Creek, where the beds fall off to the west at an angle of 5°. The thickness of the alluvial deposit ranges from a minimum of less than 1 foot along the margin of the deposit to a maximum of 40 to 100 (±) feet on Daniels Creek. The character of the material also varies greatly, but clays and sands appear to dominate. Except in the old beach deposits near the coast and on some of the smaller creeks, much of the gravel is rather angular and appears to have been little waterworn. Near the mouth of Daniels Creek a well-defined beach deposit lies about 10 feet above the present beach (fig. 16). Much of this upper deposit has been removed by the mining operations, but enough remains to show the typical beach character of the gravels. The deposit dips toward the sea, contains some shingle, and is overlain by about 8 feet of clay and muck. With the alluvium just described is grouped a series of sands and sandy clays, which are the product of rock weathering and are almost in place. On Daniels Creek alone, so far as known to the writer, does the alluvium appear to reach a considerable thickness. Not

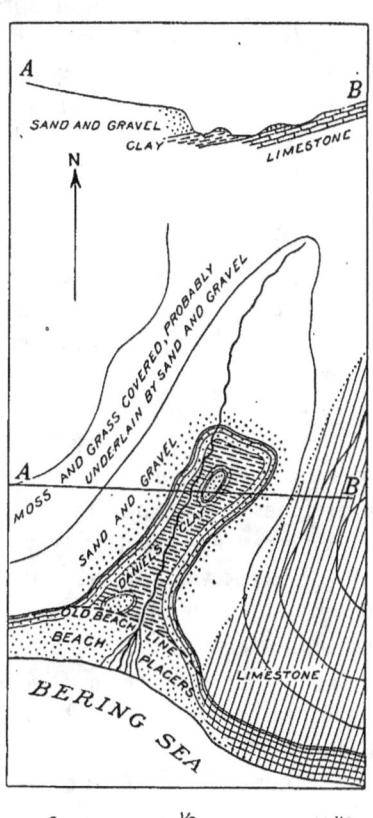

FIG. 16.—Sketch map and section of Daniels Creek placers.

more than 6 feet of gravels were found resting upon bed rock on the upper part of Eldorado Creek, and even on Daniels Creek the alluvium is only about 7 feet deep a mile from the beach. It should be stated, however, that at the time of the writer's visit (1904) the depth to bed rock had been determined at no other localities than those just described. It is reported that in the summer of 1907 excavations were made near the mouth of Daniels Creek to a depth of 60 feet. It is not clear whether these were in alluvium or in the decomposition products of the limestone, but the material is said to carry values.

THE PLACERS.

Gold is said to have been found at the mouth of Daniels Creek in September, 1899, by William Hunter and Frank Walker, but it was not until December that the first claims were staked.[a] These were located by J. S. Sullivan, George Ryan, and others. In the following January five claims were staked along the beach by H. C. Malmquist, William Hunter, and others, who organized themselves as the Black Chief Mining Company and obtained control of the Daniels Creek claims. In March the news of the discovery was published at Nome and many men found their way to Bluff. By this time the beach also had been found to be auriferous and, in accordance with the ruling of a miners' meeting, it was thrown open to all comers for a distance of 60 feet back from high tide. The exploitation of the beach placers by men who had gained their experience at Nome was energetically pushed till July 8, 1900, when military intervention put an end to beach mining. By that time, however, about $600,000 in gold had been taken out of the sands from a strip of the beach less than 1,000 feet long—probably the richest deposit of this kind ever found in the world.

Meanwhile the two lowest claims on Daniels Creek had been opened up and in 1900 yielded probably $200,000 in gold. Most of this, however, was taken not from the stream bed proper but from an old beach that stretches across the present mouth of Daniels Creek.

The beach placers were practically exhausted in 1900, yet the auriferous gravels of the lower part of Daniels Creek continued to be exploited during the following two years as far as the scanty amount of water available for sluicing would allow. Most of the gold produced here in 1901-2 came from Discovery claim, at the mouth of Daniels Creek, which was worked by the aid of a gasoline engine that pumped a sluice head of water from the sea. In the meantime gold had been found on Eldorado and Ryan creeks and on Swede Gulch, but lack of water prevented operations except on a small scale.

In 1902 a strong company called the Topkok Ditch Company began the construction of a waterway from the head of Klokerblok River to Daniels Creek. This work was completed late in the summer of 1903, and sluicing began. In 1904 the company had about 16 miles of ditch in operation, and in 1906 extended the conduit about 4 miles. (See Pl. VIII, in pocket.) This enterprise is a most successful hydraulic mining operation and demonstrates what can be done under favorable conditions and with intelligent and economical management. When the heavy gravel deposits of Daniels Creek

---

[a] Brooks, A. H., Richardson, G. B., and Collier, A. J., Reconnaissances in the Cape Nome and Norton Bay regions, Alaska, in 1900, a special publication of the U. S. Geol. Survey, 1901, p. 104.

have been sluiced off, the water of the Topkok Ditch Company can be utilized to mine the shallower deposits of Eldorado, Ryan, and other smaller creeks.

The accompanying sketch map (fig. 16) shows the known distribution of the auriferous gravels in the vicinity of Bluff. The valleys of Swede Gulch and Koyana Creek should be regarded as part of this region, but as they contain relatively shallow deposits and have not proved very rich, they are probably of no great commercial importance.

The district embraces only two types of auriferous deposits—the beach placers and the creek placers. The former, however, include an old beach placer as well as that of the present beach. These beach placers have been described as follows in the previous report: [a]

The rich part of the beach extended about 500 feet west of the mouth of Daniels Creek and about 50 feet east to a cliff that projects into the sea. The beach pay streak was 3 to 4 feet thick, overlain by a foot of barren sand and underlain by clay of undetermined thickness. Gold occurred disseminated in schist gravel, in which were scattered bowlders of mica schist averaging 10 to 15 pounds in weight and smaller pieces of limestone. The "ruby sand" so prominent about Nome is inconspicuous here, though some garnets are present in the gravels. The heavy minerals associated with the gold are magnetite, nodules of limonite, small pieces of ilmenite, and bits of cinnabar. Cinnabar is fairly abundant in the tailings, ranging from specks to rounded pebbles the size of marbles, but it has not been found in place. The beach gold is rather coarse, much coarser than that from the Nome beach. It averages about 12 pieces to the cent and assays show a purity of 0.870.

During the height of the excitement of beach mining it was common for a man to take out $100 to $300 a day, and three men are said to have taken out $10,000 in three days. It is estimated that the gold tenor of much of the pay streak must have averaged $150 to the cubic yard, or about $1 to the pan. This is far richer than the best part of the Nome beach sands, and, in fact, is the richest marine placer ever found. In gold content it has been equaled by only a very few claims in the peninsula.

Before considering the origin of the beach gold it is well to describe the other auriferous gravels of the vicinity. An uplifted beach placer which lies athwart the course of Daniels Creek has been mentioned. This deposit is less than 10 feet above the present beach, and the character of the material in both deposits is the same. It is believed, however, that the gold content of this older beach is not so great as that of the present beach. The sands of the elevated beach are about 6 feet thick, include some gravel layers, and are

---

[a] Brooks, A. H., Richardson, G. B., and Collier, A. J., Reconnaissances in the Cape Nome and Norton Bay regions, Alaska, in 1900, a special publication of the U. S. Geol. Survey, 1901, pp. 104–105.

overlain by about 8 feet of sandy clay with fragments of limestone. The relation of the two deposits is well exhibited in the section (fig. 16). The overburden of sandy clay carries but little gold. The old beach has been recognized only at the mouth of Daniels Creek, and here much of it has been removed by mining operations.

Inland from this elevated beach there is a gradual transition to the gravels of Daniels Creek, which appear to lie in a broad, rock-floored basin with some corrugations parallel to the water courses. The sides of this basin slope toward the axis, and the whole is tilted toward the sea with a grade of probably 300 feet to the mile. The bed rock, so far as exposed, is a rather massive limestone, everywhere more or less crystalline. The surface is much pitted, and the original rock floor must have been one of great irregularity. Mining operations since the writer's visit have shown that there is more than one channel. Deep crevices penetrate the limestone and in some places make it appear like an aggregate of bowlders. So far as observed, however, all these so-called bowlders are practically in place.

The alluvial material is of two general types—that in which clay predominates and that which is chiefly sand. It ranges, however, from a micaceous but very stiff clay to a loose sand with gravel layers. In many places no structure can be made out in the clays; the bedding of the sands and gravels is of the greatest irregularity, locally changing its character every few feet. The indications are very strong that the layers of clay, which in general lie near the bottom of the deposit, are formed almost in place, whereas the sands and gravels appear to have been laid down in swiftly flowing water. Many fragments of angular limestone are found embedded in the clay, and in places the whole may be traced by gradual transitions into the limestone underneath. A pit about a mile from the beach showed the following section:

Section 1 mile from beach near Bluff.

	Feet.
Muck and vegetation	1½
Sand and clay, some stratification	1
Clay, with fragments of limestone, probably lying upon bed rock	5

A section about a quarter of a mile from the beach is approximately as follows:

Section one-fourth mile from beach near Bluff.

	Feet.
Muck and vegetation	1– 2
Fine sands, some clay	10–12
Sand and gravels, with carbonaceous layer	10
Micaceous reddish clay, derived from limestone	10–12
Limestone fragments, probably resting upon bed rock.	

This section shows that the alluvial deposit thickens very rapidly toward the sea. Near the head of the creek the surficial deposits consist chiefly of clay, but near its mouth sands and gravels predominate. No adequate data are at hand for determining the amount of the alluvium in this basin, because, except at the mouth of the creek, there are but few places where measurements can be made. Moreover, the bed-rock surface is so irregular that any generalization of the average thickness must leave a large range to chance.

The clays are mostly reddish, in some places blue, and generally contain considerable mica. Their tenacity is dependent on the absence of sandy material. By the addition of arenaceous matter they pass into sandy clays and thence into sands. The typical sands are dark gray and carry much mica. Many of the gravels are strikingly angular. Lithologically, they are made up of mica schist and vein quartz. They are as a rule iron stained and in many places carry sulphide minerals. Although there are great local irregularities in the bedding of the gravels, yet in general the planes of stratification dip away from the high eastern ridge. The maximum dip observed was 5 degrees.

Little can be said of the distribution of the pay gravels. The managers of the hydraulic mine report that in general the clay carries higher values than the sand and gravel. This is very significant, for it appears to be established that the sand and gravel have been far more sorted than the clay. It would indicate a large gold tenor for the rock from which the clay has been derived. In any event there can be no question that the gold is very near its bed-rock source, which appears to be at the contact of mica schist and limestone. Since the writer's visit high values are said to have been found at depths of 50 and 60 feet, near the mouth of the creek. This would suggest that there was either a deeper channel below the present stream or that the material in which this gold was found was a product of weathering, in part derived from the hill slope to the east. The facts available are inadequate to decide this question.

Generally throughout the peninsula the richest placers, except where there has been secondary concentration, occur at the contact of limestone and schist. These contacts appear to be zones of weakness where, in consequence of stresses, fissuring has developed and thus given opportunity for the ore-bearing solutions to penetrate. The Daniels Creek deposits are no exception to this statement, for the location of the placers with reference to the limestone and schist, as well as the character of the gravels, points to the conclusion that the zone of mineralization lies at the contact of the rocks of the two types.

An impregnated zone is well exposed along the sea cliff about three-fourths of a mile east of the mouth of Daniels Creek, and, as this is

very likely the type of lode deposit from which the placer gold is
derived, it merits a detailed description, though its commercial value
has not been established. At this locality a belt of mica schist about
60 feet wide is more or less impregnated by pyrite-bearing quartz
stringers. The belt, including some irregular limestone masses, is
bounded by graphitic limestone walls, which dip away from the
schists and form a small anticline much broken by faults. The schist
is silvery gray and appears to consist essentially of sericite and
quartz. Its relations to the limestone suggest that it may be an
altered intrusive. The deformation of the series has upturned and
fractured the limestone beds and has rendered the intruded rock
rather schistose. At the west contact a band of schist 20 feet in
cross section lies between one of the included limestone masses and
the country rock. In this band the mineralization is more intense
than in the rest of the schist. Here a series of gash veins, the largest
of which is 18 inches in width, cuts the foliation of the schist (fig. 17).

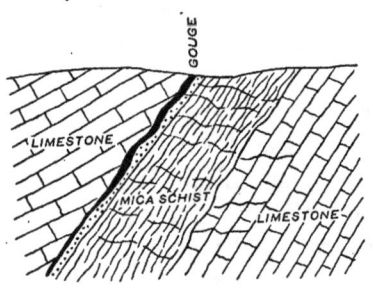

FIG. 17.—Diagrammatic representation
of character of impregnated zones,
Bluff region.

A mass of crushed material or gouge
forms the hanging wall of the deposit, and along this zone, which has
been a plane of movement, the quartz
veins are cut off abruptly. Stringers
of quartz do, however, occur in the
limestone on both sides of the schist.
The ore appears to be chiefly iron
pyrite and mispickel, with some
chalcopyrite; the gangue is mostly
quartz, with some calcite. It is reported that values as high as $3 to
the ton in gold have been found in the ore, but the writer did not learn
how the assay sample was taken.

As the source of the placer gold is at the contact of the schist and
limestone, which is also the position of the above-described impregnated zone, all the gold may have its origin in one type of lode deposit. Though the existence of workable lode deposits must remain
in question until systematic prospecting has been done, yet the finding
of rich placers like those of Daniels Creek, whose gold appears to
have been moved but little from its bed-rock source, makes it seem
possible that a workable lode deposit may sometime be discovered.

The only locality in the district where an attempt at lode development has been made is 3 miles east of Bluff, near the shore. Here a
ledge is said to be 14 feet wide and to yield up to $30 in gold to the
ton. The ore is iron pyrite and mispickel. This occurrence is believed
to be similar to the one above described, but it was not examined by
the writer.

The Daniels Creek placers are the only ones in the Bluff region sufficiently opened up to permit study. Here several facts are evident: First, the source of the gold is entirely local; second, where it is richest, as in the red-clay deposits, there appears to have been little sorting action by water; third, the gold is so intimately associated with mica-schist débris that most probably the schist has a close connection with its origin. It has been pointed out that the gold in the district is probably derived from the impregnated zones in mica schist and limestones near their contact. If this mica schist is an altered intrusive, its distribution may be very irregular. It is probable that if the surficial deposits could be stripped off, the impregnated zones would be found not far from the present placers, but there is no evidence yet that these zones will be found to carry commercial values. This statement is supported both by the fact that the auriferous clays have not been moved far from their original position, where they were formed in place by the weathering of the bed rock, and by the fact that as a rule the gold-bearing gravels are angular. It is evident that the stream gradients must have been low during the period of the formation of the clay, or the water would have quickly removed it. The area was probably exposed for a long time to the agencies of weathering, and an irregularly pitted land surface was produced. The bed rock was probably deeply buried by the clay and other products of weathering. An uplift followed, as a result of which the carrying power of the streams was increased and the deposits of sands and gravels were laid down. At Daniels Creek this uplift gave the former level a slight westward tilt, as is shown by the bedding of the gravels. The elevated beach deposit appears to have been formed prior to this uplift, but it would require a very detailed survey to establish this fact. The gold, however, was certainly concentrated by wave action in the older beach, as in the present beach. The presence of gold at a depth of 60 feet near the mouth of Daniels Creek, reported in 1907, may indicate either a deep zone of weathering or a buried ancient beach line. Subsequent uplift, which probably did not exceed 8 feet at the coast, exposed this older beach in part to wave action, and thus led to a reconcentration of the gold in the gravels of the present beach.

The other creeks of the district besides Daniels Creek have been but little developed, for none of them carry a sluice head of water except in the early spring or during heavy rains. Eventually, however, they will all be hydraulicked with water from the Topkok ditch. Sluicing has been done on about half a dozen claims on Eldorado Creek, and some work has been done on Ryan and Little Anvil creeks. So far as the scanty exposures show, the mode of occurrence of the gold on these streams is similar to that of Daniels Creek, but the deposits are probably not so rich and the auriferous gravels not so extensive.

THE KOUGAROK REGION.[a]

By Alfred H. Brooks.

INTRODUCTION.

The name "Kougarok district" [b] is generally given to a gold-placer-bearing region in the central part of Seward Peninsula, drained mostly by Kougarok River. This paper describes both the drainage basin of this river and the other gold-bearing streams tributary to Kuzitrin River from the north. Kruzgamepa River, whose drainage basin lies in the Kougarok precinct (see fig. 18), is considered elsewhere in this volume (p. 266). Investigations were begun in this field in 1900 by the writer, assisted by A. J. Collier, soon after the first actual discovery of workable placers. This work was extended by Mr. Collier in 1901. In 1903 the district was again examined by Messrs. Collier and Hess, who prepared a paper for this volume, but the delay in publication enabled the writer to visit this field again before the volume was submitted for publication.[c] As many new facts have been developed by the mining operations since Mr. Collier's studies were made, it has seemed best to entirely rewrite this section of the report. The notes of Messrs. Collier and Hess have been freely drawn on, but the writer alone is responsible for the conclusions here advanced. Descriptions of the creeks not visited by the writer have been quoted directly from their manuscript. All the surveys thus far made have been preliminary, and the data obtained leave much to be desired as to the details both of the geology and of the distribution of the placer gold.

---

[a] An abstract of this paper is contained in Bull. U. S. Geol. Survey No. 314, 1907, pp. 164–181.

[b] The "Kougarok precinct" includes the entire drainage basin of Kuzitrin and Kruzgamepa rivers.

[c] The writer's task was rendered pleasanter and the work was much accelerated by the hospitality and courtesy shown him by the operators and prospectors in the region during his ten days' visit. Among the many that deserve special mention are the Irving Mining Company, the Northwestern Development Company, J. M. Davidson, Al Garvey, Joe Turner, Nels Leding, E. Anderson, R. Anderson, and Andrew J. Stone.

294

TOPOGRAPHY.

The area here considered is made up of three topographic provinces which, named from south to north, are (1) a mountain mass, (2) a

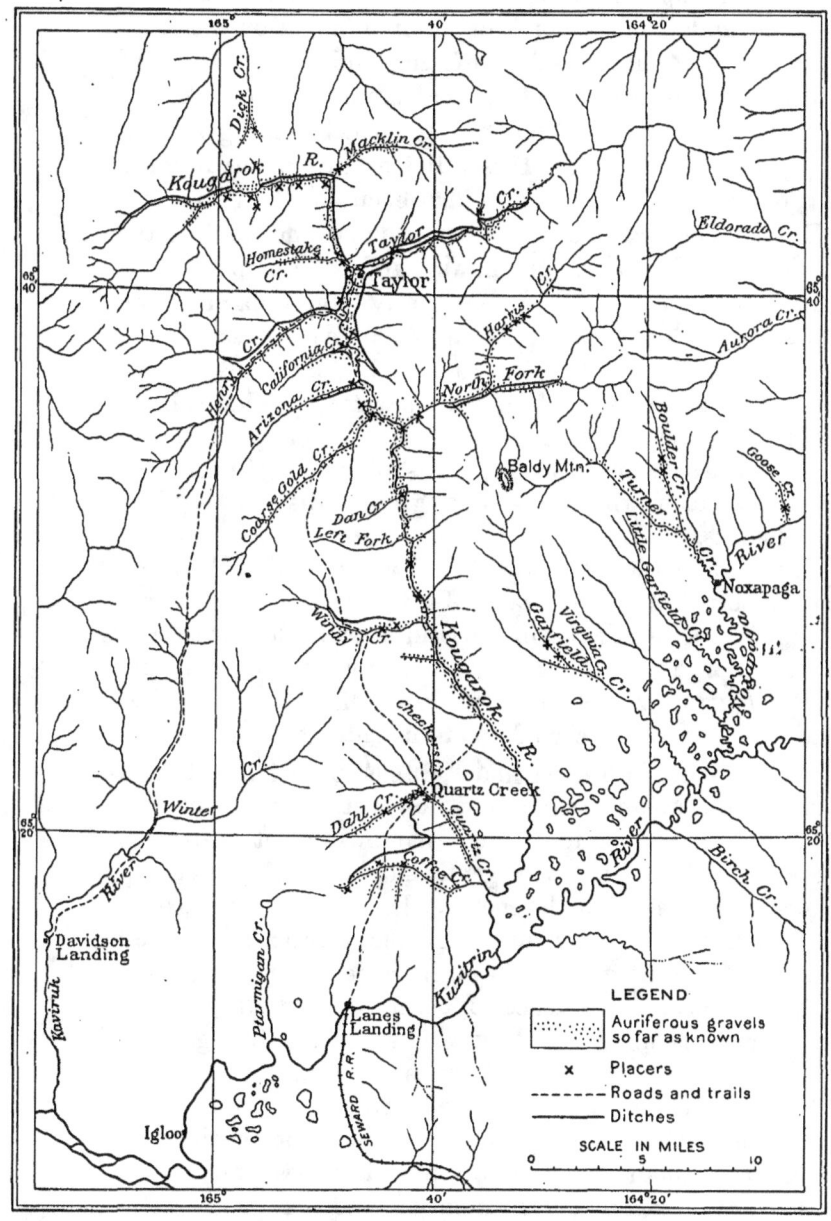

FIG. 18.—Sketch map of Kougarok region.

lowland, and (3) a dissected upland. It is bounded on the southeast by the Bendeleben Mountains, which stretch eastward from lower Kruzgamepa Valley and form with the Kigluaik Mountains the backbone of the peninsula. These mountains are sharply serrated, but attain no great height, the highest peak, Mount Bendeleben, being only 2,700 feet in altitude. They are deeply incised by streams, which occupy steep-walled valleys.

The northwestern slope of the Bendeleben Mountains descends to a lowland basin bounded on the north by an upland and drained into Imuruk Basin by Kuzitrin River. This lowland has a width of nearly 10 miles at its upper end, but gradually narrows to the southwest as the upland and the mountains approach, becoming a well-defined valley about 3 miles wide. This valley extends about 10 miles and broadens out again to a lowland which encircles the upper end of Imuruk Basin. The margins of the lowlands are indented by the low spurs which extend from the mountains and the upland.

To the north the Kuzitrin lowland rises gently to an upland, whose flat summits stand at altitudes of 800 to 1,600 feet. Here broad, interstream areas, with flat hilltops, diversified by some higher domes reaching altitudes of 2,500 feet, are separated by wide valleys. This upland level, as elsewhere in the peninsula, marks a former stage of erosion, when the entire region was planated. Subsequent uplift formed a plateau which has been greatly dissected by the present water courses.

Kuzitrin River drains the district southwestward to Imuruk Basin, a tidal inlet connected with the sea at Port Clarence. This river meanders sluggishly across the basin lowland already described, and receives numerous large tributaries from the north, the longest being Noxapaga and Kougarok rivers and Garfield Creek. Several smaller streams heading in the Bendeleben Mountains enter from the south, but these are outside of the region under discussion. The northern tributaries rise in the upland region and flow with tortuous courses through valleys of irregular form, here narrowing between steep rock walls, there broadening out with gentle slopes, which are broken at many places by gravel-covered benches. Heavy talus and extensive masses of ground ice cover many of the gentler slopes and obscure the bed-rock form of the valleys, in some places completely obliterating all topographic evidence of benches. Tributary to the larger water courses are many smaller streams flowing through narrow, steep-walled valleys or V-shaped gulches. Where the valleys are broad the floors are, in the main, gravel covered; where they are narrow, bed rock is exposed at many points. The characteristic topographic forms are smooth slopes and rounded domes and ridges, which, as a rule, are broken by outcrops only at their crests.

As elsewhere in the peninsula, the dominant vegetation is moss. Timber is entirely absent, but thick growths of alder and willow are present along the water courses. Grass, though not abundant, occurs in favorable localities along the valley floors. The hill slopes are, for the most part, moss covered, with here and there some grass. As bed rock is exposed only on the highest summits and in the sharply cut valleys, it is exceedingly difficult to decipher the geology.

GEOLOGY.

BED ROCK.

The two main stratigraphic units described on pages 66–70 by Mr. Collier—the Kigluaik group and the Nome group, with its subordinate member, the Port Clarence limestone—are represented in the Kougarok region. The limestone schists and granites of the Kigluaik group make up the Bendeleben Mountains and stretch along the southern margin of the region here described. These rocks, so far as known, are not gold bearing in this district, and need no further description here. This older group is separated by the broad belt of alluvium that floors Kuzitrin Valley from the schists and limestones of the Nome group, which form the country rock of the uplands and are the source of the placer gold. Here the Nome group is clearly divisible into two members—(1) the Port Clarence (?) limestone and (2) a succession of graphitic phyllites and quartz schists,[a] mica and greenstone schists, and some beds of semicrystalline limestone. The Port Clarence limestone of this region is much more crystalline than that in the type exposures to the west. (See p. 73.) The schist series is closely folded and faulted, and its stratigraphic relation to the massive limestones has not been definitely established.

Whatever the stratigraphic relations are, the fact of the occurrence of two series, one essentially schistose and the other a massive limestone, is well established. The limestone forms the country rock in two large areas. One of these with an oval outline lies between Kougarok and Noxapaga rivers. The other, more irregular in form, lies to the west, partly in the Kougarok and partly in the American basin. Between the two is a belt of closely folded and faulted schists.

Besides these sediments there are several types of igneous rocks in this province or immediately adjacent. Greenstone schists, probably altered intrusives, occur with the schistose rocks. Dioritic rocks, some massive and some more or less schistose, are common among the schists as dikes and small stocks. A large stock of granite outcrops a few miles northeast of the Kougarok-Arctic divide. A noteworthy hot spring occurs near the margin of this granite mass. In upper Kuzitrin Valley a large area is occupied by a basaltic lava stream of recent age. (See Pl. XI, in pocket; also p. 101.)

---

[a] In former reports the graphitic schists, which form a well-defined east-west belt about 2 miles wide and parallel to lower Kuzitrin River, were mapped as a stratigraphic unit under the name "Kuzitrin series." These rocks are here included in the Nome group. See p. 70.

As in other placer districts of the peninsula, the schistose rocks appear to be the source of the placer gold. Quartz seams and small veins are common in the schists, and many are iron stained. Prospectors say that some of these veins carry gold, and it is reported that a copper-bearing lode has been found near the divide between Kougarok and Serpentine rivers, but, so far as known to the writer, no lodes of proved commercial value have yet been discovered. There appear to have been two generations of quartz intrusions. The earlier preceded the extensive deformation of the schists, for its veins are crushed and sheared. The later intrusion, which cuts the first system of veins and is comparatively little deformed, appears more mineralized than the first. The presence of a massive granite intrusion near the later quartz suggests a genetic relation between this quartz and the granite, but of this there is no proof.

The evidence in this district lends additional support to the opinion set forth elsewhere (p. 122), that the locus of mineralization is at or near the contacts between the schists and the limestones. The geologic map (Pl. XI, in pocket) shows a close correspondence between the contacts of the limestone and schist and the known distribution of the placer gold.

ALLUVIUM.

The genesis and correlation of some of the alluvial deposits of this province present problems whose solution must await more detailed surveys than have yet been made. The gravels of the present streams and associated benches are of simple genesis, but the origin of the deeper deposits of the Kuzitrin lowland and of Quartz Creek is less easily explained. The alluvial deposits fall into three groups, which in order of age are as follows: (1) The gravels, sands, and clays that floor the basin lowlands and underlie the modern gravels of a part of the Quartz Creek basin; (2) the bench gravels, such as occur along Kougarok Valley; and (3) the present stream gravels.

The gravels, sands, and clays of the basin lowlands—the oldest alluvium—are known only where exposed by river erosion. They consist chiefly of gravels and sands that are somewhat finer than those of the present streams. Fragmentary evidence from drill records indicates that much of the basin is filled by clays, which were probably deposited in a large body of water, such as a lake or an estuary. However, the surface deposits of gravel and sand in the Kuzitrin basin probably do not as a rule exceed 20 or 30 feet in depth. But outside of the basin, in the more constricted part of the valley, the gravel deposits are probably much deeper.

Bench gravels, the second type of unconsolidated deposits, are common in this region. The best known are along Kougarok River and range from 10 to 75 feet above the present stream level. Washed gravels in benches at still higher altitudes are reported, but have not been examined. These bench gravels are of the same character as the alluvium of the present streams and appear to contain no material foreign to the basins in which they occur. Quartz pebbles usually predominate among the gravels, and much of the material on and close to bed rock is rather angular. A well-defined gravel terrace lies west of the Kougarok, extending from the mouth of Quartz Creek to the point where the river emerges from the upland. East of this point the terrace is traceable along the southern margin of the upland as far as Turner Creek by remnants of a gravel sheet, much of which has been removed by erosion. At the mouth of Quartz Creek the gravels are about 75 feet thick and can be traced up that stream nearly to Dahl Creek. This gravel bench on Quartz Creek includes at least two unconformable layers, and the uplift which raised it to its present position was differential. The genesis and physiographic significance of this deposit are considered on page 305.

The stream gravels constituting the third and latest type of alluvium are better known than those of the other two types. They vary in character according to their occurrence. Nearly all the developed placers are in these gravels. So far as known, they carry only material derived from the basins in which they are found. Much of the surface material is well rounded and very coarse, but many of the pebbles next to bed rock are subangular. Quartz is usually the predominant constituent of the pebbles. In some streams, as Kuzitrin River, the gravel bars are made up of iron-stained quartz, almost to the exclusion of other material. Sands and some clay are interbedded with the stream gravels, forming, however, but a small percentage of the bulk of the alluvium. In all the smaller streams and in parts of the larger ones a bed of clay or sandy clay, in which more or less vegetable matter is intermingled, forms the topmost layer. This surface bed, which is called "tundra" by the miners, ranges in thickness from 2 to 30 feet and appears to be a subaerial accumulation, due in part to the decay of vegetable matter and in part to the deposition of silt during the rainy seasons. Though sometimes explained as a lake deposit by the miners, its distribution and character seem to preclude lacustrine origin.

Another type of alluvial deposit which deserves mention is the ground ice, which occurs more extensively than in the Nome region. Along the northward-facing slopes of the valleys this ice forms in many places almost continuous layers for several miles. It ranges in consistency from a frozen mud to almost pure ice. Fragments of beaver-gnawed wood have been found in it at a number of places. The ice beds as a rule slope with the valley wall and in places extend up the hillside to a height of 100 feet above the stream. This ice can probably be explained best as the accumulation and subsequent solidification of winter snow, which has become buried by the talus and alluvium. The thick coat of moss, once established, effectually prevents its thawing. This ground ice is a constant source of trouble and expense to the ditch owners.

RECENT GEOLOGIC HISTORY.

INTRODUCTION.

The genesis of the present land forms, as well as of the alluvial deposits, is so closely related to the occurrence of the placers that reference to this phase of the geology in discussing the placers is unavoidable, though many of the conclusions advanced must be regarded as only tentative. In the foregoing description of the topography emphasis has been laid on its simplicity, yet when it is submitted to final analysis in terms of physiographic development, the succession of earth movements by which the present land forms were evolved is found to be complex.

Collier[a] has recognized four periods of erosion in this and the adjacent provinces. The topographic features constituting the evidence for these periods he has termed (1) the Nuluk Plateau (2,000–2,700 feet), (2) the "Kugruk"[b] Plateau (1,000–1,200 feet), (3) the York Plateau (100–700 feet), and (4) the lowland plains (0–200 feet). Collier's type localities for these topographic features are mostly west of the area here under discussion. Moffit,[c] who has studied the region to the northeast, recognized only two of these cycles of erosion, represented by remnants of an upper plateau standing at an altitude of 1,000 to 1,400 feet, which he correlates with the "Kugruk" Plateau, and a lower, which represents the same cycle as Collier's most recent—that of the lowland plains.

The writer was unable to find definite proof that four well-defined cycles of erosion have occurred in the Kougarok district. The indications are that most of the level and flat-topped ridges and domes are the product of one extended cycle of erosion and that their present differences of altitude are due to differential uplift. The suggestion made in a previous report[d] by the writer, that certain rock benches found at altitudes of 1,000 to 1,700 feet were of marine origin, has not been confirmed by later work, but the genesis of these land forms is still unexplained. However, the presence of well-defined stream

---

[a] Collier, A. J., Reconnaissance of the northwestern portion of Seward Peninsula, Alaska: Prof. Paper U. S. Geol. Survey No. 2, 1902, pp. 34–43.

[b] At the time Collier wrote Kugruk was the accepted spelling for the river now known as the Kougarok.

[c] Moffit, F. H., The Fairhaven gold placers, Seward Peninsula, Alaska: Bull. U. S. Geol. Survey No. 247, 1905, pp. 42–47.

[d] Reconnaissances in the Cape Nome and Norton Bay regions, Alaska, in 1900, a special publication of the U. S. Geol. Survey, 1901, pp. 57–58.

benches in the existing valleys proves that the uplift after planation was intermittent, and it is possible that some of these benches were formed during one of the later cycles described by Collier. The problem is intricate, for if there were several cycles of erosion it is evident that the last extensive one must have largely obliterated the topographic evidence of any that preceded it.

Collier has correlated a plain of denudation, which he recognized in the flat-topped spurs of Midnight and Kougarok mountains, with a plain marked by the summit level of the York Mountains and associated domes, which he called the Nuluk Plateau, but the evidence hardly seems conclusive. It appears equally probable that there may be uplifted portions of the peneplain which is strongly marked in this district. In the opinion of the writer the only well-defined peneplain is recorded in the general summit levels of the upland that now stands at altitudes of 800 to 2,000 feet, and this he would correlate, as Moffit did, with Collier's " Kugruk " Plateau. At the west end of the peninsula a well-marked elevated surface of erosion, called the York Plateau,[a] was traced eastward by Collier along the seaward front of the York Mountains. With this he correlated certain low benches and flat-topped spurs in the Kougarok district. In the opinion of the writer the facts bearing on this point admit of different interpretation. The York Plateau throughout much of its extent is an uplifted plain of marine denudation, and no evidence appears that any considerable inland areas were base leveled at the time that it was formed. Therefore it would seem that the inland equivalent of this marine benching will be found among the benches along the present valleys rather than as a peneplain. If, then, the York Plateau can be correlated with any of the topographic features of the Kougarok district, the proper correlation is with the stream benches of the present drainage channels.

Collier's fourth epoch of erosion left its record in the coastal plains and inland basins, and these he grouped under the name, " lowland plains." There is but scant evidence as to the form of the rock floor of the lowland areas, but probably it is in general flat, with no great thickness of alluvial covering. The bed-rock floor of the coastal plains, in the lack of evidence to the contrary, might be ascribed to marine denudation; the extensive inland basins, however, such as that of the Kuzitrin lowland, must be the product of a long period of erosion. Though, as will be shown, the sea probably invaded this basin, yet this was but a minor event in its history and throws no light on the origin of the depression itself.

The relation of the northwestern margin of the Kuzitrin lowland to the upland is obscured by the presence of a heavy bed of gravel,

---

[a] Brooks, A. H., Reconnaissances in the Cape Nome and Norton Bay regions, Alaska, in 1900, a special publication of the U. S. Geol. Survey, 1901, p. 52.

already referred to (p. 298). It appears, however, quite possible
that the rock floor of the basin merges with the gentle slopes of the
upland, which has been described as a planated surface. Such a
relation was suggested by the character of the topography along the
trail from Dahl Creek to Kougarok Valley. Here the upland falls
off to the north with a gentle slope from an altitude of about 800
feet to about 200 feet and then passes underneath the gravel terrace
that bounds the Kougarok Valley. An extension of the same slope
would carry it underneath the gravel-floored Kuzitrin basin. The
writer explains this feature as an erosion surface that has been
warped. If the basin lowland has such a structural origin, it is probable that its margin may be faulted in places. This would account
for the abrupt local transitions from the valley floor to the upland.
Warping may have influenced the formation of these basins also by
causing local barriers that brought about headwater erosion. The
evidence presented in Collier's reports as to the warping of the old
land surfaces is conclusive, and only the extent and time of this warping can be called in question.

Both Moffit and Collier have called attention to the gradual transition between the coastal plains and the uplands of the northern part
of Seward Peninsula. The topographic map (Pl. IX, in pocket)
bears out this interpretation of the topography, for everywhere along
the northern margin of the peninsula this gradual transition is
evident.[a]

One feature of the rock floor of the Kuzitrin basin deserves special
mention. Near the entrance of Dahl Creek into Quartz Creek, a
tributary of Kougarok River, a shaft has been sunk to a depth of 187
feet and, although within 50 feet of sea level, has not reached bed
rock. The exposures of bed rock in adjacent areas indicate that at
this locality a depression exists in the bed-rock floor, but the data
at hand are insufficient to define its limits. It is also impossible to
account for this depression, but the suggestion may be offered that it
is of structural origin. If it was a channel of erosion its outlet is
not apparent, because so far as known it is cut off from the sea on
all sides by higher bed rock. It may, however, be the remnant of a
channel which formerly discharged into Norton Bay through Koyuk
Valley and which was blocked by the recent lava flow at the upper
end of the Kuzitrin lowland.[b]

---

[a] The writer has elsewhere discussed similar warped peneplains and their relations to
coastal plains. See Geography and geology of Alaska: Prof. Paper U. S. Geol. Survey No.
45, 1906, pp. 276–280, 286–290.

[b] Compare Pl. XI, in pocket; also Pl. III of Bull. U. S. Geol. Survey No. 247, 1905.

SEQUENCE OF GEOLOGIC EVENTS.

This section treats only of the recent geologic history. The various periods of sedimentation, deformation, intrusion, and metamorphism from which the hard rocks have resulted have been discussed by Mr. Collier elsewhere in this report (p. 60).

In the opinion of the writer the present topography may be traced through four chief epochs—(1) planation; (2) general uplift, with local depressions permitting the invasion of the sea; (3) stream dissection, marine benching along the coast, and sedimentation in local areas; and (4) uplift, with dissection.

PLANATION.

It has been pointed out that certain high, flat-topped domes and spurs, ranging from 2,000 to 2,700 feet in altitude, may be the remnants of a dissected peneplain, as shown by Collier. Within the province here under discussion, however, the evidence of this old surface of erosion is too fragmentary to prove or disprove this theory. But it can be definitely asserted that if there was an earlier epoch of planation little indication of it is left in the present topography.

Throughout the peninsula the flat-topped ridges and domes, ranging in altitude from 800 to 2,400 feet, are among the most striking features of the topography, and they have been generally interpreted as the remnants of an uplifted and dissected peneplain. Some of the best examples of this feature are found in the Kougarok district, where extensive smooth, flat-topped interstream areas characterize the topography. To an observer standing on one of these prominences the old plain, represented now by the summit levels, appears to sweep off toward the horizon gently rolling and unbroken, save for an occasional dome or rocky crest which stands above it. In such a view the valleys, unless close at hand, are almost completely lost sight of, for their bounding ridges coincide with the general level. There can be little doubt that this is an uplifted erosional surface which had probably been reduced nearly to base-level, or at least to a mature state of dissection. The form of the valleys furnishes additional evidence of the base-leveled character of this upland. Their tortuous courses have evidently resulted from incised meanders inherited from the drainage system of a previous period.

This peneplain, for such it may be called, is gently rolling and contains many minor ridges and depressions. Some of these irregularities are without doubt an inheritance from the former cycle, but they are believed to have been largely caused by warping after the peneplain was formed.

DIFFERENTIAL UPLIFT.

After this period of planation came differential uplift with local depressions. The central axis of the peninsula, lying north of Kuzitrin River, was elevated to at least 1,000 feet, while depression probably took place along its border. The writer is not personally familiar with the coast along the Arctic Ocean, but the published descriptions and the topographic maps suggest at least that the peneplain surface lies underneath the gravels of the coastal plain. It has been shown that along the southern margin of the upland the old peneplain appears to pass underneath the alluvial filling of the Kuzitrin basin.

As this peneplain was formed under subaerial conditions, it is evident that the part of it now buried under alluvium must have been depressed when the rest of the area was uplifted. In other words, the movement was differential and the result was a warped land surface. In at least one locality (Quartz Creek valley) a sharp flexure of the old land surface is indicated.

But this movement did not take place at once, for the benches, both marine and stream, as well as the character of the sediments, indicate several stages of active deformation separated by periods of quiescence during which erosion and sedimentation took place. Because of these facts no sharp line can be drawn between this and the succeeding cycle. Some movement took place, but the first was the period of the more far-reaching orographic disturbances. It has been stated that the Kuzitrin lowland was at about this time invaded by the sea; this implies a submergence of the lowland surrounding Imuruk basin and of a portion at least of Kruzgamepa Valley. The evidence of this submergence lies both in the form of the basin and of the adjacent territory and in the character of some of the sediments which it contains. A log about 60 feet long, which must have been brought in by ocean currents, has been found in these sediments. The area submerged possibly included the upper part of the Kuzitrin basin, which is now covered by recent lava flows, and this inland sea may have been connected with Norton Sound by a strait through the depression now occupied by Koyuk River.

STREAM DISSECTION AND SEDIMENTATION IN LOCAL BASINS.

It is believed that at the close of the first and more intense period of deformation the land stood at a somewhat lower level (200 to 500 feet lower) than it does now and that a period of quiescence followed, permitting extensive stream dissection and marine benching along the coast. During this period most of the present valleys were carved out, and probably also the marine bench was formed which was later elevated and is now called the York Plateau.

While the drainage channels were being incised sediment was deposited in the Kuzitrin basin, then an arm of the sea. It has been shown that the character of this sediment is but imperfectly known. Underneath Quartz Creek valley there are nearly 200 feet of white quartz gravels, which would appear to be the result of the denudation of the old planated land surface. Such quartz gravels, many of which are not greatly rounded, are most likely to have come from an area which had been long subjected to subaerial decay. Their accumulation was probably brought about by rapid erosion and sedimentation, which would indicate rapid uplift. It has been stated that both rounded gravels and sands occur on the lower part of Quartz Creek and in a terrace along the west bank of Kougarok River, and that these are believed to be of about the same age as the white gravels described above. They probably represent a somewhat later period of deposition, when the materials had become more waterworn by stream action and more or less sorting had taken place.

The explanation of the occurrence of the clay deposit underneath the Kuzitrin lowland proper is still more difficult. Perhaps an attempt to explain this deposit is useless when so few facts regarding it are available. If, as is probable, this clay bed is extensive underneath the Kuzitrin lowland, it is probably a deposit in deeper water, laid down at the same time as the gravels described above. In other words, it is the result of sedimentation in the central part of the embayment along whose margin the gravels and sands of fluvial origin were laid down. This clay may, however, be a glacial mud, derived from the valley glaciers of the Bendeleben Mountains, which were the locus of an ice accumulation in recent time. Opposed to this genesis for the clays is the fact that the glaciation, which was confined to a very small area, occurred at a somewhat later period. One additional fact deserving mention is that the few borings made in these clays encountered several layers of vegetable matter which is in every way similar to the present tundra growth. This indicates that deposition in that part of the basin where the clay has been found has been interrupted by periods when the land stood above water long enough to permit a covering of vegetable growth.

The gravels of Quartz Creek comprise two unconformable series. This indicates a period of erosion which interrupted sedimentation and must have been brought about by orogenic movement. Though the fact is important from the standpoint of the placer miner, as the upper gravels alone have thus far been found to be gold bearing, it is probably of no great physiographic import. It simply shows that there were repeated oscillations of the land. More detailed studies will probably disclose the identity of age of this uncomformity with one of the layers of vegetable matter in the clay deposits of the central part of the basin.

UPLIFT WITH DISSECTION.

The present topographic cycle was begun by general elevation, which through a series of intermittent uplifts brought the entire province to its present position relative to sea level. Like the preceding orographic disturbances, these uplifts were differential and in some localities were accompanied by downward warping. The best example of sinking is the Tuksuk Channel, a tidal waterway which connects Imuruk Basin with Bering Sea. Collier [a] has shown that in the York region there is indisputable evidence of warping since the York Plateau was cut.

It is shown in the description of the Quartz Creek placers that the fluvial deposits of this area have been locally depressed. The best evidence of depression is the northwest dip under the valley shown by the bench gravels on the southeast side of Dahl Creek. As a result, gravels of the same horizon are found in both the bench and the creek.

The presence of stream benches at various altitudes in Kougarok Valley and some of its tributaries is interpreted as signifying that the last uplift was interrupted by intervals of tranquillity long enough to permit considerable stream cutting. A discussion of this phase of the physiographic history will have to await more complete field studies. The relation of the stream cutting to the gold values in the placers is important, for in some places the dissection of the auriferous bench gravel has led to an enrichment of the placers of present stream gravels.

This last cycle resulted in the present topography. The streams are still lowering their grades and are actively at work extending their drainage basins.

THE GOLD PLACERS.

MINING DEVELOPMENT.

This district was probably visited by prospectors as early as the summer of 1899, but the first locations were not made until the winter of 1899–1900, and there was no actual discovery of gold until the following summer. A rush from Nome to the new field took place in March, 1900, and another in July of the same year. The first claim staking in March was on Harris Creek, and in July gold was found on Quartz and Garfield creeks. In August and September considerable gold was taken out of the shallow placers of these two creeks. Meanwhile gold had been found on the Kougarok and on many of its tributaries, but no claims were opened up. In 1901 there was a decrease in the gold output, for the shallow diggings were

---

[a] Prof. Paper U. S. Geol. Survey No. 2, 1902, p. 39.

rapidly exhausted and no very rich gold had been found on other creeks. The remoteness of this field from transportation facilities discouraged prospectors, unless they were heavily backed financially, and there were no bonanzas to give an impetus to mining. Probably the most important discoveries were those of Kougarok River, but these could be exploited only by individual miners during low stages of flow, and sudden freshets often destroyed the work of weeks of preparation. So mining interests in the Kougarok district may be said to have lain dormant for several years, though some gold, chiefly from Dahl Creek, was produced every year. When ditches were successfully constructed at Nome, a renewed interest in these outlying placer fields arose. In 1903 T. T. Lane constructed the first long ditch in the district, from the head of Coffee Creek to a bench at the mouth of Dahl Creek. In the following years many more ditches were planned and surveyed. In 1905 and 1906 ditch construction went on with feverish activity, and by the end of the summer of 1906 upward of 100 miles of ditch were planned, about half of which was completed.

One of the larger ditches is the North Star, which extends from Arctic Creek, on the east side of the Kougarok, to the mouth of Taylor Creek, and up that stream about 10 miles, a total length of 15.2 miles. The Cascade ditch takes water from Taylor Creek about 6 miles upstream and discharges at the mouth of the creek. Both of these ditches were completed in 1906. The Kougarok Mining and Ditch Company had one ditch in operation in 1906 and two more partly constructed. Of these the Homestake ditch, which heads on the Kougarok, $3\frac{1}{2}$ miles above Macklin Creek and discharges at the mouth of Homestake Creek, with a head of 172 feet, was completed in 1905. Work has been begun on the Altoona ditch, which heads $1\frac{1}{4}$ miles above the mouth of Washington Creek and discharges at the intake of the Homestake ditch. A third has been located which is to be built up Macklin Creek, to take water from Schlitz and Reindeer creeks north of the Arctic divide. T. T. Lane has completed a ditch from Henry Creek, which discharges at Homestake Creek. All the above-mentioned ditches discharge within a few miles of each other on Kougarok River and represent a large aggregate outlay.

The Irving Mining Company has constructed a ditch from Washington Creek along the north slope of Kougarok Valley nearly to the mouth of Mascot Creek. Another ditch has been built by the Northwestern Mining Company, on North Fork, from the junction of Alder and French creeks; it is to be continued to the Kougarok, about 7 miles being complete in 1906. The Lane ditch, from Coffee Creek to the mouth of Dahl Creek, has already been mentioned. Smaller ditches have been built or surveyed at various places, includ-

ing Arizona, California, Coarse Gold· and Windy creeks.   Besides
these there are many other schemes for ditch building, which in 1906
had not gone far enough to deserve individual mention.

The summer of 1906, being abnormally dry, was especially favorable for ditch construction, but was very unfavorable to those who
were prepared to sluice.   It is perhaps well, however, that the managers of the large companies should know what to expect and be able
to include an allowance for a dry season in their estimate of cost.
The records show that in the last seven years there have been two
very dry summers, 1900 and 1906, and that therefore the last season.
is not by any means so abnormal as some promoters try to make the
public believe.

The Kougarok district up to 1906 could be reached from Nome only
by an overland journey of about 100 miles or by a very circuitous
water route via Teller, Imuruk Basin, and Kuzitrin River.   From
Lanes Landing, at the head of scow navigation on the Kuzitrin,
freighting by wagon to the creeks costs 6 to 15 cents a pound in summer, but the winter rates are much lower.   J. M. Davidson and
Andrew J. Stone, who are among the largest operators in the district, have established a transshipping point on Kaviruk (Marys)
River called Davidson Landing and have built a road from that
point to the upper Kougarok region, a distance of 40 miles.   Small
lighters can be towed directly from the ship's side at Port Clarence
to Davidson Landing, so that at least one handling of freight is
avoided.

In 1906 the Seward Peninsula Railway was extended northward
to the head of Nome River, and thence down the Kruzgamepa to
Lanes Landing.   Surveys have been made looking to an extension of
this line farther up Kougarok Valley.   This railway will bring the
district into close communication with Nome and will do much to
accelerate its development.   Two telephone lines connect many of
the camps with Nome.   The recording office is at Lanes Landing.

Mining operations in 1906 may be summarized as follows: One
hydraulic plant was operated for a part of the season on a bench
claim on Dahl Creek, and two plants on Kougarok River above the
mouth of Taylor Creek.   The two on the Kougarok removed the overburden and part of the pay streak by hydraulic power and cleaned
the bed rock by hand.   In both plants hydraulic lifts were operated.
Considerable work was done on the lower four claims on Dahl Creek
by shoveling into sluice boxes.   Ground sluicing was done by a number of operators, notably on Windy Creek and on Solomon Creek, a
tributary of Taylor Creek.   Several claims were worked in a small
way on Coffee Creek and on some of the tributaries of the Kougarok.
A dozen outfits were engaged in mining the river gravels and some
of the tributary gulches of the Kougarok above Macklin Creek, but,-

they were handicapped either by lack of water or by an excess of water that flooded them out. Below Taylor Creek on the main Kougarok attempts were made to exploit the bench gravels either by sinking shafts and drifting or by the aid of small hydraulic plants, but in most cases the equipment was insufficient to produce anything but meager results. Probably the most successful of these attempts consisted in the drifting on some benches on the west side of the Kougarok near the mouth of Taylor Creek. Harris, Garfield, and other creeks received some attention. In all, probably not more than 150 or 200 men were engaged in productive mining, chiefly because of the inadequacy of the water supply.

The amount expended in ditches and the purchase of claims during 1905 and 1906 probably exceeds $1,000,000—a sum hardly justified by the extent of placer ground actually proved. The total gold output to the end of 1905 is estimated at about $600,000. This amount is, however, only an approximation. The production of 1906 [a] appears to have been small, owing to the lack of water.

DISTRIBUTION OF THE AURIFEROUS GRAVELS.

Broadly speaking, the auriferous gravels thus far discovered in the Kougarok district fall into two zones, which converge toward the lower part of Kougarok River. (Pl. XI, in pocket.) The larger zone stretches northward from the lower Kougarok, embracing much of the Kougarok basin, and is here termed the "northern belt." The smaller zone, which appears less well defined, extends eastward from the Kougarok to the Noxapaga, embracing the streams tributary to Kuzitrin River. This second zone will here be called the "southern belt." The northern belt lies in a zone of schistose rocks, bounded on both sides by massive limestones. It is therefore consistent with the general rule that the gold has its source at or near the limestone and schist contacts. Nor does the second belt, so far as known, offer an exception to this rule. The types of placers in each of the two belts are described below.

SOUTHERN BELT.

GENERAL DESCRIPTION.

The auriferous gravels forming a broken fringe along the southern margin of the highlands that bound the Kuzitrin basin on the north and west have certain features in common, which justify describing them as a unit. This belt includes the placers of Quartz and Garfield creeks, as well as those of the Noxapaga basin. The bed-rock geology of the belt is obscured both by the extensive alluvial deposits

---

[a] Not a single operator in the district responded to a request for information in reference to production.

and by the products of deep rock weathering. However, a belt of graphitic phyllites and schists, including some calcareous beds, appears to stretch across the upland between Kaviruk and Kuzitrin rivers. Many of these rocks carry quartz veins, locally stained with iron. Schists occur north of these graphitic rocks, and farther north are succeeded by limestone. Though these formations can not be exactly delineated, because of the deeply weathered character of the rocks and the absence of outcrops, yet most of the gold-bearing creeks appear to cross the contact of the limestone and schist.

The unconsolidated formations embrace (1) the present stream gravels, (2) the deposits flooring the Kuzitrin lowland, and (3) the bench gravels. The first group, which embraces most of the working placers of the district, is fully described on page 299. Little can be added to the description of the second group already given. The bench gravels merit closer consideration.

The upland region descends by a gentle slope, here and there broken by a well-marked terrace, from an altitude of about 1,100 feet to the Kuzitrin Valley floor, 100 feet above sea level. The best-defined terrace is about 100 feet above the present water level and is traceable from the mouth of Quartz Creek northward along the western side of Kougarok River to the point where the valley of that stream emerges from the upland. A similar terrace is present along the northern margin of the Kougarok, and the lower part of Turner Creek and some of the tributaries of the Noxapaga are reported by Collier to be incised in deep gravel deposits; a fact which indicates an easterly extension of this terrace. The alluvium of these terraces in nearly every exposure is made up of the same kind of material—well-rounded and stratified brown sands and gravels. Certain exceptions to this are noted below. These benches are without doubt the remnants of an extensive gravel sheet, as is indicated by the hillocks of stratified gravels which stand here and there above the floor of the Kuzitrin lowland.

Near the mouth of Quartz Creek the top of the terrace is about 125 feet above the water, and the gravels rest upon clay of unknown thickness about 15 feet below water level. The exposed material consists of well-rounded gravel and sand. Along Quartz Creek the surface of the gravel dips with about the grade of the stream, and 1 mile below Dahl Creek about 100 feet of gravel and sand are exposed in the valley wall. Above this point the bed was not definitely recognized, but it is believed to be represented by a white quartz gravel exposed on Quartz Creek just below the mouth of Dahl Creek. On the north side of Dahl Creek valley a shaft was sunk to a depth of 187 feet entirely in this white gravel and did not reach bed rock. These white gravels are probably a phase of the bench gravels of lower Quartz Creek and the Kougarok described above. The surface of the white gravels dips to the northwest under the trench occupied by Dahl Creek. In other words, the gravels underlying the pay streak at the Lane hydraulic mine and those on Dahl Creek are parts of the same bed. The surface of the same gravel deposit is believed to be exposed near the mouth of Joe Creek, a tributary of Quartz Creek. These relations are too complex to permit detailed analysis here, but they point to the following conclusions: (1) The auriferous gravels of the Lane hydraulic mine, Dahl Creek, and of Joe Creek belong to the same horizon; (2) they are underlain by barren alluvium, from which are formed the white gravels of Dahl Creek and the bench gravels of the lower part of Quartz Creek, of the Kougarok, and of the front of the upland near Garfield and Turner creeks; (3) this older gravel series is believed to carry no values, though it is known to be more or less auriferous. The last conclusion is supported both by the results of prospecting and by theoretical considerations. In general the rich placers of the peninsula occur in alluvium that was deposited under conditions of subaerial decay rather than during floods. Much of this bench gravel, however, is well rounded and stratified and appears to have been laid down during periods of flood, which are not favorable to a concentraton of values. Still, so far as known, these gravels next to bed rock have never been prospected. They may be gold bearing, and in the opinion of the writer the chances of finding gold at depth warrant the outlay of the cost of prospecting to bed rock.

QUARTZ CREEK BASIN.

The basin of Quartz Creek includes a dendritic drainage system lying in an elbow formed by the lower Kougarok and Kuzitrin rivers, and is incised partly in the upland and partly in the broad gravel terrace already described. The valleys cut in the gravels are steep-walled trenches; those cut in bed rock are broader, with gentler slopes. Placers have been found chiefly in the bed-rock portion of the valleys. So far as known this bed rock is phyllite or mica schist.

Throughout the basin the stream gradients are low, averaging probably less than 10 feet to the mile. The amount of water ordinarily available is very small, and in dry seasons, such as that of 1906, the supply fails entirely. Gold has been mined on Quartz Creek and on its tributaries Dahl, Joe, and Coffee creeks, but of these only Dahl Creek has been a considerable producer. The terrace gravels of the lower basin carry colors of gold, but no values have been found in them.

In 1900 and 1901 half a dozen claims were operated on Quartz Creek near the mouth of Dahl Creek, but these were quickly worked out. The flood plain here is not more than 50 to 75 feet wide, and the pay streak, which comprised only a part of this plain, was 2 to 3 feet thick. No bed rock was found in this part of the basin, the values being concentrated on a layer of blue, sticky clay. The gold is medium coarse and probably came from the Dahl Creek basin.

Dahl Creek has been the center of mining interest since 1900, for the lower six claims have been the largest group of producers in the district. This stream flows in a shallow trench cut for about 15 feet in muck. Above this is a bench a quarter of a mile wide, from which the valley walls rise gently. From the mouth of the creek up to claim No. 6 the gold rests upon a clay seam, but on No. 6 the pay streak was found 10 to 20 feet below the surface, just above a mica-schist bed rock. Underneath the clay floor upon which the gold-bearing gravels below claim No. 6 lie are the white quartz gravels, already described; these have been penetrated by a shaft, which goes to a depth of 187 feet without reaching bed rock. The following section was measured on Dahl Creek about one-fourth mile above its mouth:

Section on Dahl Creek one-fourth mile above its mouth.

	Feet.
Frozen muck with impure ice lenses	15–20
Ferruginous gravels and sands (pay streak)	3– 4
Blue clay, carrying some sand	1– 2
White quartz gravel.	

Near the mouth of Dahl Creek the bench just described is about 20 feet above the present stream bed and has been found to carry values. A measured section at Lane's hydraulic mine is as follows:

Section of bench at Lane's hydraulic mine, Dahl Creek.

	Feet.
Frozen muck with lenses of ice	15–20
Ferruginous, gold-bearing gravel	3– 4
Sticky blue clay	½– 1
White quartz gravel.	

These deposits dip under the creek bed, and both bench and creek placers are of the same age, their difference in position having been caused by local warping. Similar bench gravels will probably be found to underlie the talus and muck that cover the lower slopes of the Dahl Creek trench, but so far values have been found only at Lane's mine. The possibility of finding values on bed rock underneath the white gravel has already been noted.

Placer gold has been found near the mouth of Joe Creek, on Coffee Creek, and on a number of other smaller tributaries of Quartz Creek. No gold is reported from Checkers Creek, a northern tributary of Quartz Creek. The course of Checkers Creek appears to be along the old planated surface, which dips under the gravels and from a theoretical standpoint should be a favorable place for gold concentration.

Considerable prospecting has been done throughout Coffee Creek basin and a little gold has been mined in its upper part. The lower part of the creek is incised in the gravel terrace, and here also a little gold has been found. Prospecting on the upper part of the creek is expensive because the gravels are everywhere buried under 10 to 25 feet of muck. This is true not only of the floor of the valley, but also of the slopes and the tributary gulches. This overburden is too heavy to permit open-cut mining, and the cost of fuel has so far been too great to encourage drift mining or even prospecting. As a result much of the prospecting on Coffee Creek has been confined to cuts made by ground sluicing at favorable localities during floods, and few careful tests of claims have been made.

In the winter of 1906 some rich placer ground was found in the talus of the valley slope near the head of Coffee Creek. This peculiar auriferous deposit appeared to be almost in place. The gold occurs in 4 to 7 feet of angular schist and quartz débris and weathered schist bed rock covered by 18 to 20 feet of muck. The quartz is iron stained, but does not appear to be auriferous, and the gold probably came from the associated schist. The gold is angular, spongy, and bright colored. All these facts point to the conclusion that the material mined is a weathered part of a mineralized zone. The deposit has been traced about 1,000 feet, but is buried so deeply that its boundaries are not well known. It is indicative of the source of the gold and suggests the possibility of finding lode deposits that may carry values.

The other placers of the upper part of Coffee Creek are as a rule buried under an overburden of muck 10 to 20 feet thick. The gravels are chiefly schist and quartz, and range from 3 to 7 feet in thickness.

GARFIELD CREEK.

Garfield Creek heads near Baldy Mountain, flows in a rather broad valley through the upland, and in the last 5 miles of its course meanders across a gravel plain to join Kuzitrin River. The gold placers are confined to the upland parts of its valley, and may be reached by a trail crossing the tundra from the Kougarok, but such traveling is difficult and transportation is very expensive. The discovery was made here in 1900,[a] and considerable gold was produced in that year from a shallow pay streak confined to the creek bed and resting upon a white clay.

The valley of Garfield Creek is broad, the sides are benched, and gravels and talus with a thick layer of moss and muck cover the bed

---

[a] Brooks, A. H., Richardson, G. B., and Collier, A. J., Reconnaissances in the CapeNome and Norton Bay regions, Alaska, in 1900, a special publication of the U. S. Geol. Survey. 1901, p. 122.

rock.   The gold occurs under about 2 feet of gravel in the creek bed and on a false béd rock of clay.   It is coarse, dark colored, and of irregular outline.   Grains of average size were worth about 1¼ cents. Except on two or three claims the prospecting since 1900, either in the benches or on the bed rock, has failed to locate placers that can be profitably mined under present conditions.   The creek has been almost abandoned, though it is reported that one claim is worked in a small way every year.

BOULDER CREEK.

Boulder Creek is a tributary to Turner Creek a short distance above the junction of the latter with the Noxapaga, a tributary of the Kuzitrin.   The following description is by Mr. Collier:

The placers on this creek are estimated to have yielded $7,000 in 1901, but since then production has been small and the creek as a whole has been a disappointment to miners.   In 1903 there were, however, still a few men at work here.   The creek occupies a straight, rather broad valley about 5 miles long, which is deeply cut in the higher upland about the headwaters, but has less relief in the lower country where Boulder Creek empties into Turner Creek. The bed rock consists of calcareous schists and limestones, with some beds of graphitic schist.   The rocks are deeply covered with moss, muck, and gravel, being exposed only in mining excavations.   Sluicing was begun in August, 1901, and some gold was taken out with the rockers previous to that time.   A mile from the mouth, excavation revealed about 2½ feet of moss and muck overlying a gravel consisting of angular fragments of graphitic schist and limestone. This gravel was reported to carry values to a depth of 4 or 5 feet, below which it was frozen and had not been tested.   Colors of gold were found through a width of 56 feet.   About three-fourths of a mile above this mine a prospect hole through the gravel to a depth of 11 feet failed to reach solid bed rock.   The gravels there are chiefly limestone pebbles.

The gold on Boulder Creek is comparatively coarse, and nuggets ranging in value from 25 cents up are not uncommon.   They are as a rule rounded and dark colored.   The largest nugget found weighed over 2 ounces and was valued at $36.   It was a well-rounded piece of gold, stained with iron.

SMALLER TRIBUTARIES OF THE NOXAPAGA.

Above the mouth of Turner Creek the Noxapaga receives a number of small tributaries, the gravels of several of which are said to be auriferous.   These gold-bearing creeks, from south to north, are Grouse Creek, Black Gulch, and Buzzard, Goose, and Frost creeks. Gold was found on Goose Creek in 1900 and a little mining has been done.   All the mining operations on these streams have been seriously hampered both by their remoteness from transportation routes and by scarcity of water.

NORTHERN BELT.

GENERAL DESCRIPTION.

The auriferous gravels of the main Kougarok above the flats and of its tributaries form the northern belt of placers. This belt embraces stream placers and bench placers. Up to the present time the first type has yielded most of the gold. There are two forms of the stream placers—those of the smaller gulches and creeks and those of the main river. The gulch and creek placers are in general of small extent, but many are so situated that they may be profitably mined by pick and shovel. On the other hand, the placers of the main river, though many are of greater extent, are difficult to exploit except with equipment that is capable of handling a large amount of material and is adapted to both high-water and low-water conditions.

Bench gravels have been reported at many localities, but those of proved economic importance are confined to the main Kougarok and some of its larger tributaries. These gravels are chiefly within 25 feet of the present water level, but some higher auriferous benches have also been reported. The geologic map (Pl. XI, in pocket) shows that the main valley of the Kougarok lies in a belt of schists, which is flanked on the east and west by broad areas of semicrystalline limestones. These limestones are cut by all the larger tributaries of the Kougarok. The schists are predominately argillites, some altered to phyllites, others to mica schists. Graphite is very commonly associated with these rocks, and many thin limestone beds are present. A dark-green dioritic rock occurs as dikes, stocks, and sills, in some places massive, in others schistose. Pyrite-bearing quartz veins are common in the phyllite and are probably the source of the gold. On Mascot Gulch, described on page 320, the placer gold may be traced to its source in a mineralized shear zone in the phyllites.

In this discussion Kougarok River and its smaller tributaries are considered under one heading and the larger tributaries are treated separately. The descriptions are given in order from south to north.

KOUGAROK RIVER.

Kougarok River joins the Kuzitrin about 17 miles above Igloo, the head of steamboat navigation, and has a length, exclusive of meanders and smaller bends, of about 60 miles. Through the upper 10 miles it flows nearly due east, then it turns southward and flows about 40 miles to its confluence with the Kuzitrin. In the lower 10 miles it meanders across the Kuzitrin lowland. Above this portion its valley is incised in the upland plateau, which here has an elevation of 1,000 to 1,500 feet above the sea.

Though the general trend of the different parts of the valley is in one direction, its course in detail is tortuous. Within the valley walls the river flows in a very irregular channel, and many of the meanders are separated by well-defined benches whose flat surfaces mark former stages of water level. This is especially evident above Taylor Creek, where the river flows through a continuous series of oxbow curves, which are separated by sloping benches.

At the point where the Kougarok enters the Kuzitrin lowland both valley walls show well-marked benches. Two levels are here noticeable—one 50 feet and one 25 feet above the stream. These can be traced for several miles above Windy Creek; the walls then become steeper and the river occupies a canyon-like valley up to Left Fork. From this point to Washington Creek, 20 miles above, some evidence of benching can be observed in most places, though the benches are not continuous. The individual levels have not been traced, but in the part of the valley below Taylor Creek there are at least two high-level gravels and possibly three.

Kougarok River is a swiftly flowing stream that carries at its mouth probably 5,000 to 8,000 miner's inches of water, and has an average gradient of about 20 feet to the mile. Most of the material transported is coarse, ranging from fine gravel to coarse cobblestone.

The most extensive deposits of auriferous alluvium yet discovered in the district are on the main Kougarok, both in the present stream bed and in the benches. Gold has been found in the gravels of this stream for about 40 miles of its length. It is not known how much of this stretch carries commercial values, but it is probably not more than 50 per cent.

Several placers of the present stream bed have been worked since 1901. Some gold has been taken out of the river bed with shovel and rockers near Coarse Gold Creek and at various points as far up as Taylor Creek. Much more work has been done at and above the big bend of the Kougarok, near the mouth of Macklin Creek and as far up as Washington Creek. Most of this mining was done during low-water stages. It should be stated that these placers are in no sense of the river-bar type, but carry coarse gold mingled with gravels and largely concentrated on bed rock.

The gravels are in the main fairly well rounded and stratified. The largest pebbles are not more than 1 or 2 feet in diameter, but a few larger bowlders have been contributed by the talus of the valley slopes. No general statement as to the thickness of the gravels can be made, as it varies greatly in different parts of the river. In the canyon already described bed rock is exposed throughout the river bed. In many places above the canyon, gravels are almost entirely absent; in other places the depth to bed rock is 6 to 20 feet.

The width of the alluvial floor also varies; in some parts of the valley the entire floor is buried in gravels, but in others the stream has uncovered bed rock over a part of it. The actual flood plain of the river ranges in width from 100 feet in the canyon to 800 feet at the flat. At the mouth of North Fork it is 300 or 400 feet wide, at the mouth of Taylor Creek about 600 feet, and near the mouth of Trinity Creek about 300 feet. Below the flat at the mouth of Taylor Creek the alluvium is made up almost entirely of gravel, but above this flat the gravel is in places buried under a considerable stratum of muck. So far as the writer knows, the gold found in the stream bed below Coarse Gold Creek, except at the mouth of North Fork, is chiefly fine, but the auriferous gravels at and above the mouth of this creek are said to contain a large percentage of coarse gold. This indicates that in the upper half of the river the enrichment is of local origin, and therefore that the gold has not all been brought in by the main stream from its source. The gold of the flood plain is mostly dark, but that of the smaller tributaries is bright. In the flood plains the only placers yet opened up on a commercial basis are those at the mouth of North Fork, where little has been done, and at the mouths of Taylor and Homestake creeks and between Macklin and Washington creeks, where considerable gold has been procured.

The bench deposits of the Kougarok appear to afford an attractive field for the gold miner. Their position makes them easy of access and no hydraulic lifts are required to dispose of the tailings. Between Coarse Gold and Taylor creeks at least two distinct bench levels are well defined, one about 25 feet and the other 50 feet above the water. So far as observed, the gravels are from 8 to 10 feet in depth and are in most places covered with muck. No determinations of values are known to the writer, but the fact that some of the lower benches have been worked at a profit by crude means makes it seem probable that their gold content is sufficient to assure returns if handled by cheaper methods. Bench gravels have been reported at various places above Taylor Creek, and some are known to be auriferous, but they have not been developed on a commercial scale.

The following notes on the occurrence of the auriferous gravels are the basis for the generalizations which have preceded.

At the point where the river emerges from the upland two benches were observed—a lower, which is well defined, standing about 15 feet above the present flood plain, and an upper, about 35 feet higher, which slopes toward the axis of the valley. These benches can be traced upstream for about a mile; above this stretch the valley walls steepen and become rock bound. So far as known, this part of the river has yielded no placer gold.

From the mouth of Windy Creek up to a point within a mile of Left Fork the valley is broad and the slopes are interrupted by benches, one at 20 and one at 50 feet being recognized. The gravels of Windy Creek are auriferous (see p. 320), but no mining has been done on the main river in this vicinity. Above this stretch the valley is a steep-walled canyon cut in hard diorite and there is practically no gravel on the rock floor. This topography continues nearly to Queen Creek, where the valley opens out again. Here there is a bench, covered by 4 to 5 feet of well-rounded gravel overlain by 6 to 10 feet of frozen muck. Near the mouth of Queen Creek two benches were observed—a lower one 20 feet above the river, covered by 4 to 5 feet of gravel, with an overburden of muck 10 feet thick, and an upper one 45 feet higher, carrying about 10 feet of gravel that is more or less auriferous. These benches, though not continuous, can be traced along the valley to Coarse Gold Creek, where the river makes a long curve, swinging around a bench 100 feet high, on which lies about 10 feet of river gravel. This bench, together with a lower one standing about 25 feet above the water, can be traced almost continuously between Coarse Gold Creek and North Fork. Similar benches occur along North Fork. Gold has been found in the river gravels at the mouth of North Fork and at the big bend above Coarse Gold Creek. A scheme is under way to drain the river bed at the big bend by tunneling through the neck of land. It is claimed that by building a 20-foot dam and driving the tunnel, water power with an 18-foot fall can be developed and $2\frac{1}{2}$ miles of the river bed can be laid bare. This power is to be utilized in mining the river gravels, which are said to carry values. The bed rock, a schist, is exposed along a part of this stretch of the river, but in some places it is buried under gravels at a maximum depth of 8 feet. The bench gravels along this part of the river also are said to be auriferous.

Between Coarse Gold and Taylor creeks the Kougarok Valley maintains its open character, and benches are well marked in many places. A low bench 15 to 20 feet above the river bed is the most conspicuous of these features and is usually covered with 3 to 6 feet of gravel with an overburden of frozen muck. Here, as elsewhere in the valley, the benches are obscured in many places by the heavy talus, and therefore the absence of a topographic expression of the benches where the slopes are gentle does not necessarily prove their absence. About 2 miles below Taylor Creek a bench about 20 feet above the river carries 10 feet or more of gravel covered with frozen muck. The same bench is recognizable at the mouth of Arizona Creek and at other places. Some gold has been extracted from these bench gravels.

Several lower benches occur along the valley slope opposite the mouth of Taylor Creek. Here a measured section showed 6 feet of auriferous gravels resting upon mica-schist bed rock, whose surface slopes toward the axis of the valley. The bed-rock floor of this bench is only a few feet above the present flood plain of the river, but, as the alluvium of the flood plain is 20 feet or more deep at this point, the bench deposit was originally at least 25 feet above the river bed. The slope of these benches suggests that they were formed by streams tributary to the Kougarok. They were mined during the winter of 1905-6 by drifting, and considerable gold is said to have been taken out. The largest nugget had a value of $1.80.

Considerable prospecting and some mining have been done in the flood-plain gravels between the mouths of Taylor and Homestake creeks. The bed rock is schist, and the gravels are from 8 to 10 feet deep, with an overburden of muck varying in thickness. The values are said to be chiefly concentrated on bed rock. The flood plain is several hundred feet wide, but the width of the pay streak was not learned. The largest plant at this locality includes in its equipment a hydraulic elevator.

No mining appears to have been done on the main Kougarok for several miles above Homestake Creek, though prospects are reported. In this stretch of the valley the river meanders from one side to the other, forming a series of symmetrical curves separated by sloping benches. These benches may carry gravels, but were not examined by the writer. A mile below Macklin Creek a little work was done in 1906 on the river gravels, here about 6 to 8 feet thick. An attempt to control the river by a small dam, so as to make the bed accessible to mining operations, was only partially successful.

Considerable mining has been done in the river bed about a mile above Macklin Creek, where the water has been diverted by a dam. The gravels here are about 6 to 8 feet deep, of which 18 inches to 3 feet 6 inches carries most of the gold. The pay streak is said to have in some places a width of 400 feet, but in these places the values are much less than where the pay streak is narrow. Values of 3 or 4 cents to the pan and pockets very much richer are reported. Most of the gold is dark, but a little bright gold is intermingled.

Several claims have been opened up at the mouth of Trinity Gulch. Here about 6 or 7 feet of gravel rests upon schist bed rock and is overlain by 4 or 5 feet of frozen muck. The gold from Trinity Gulch is bright; that of the river is dark. Above this group of claims the valley walls close in, forming a canyon. At the canyon an ill-defined bench lies about 20 feet above the river bed. At the lower end of the canyon the depth to bed rock is 12 feet, more than half of which is gravel. Above the canyon the gravels, which are 3 to 4 feet thick, are coarse and well rounded. The gold is dark and chiefly coarse, occurring in flat nuggets. Two miles above, at the mouth of Mascot Gulch, the river bed is about 300 yards wide, but the pay streak lies well toward the southern wall of the valley. The gravels in the river bottom are about 16 feet deep and have been mined by winter drifting. The gold includes both the dark variety of the main river and the bright variety found in the tributaries. The pay streak is 2 to 3 feet thick and is said to have carried high values on the schist bed rock.

Considerable mining has also been done in Mascot Gulch, on a deposit similar to that at the head of Coffee Creek. (See p. 313.) Mascot Gulch has a steep slope toward the river and carries very little water. The bed rock is a blue mica slate or phyllite, with many quartz stringers. This is covered by 2 to 3 feet of subangular talus and decomposed bed rock much stained with manganese. The gold is distributed in patches irregularly on the bed rock. It is bright and angular, some of it showing crystalline faces. Much of it is attached to small fragments of quartz. Considerable cassiterite occurs in the concentrates. This gold is without doubt very near its source in the parent rock and is probably derived from a mineralized shear zone in the slates. Washed gravels are said to occur 100 to 150 feet above the river level on the slope of the ridge south of the Kougarok Valley at Mascot Gulch. A test pit at this locality is reported to have shown 40 feet of muck and talus overlying 10 feet of well-rounded gravel or " river wash." These gravels were found to be auriferous, but no values were discovered.

Above Mascot Gulch the valley floor widens out between gentle slopes and is entirely gravel filled. About a mile above Mascot Gulch some mining has been done with a hydraulic plant. Here the gravels are 3 to 7 feet thick, most of the values being on or near the bed rock. The valley floor is 400 to 500 feet wide. A series of prospect holes at this place showed washed gravels on bed rock 50 feet above the present water level. The bed rock of this bench is said to slope toward the axis of the valley. Gold has been found in the gravels above this point as far as Washington Creek, but no mining has been done. It is of interest to note that the limit of the distribution of the placer gold appears to be determined here, as elsewhere, by the limestone-schist contact.

The foregoing descriptions indicate a very wide distribution of gold in the Kougarok Valley, in both the bench and the flood-plain gravels. No high values are known to have been found, but many sections of the gravels of this valley can probably be mined at a profit. The bench gravels are easily workable by hydraulic methods where water is available.

WINDY CREEK.

Windy Creek, a western tributary of Kougarok River about 15 miles in a straight line above its mouth, is about 10 miles long and carries sufficient water for ordinary mining purposes. Some mining has been done on this stream since 1901. The developed placers occur along a small tributary from the south, called Anderson Gulch, which is merely a minor depression in the valley wall. A 4-mile ditch furnishes water under a 140-foot head. The gravels exposed in cuts and forming the pay streak are 2 to 3 feet thick, and in addition 1½ to 2 feet of bed rock carries some gold and is put through the sluice boxes. The bed rock is a silvery mica schist with much iron-stained quartz. These placers have been traced for 1,500 feet along the slope of the valley of Windy Creek. The known area of workable deposits is not large. Gold has been found in other parts of the basin, but not yet in paying quantities.

NEVA CREEK.

Some sluicing has been done in shallow gravels on Neva Creek, a tributary of the Kougarok from the east about one-half mile above the mouth of Windy Creek, but the production has been small.

NORTH FORK.

North Fork joins Kougarok River from the east about 25 miles above its mouth. It is about 10 miles long and flows nearly due west in a rather broad valley. Its important tributaries, from an economic point of view, are Harris and Eureka creeks. At its mouth North Fork carries in an ordinary season more than 1,000 miner's inches of water. The basin of this stream was the scene of the first gold discoveries in the Kougarok district, and some of the placers have yielded considerable values. The benches, of which possibly three different tiers are known, are prominent features, and their gravels are said to be auriferous. The following notes are by Collier and Hess:

The original discoveries of gold in the Kougarok region were made in 1900 on Harris Creek, a northern tributary of North Fork, about 7 miles from its mouth. Since then mining operations have been continuous, but the whole amount of gold produced up to 1903 probably did not exceed $50,000. The bed rock of North Fork and its tributaries consists of schists interbedded with limestones that are correlated on fossil evidence with the Port Clarence. A small amount of gold has been mined on Eureka Creek, a tributary of North Fork, about 1 mile above its mouth. The creek valley is narrow, but it contains some gravels consisting of schist and vein-quartz pebbles. The gold is mostly fine and the largest nugget found was worth only about 7 cents. Harris Creek, from the standpoint of gold mining, is the most important tributary of North Fork. The camp located here can be reached by pack train from Lanes Landing, which is about 30 miles distant, or, during high water, freight may be brought up Kougarok River in small boats within a few miles of the camp.

The plateau is well developed in this vicinity, but several limestone buttes rise above it. The largest of these buttes, known as Harris Dome, lies between Harris Creek and the main North Fork. The valley of Harris Creek is comparatively broad in its upper part, but within 2 miles of its mouth narrows to

15604—Bull. 328—08——21

a canyon. Through the upper 7 miles of its course the creek runs nearly southwest, then it turns somewhat abruptly and flows due south for nearly 2 miles to its mouth. It ordinarily carries from 200 to 300 miner's inches of water, but about 2 miles above its mouth the water sinks, and only after a heavy rain, when the creek is unusually full, does it carry water through its whole length.

The bed rock for 2 miles above the mouth of the creek is massive limestone, from which Silurian fossils have been obtained. It was impossible to make out the bedding of this limestone. Above the limestone area the creek flows for a number of miles over siliceous mica schist, which strikes to the northwest and dips to the northeast at a high angle, and which contains many small stringers of quartz. The headwaters of Harris Creek are in limestone somewhat similar to that near the mouth of the creek.

All the working gold placers are located within 3 miles of the mouth. For about 1 mile up the creek the valley is very narrow and, although the claims here have nearly all been worked, it is reported that they have yielded scarcely more than wages for the men employed. Above this stretch the gravel deposit is 4 or 5 feet thick and about 100 feet wide. A pit, 40 by 300 feet, 1 mile from the mouth, is said to have yielded $12 to $15 to the shovel. The bed rock is limestone, more or less shattered in some places, and the gold-bearing sediments extend downward in the crevices. Two miles from the mouth of Harris Creek the workings show a section about as follows: Brown sandy muck, 6 feet; gravel with values in the lower half, 4 feet, below which was broken limestone, in place, more or less weathered. It is reported that in one place this broken limestone material yielded on an average 2 cents to the pan from the surface down to a depth of 10 feet. When the claim was examined a strong stream of water was disappearing into a hole in the bed rock, indicating an underground channel. The fragments of limestone taken from this hole had a diameter ranging from a few inches to about 1 foot and the crevices between them were filled by a small amount of yellow clay sediment which carried the gold. The preparations for working claims near this point consist of two ditches, one from a small tributary called Left Fork and another from the main stream.

About 3½ miles from the mouth the gravels have a considerably greater extent. The creek valley here is wider, and on the north side there is a gravel bench, slightly above the level of the creek, that has been prospected by a series of holes extending northward and is known to have a width of about 1,500 feet; throughout this distance good prospects of gold are reported. The workings at this point show 2 feet of muck above 6 feet of gravel resting upon 4 feet or more of broken limestone bed rock. The whole deposit from the surface down is reported to average 3 cents to the pan, which is equivalent to $4 or $5 a cubic yard. The best pay is not found in the present creek bed, but extends under the right bank. The claim here is worked by a short ditch that takes water from Harris Creek, and about 10 men were employed in 1903. Above this claim the bed rock of Harris Creek changes to a mica schist, and there are no more working claims. It is the common report that the gravels above this point do not carry gold in paying quantities, and certainly no pay gravel has yet been found.

To explain the origin of the gold placers on Harris Creek the following facts must be considered: The gold placers are confined to the part of the creek in which the bed rock is limestone and they have been found richest where the bed rock is broken up by fissures and crevices. A possible explanation of the distribution of the gold is that it was derived from small quartz veins contained in the limestone or from the limestone itself, but no evidence of such occurrence was obtained by the writers, and no distinct veins in the limestone were seen, though they may have been overlooked. A second possible explanation is that an old channel crossed the ridge from Taylor Creek and entered the present valley of Harris Creek at the upper contact of the limestone area, where the present richest claim is found. A study of the topography fails to indicate that such a channel has existed, and no gold placers, which should exist in such a channel, have been found. A third and probably the true explanation of the occurrence of gold on Harris Creek is that it is derived from small quartz veins and stringers in the schist belt which crosses the creek above the limestone, and that it has been washed down the creek and concentrated in the broken bed rock of the limestone belt. This hypothesis is borne out by the facts that small quartz veins have been observed within these schists, that all the gold of Harris Creek is well rounded and waterworn, and that no large nuggets have been found. The failure to find gold in paying quantities within the schist belt may be accounted for by the fact that the schists do not form proper riffles for holding and concentrating fine gold.

The largest nuggets do not exceed $1 in value. The average value of the particles is less than one-tenth of a cent.

Gold in paying quantities has also been discovered in the gravels of North Fork itself, and in 1903 mining was in progress about a mile above the mouth of Harris Creek. North Fork here cuts across the massive Silurian limestone already described in connection with Harris Creek. In crossing this limestone belt the valley of North Fork is constricted, and a large part of the water sinks, but 300 to 400 miner's inches still flow on the surface. The stream gravel here consists mainly of limestone in more or less angular pieces. The workings are in the bed and under the north bank of the creek, and the gold is found partly in the gravel, but mostly in crevices between angular blocks of the limestone bed rock. These angular blocks are called "slide" by the miners, but are more probably residual bowlders from the weathering and solution of the limestone bed rock in place. This bed rock is mined to a depth of about 3 feet, below which there are still some colors of gold, but not enough to pay under present conditions. The gold is bright, clean, and well rounded. It is rather coarse, probably averaging 2 or 3 cents to the piece, and is easily saved in the sluice boxes. The concentrates consist mainly of hematite and magnetite.

A gravel bench a few feet above the present creek bed, on the north side, has also been prospected. The gravel consists of rounded schist and quartz pebbles and angular pieces of limestone, and is reported to yield fair prospects of gold, but it has not been mined. The occurrence of the gold here seems to be similar to that on Harris Creek, and it is more probably derived from small quartz veins in the schists, which outcrop farther up the creek, than from the limestones in whose crevices it is found.

COARSE GOLD CREEK.

The alluvium of Coarse Gold Creek is auriferous, but as yet little gold has been extracted. A hard diorite forms the bed rock of a part of the creek and does not afford a favorable surface for the concentration of values. The lower part of the creek is in schist and deserves the attention of prospectors.

ARIZONA AND CALIFORNIA CREEKS.

Arizona and California creeks are small streams near whose mouths are heavy gravel deposits. Both the flood plain and bench deposits have yielded gold. Considerable mining has been done on benches near the mouths of these creeks about 20 feet above the present water level.

HENRY CREEK.[a]

Henry Creek, which joins Kougarok River from the west about 35 miles from its mouth, is only about 15 miles long, but on account of its many tributaries it has a large drainage area and at its mouth carries between 1,000 and 2,000 miner's inches of water. One mile above its mouth the flood plain is more than 1,000 feet wide, and the creek flows in a very tortuous course. The bed rock consists mainly of graphitic and chloritic schists, containing many bodies of greenstone whose intrusive origin is evident. Limestone outcrops near the head of the creek. The gravels are made up of graphitic, chloritic, and calcite schist and greenstone pebbles.

Although there is evidence of some prospecting all along the creek, mining has been attempted at only one or two places. Two men were sluicing the gravel from the creek bed about 1 mile from the mouth in August, 1903. The gold is flat and comparatively coarse, the pieces probably averaging 5 or 6 to the cent, but it contains no nuggets worth more than 3 or 4 cents. All the gold is well rounded and iron stained. Half a mile farther up the creek three men were sluicing, but probably were not saving much gold, as work was suspended before the end of the season. From present indications it seems probable that there are low-grade placer deposits on Henry Creek which may be made to yield profits by more economical methods of mining than were in vogue at the time the examination was made.

DREAMY GULCH.[a]

Dreamy Gulch is a small depression, carrying little water at any season, which cuts across the hillside and benches on the east side of Kougarok River, about one-half mile above the mouth of Henry Creek. The whole hillside, as well as the head of the gulch, is gravel. Beneath the tundra, gravel from 3 to 5 feet thick rests upon a bed rock of dark, calcareous mica schist, 30 to 40 feet above the river bed. The gravel contains pebbles of mica and chlorite schist from one-half to 3 inches in diameter and large bowlders of well-rounded quartz up to 18 inches in their longer dimension. A few men worked here in 1903, and, although they had water for sluicing only when it rained, they probably averaged from $7 to $10 a day to the shovel. The gold is coarse, rusty, and well rounded. The largest nugget taken out was valued at $1.10.

TAYLOR CREEK.

Kougarok River forks about 38 miles from its mouth in two branches of about equal size. The eastern branch, called Taylor Creek, rises southeast of Midnight Mountain and flows westward to its junction with the Kougarok. The bed rock at its mouth is a dark

---

[a] Description by Messrs. Collier and Hess.

schist that extends northeastward to Midnight Mountain, but immediately south of that mountain, about 6 miles above the mouth of the creek, there are exposures of a massive limestone that is correlated with the Port Clarence. The creek cuts across the strike of the schists and limestones. Colors of gold are known to have been found on this creek and some of its tributaries, but not much mining has been attempted.

Some mining has been done near the mouth of the creek, where the placer deposits are similar to those of the flood plain of the Kougarok, of which they form an extension. Besides this the only mining attempted in this basin is on a small tributary called Solomon Creek. At the mouth of this stream there is a sloping bench, on which lies 3 to 7 feet of gravel covered by 8 to 10 feet of muck. The gravel is auriferous and has been mined in a small way, as have also the stream gravels of Solomon Creek one-half mile above its mouth.

HOMESTAKE CREEK.[a]

Homestake Creek enters Kougarok River from the west about 35 miles above its mouth, and is 4 miles long. It heads in the plateau at an elevation of 1,600 feet, and in the 4 miles of its length falls approximately 800 feet, averaging about 200 feet to the mile. Although gold was probably discovered here as early as 1900, active efforts at mining were not begun until 1902. In that year several thousand dollars in gold were produced along the creek. In general, the workings have yielded scarcely more than wages to the men employed.

The valley of the creek is broad. Nowhere along its course is a definite canyon developed. The bed rock consists of dark-colored graphitic and calcareous mica schist similar to that which prevails along this part of Kougarok River. In the creek bed and extending out from it for an undetermined distance are deposits of more or less stratified gravel, which in the bank are overlain by nearly pure ice covered by muck and moss. The creek bed is practically incised in the gravel deposits that cover the west wall of the Kougarok Valley. Some gold has been found in the gravel floor of Homestake Creek from its mouth to its head. One-fourth mile from the river the deposit consists of 4 to 5 feet of muck and fine sand resting upon bed rock. A pit about 20 by 50 feet in the creek bed has been mined out. The gold obtained is generally in larger and thicker pieces than the river gold and included one nugget worth $14.40.

Half a mile from the mouth a crosscut ditch showed about 5 feet of gravel resting upon bed rock. It is reported that on this claim some pans yielded as much as 25 cents each, and further tests show that all pans carried some gold, but that no continuous body of pay gravel was found.

One mile from the mouth a body of pay gravel 8 feet wide and 40 feet long has been worked out. The values in some places extended under the banks, beyond the limit of the workings, and probably some values have been left in the tailings. The owners of the claim estimated that the pay streak is possibly 40 feet wide. It is reported that near the head of Homestake Creek a pay streak 25 feet wide has been followed for a distance of 65 feet along the creek. The gold was coarse and the highest values were near the creek bed. It was not mined at the time of visit, as to work the claim economically a long drain ditch is needed.

---

[a] Description by Messrs. Collier and Hess.

In general, the pay streak along Homestake Creek is from 8 to 40 feet wide, and is confined to the valley floor, but not necessarily to the present stream bed. The gold is of a little lower grade than that from Kougarok River, and is in the main darkly stained.

MACKLIN CREEK.[a]

Macklin Creek joins the Kougarok at the great bend 40 miles above its mouth. The two valleys have the same axis, running about east and west, and the combined streams run southward, thus forming a T. An attempt to work placers along this creek in 1901 is described in a previous report,[b] but no work was in progress in 1903. Some spots of comparatively rich gravel have been found, but their development has been stopped by lawsuits. It seems probable that some of the gravels of Macklin Creek carry sufficient values to yield fair returns if worked economically.

COLUMBIA CREEK.[a]

Columbia Creek is another small tributary of Kougarok River about 6 miles above the mouth of Macklin Creek. It heads near Homestake Creek and flows northward. It is reported that twenty claims have been located along the creek, and that during 1903 one was being developed. The gold is said to be coarse, the largest nugget found being worth $10. Some of it is stained brown and black, but the rest is bright yellow. It is all more or less angular and some of it has quartz attached. Two men working on this claim obtained $220. There is probably considerable low-grade gravel along this creek that can not be worked profitably under present conditions.

CONCLUSIONS.

The investigations on which this report is based were entirely too inadequate to permit a final word on the value of the auriferous gravels of the district. That there are extensive alluvial deposits which may carry sufficiently high values to yield returns for economical mining no one can deny who has studied the field carefully. It is equally well known that as yet, with the exception of a few claims, no gravels of very high grade have been developed. Certain conditions already referred to are favorable to the probable extension of the placer-mining industry. One of these is the wide extent of the mineralization. If, as stated elsewhere in this report (p. 122), the zones of mineralization of the peninsula are most commonly found along or near the contacts of mica schist and limestone, then placers should be expected in the Kougarok region. As in other mineral-bearing districts of the peninsula, the bed rock is closely folded, faulted, and fractured, and mineralized quartz stringers are not uncommon, but have not been tested as to their gold content. In at least two localities the gold has been traced to its bed-rock source in the schists. So far as the present studies can determine, the bed

---

[a] Description by Messrs. Collier and Hess.
[b] Collier, A. J., Reconnaissance of the northwestern portion of Seward Peninsula, Alaska: Prof. Paper U. S. Geol. Survey No. 2, 1902, p. 64.

rock is no less favorable for the occurrence of gold here than in other districts of the peninsula.

The history of this province since it was last elevated above the sea, interpreted according to theories elsewhere presented, points to the concentration of gold in the alluvial deposits. The various epochs of erosion indicated by the bench deposits would promote the concentration of the heavier materials in the gravels. In several localities on Kougarok River the gold was probably derived by reconcentration from older elevated placers. Yet it must be said that, in spite of this reconcentration, the resulting placers have not been found to be as rich as those of similar origin in other parts of the peninsula. This fact points toward the conclusion that the bed-rock source is not as heavily mineralized as in some of the other districts. The lower bench gravels of the Kougarok and of some of its tributaries are undoubtedly among the most important deposits of the district, if only because of their favorable position for cheap mining. The highest gravels (i. e., those above 50 or 60 feet) reported at various places have now little prospect of development unless they are far richer than any of the other deposits. Their topographic position makes it difficult, if not impossible, to hydraulic except at great cost. Experience has shown that the abundance of ground ice, the limestone masses, and the heavy talus all combine to make ditch construction and maintenance expensive.

The writer is unable to make a definite statement as to the gold tenor of the gravels in this field, for the results of the little prospecting that has been done are not available for this report. When the meager evidence is carefully weighed, it seems probable that $1 to the cubic yard must be considered high value for most of the placers of the district. Whether or not there are considerable bodies of gravel which carry such values, the writer is not prepared to state. While a gold tenor of $1 would be considered very rich in most placer camps, it is low compared with that of some of the auriferous gravels of Anvil and Ophir creeks. Nevertheless it is probable that gravels can be profitably mined at but a fraction of this amount in many places in this district.

The two dry summers of 1900 and 1906 make it evident that such climatic conditions must be reckoned with in counting cost, especially where large investments of money are made. Though during a wet season water is abundant, nevertheless the Kougarok region has no such reservoirs to draw on as the Kigluaik Mountains, which are being tapped by the Nome ditches, and this fact is emphasized by a dry season like that of 1906. At the present rate of ditch building every possible source of water supply will soon have been utilized. Here, as elsewhere, more careful prospecting of the ground would probably have curtailed some of the ditch building. It appears that some operators have been too ready to believe without adequate prospecting that the values in the ground were sufficient to warrant large expenditures for ditch construction. This hit-or-miss style of mining has fewer odds against it in regions where the hope of finding bonanzas is better than in the Kougarok. It certainly can find no place in a region where the question of costs has to be carefully considered.

The Kougarok does not appear to be an inviting field for the miner without capital. Though considerable gold has been recovered by pick and shovel, on the whole the values thus far developed are not high enough to yield profits by such simple methods of mining. This is certainly true now, but conditions may alter with the reduction of costs of labor and supplies.

To recapitulate briefly, the following facts appear to be established: (1) Prospecting up to the present time, so far as known to the writer, has not established the existence of many bonanzas. (2) There are some extensive deposits of heavy auriferous gravels, yet it appears that but few of them have been sufficiently prospected to prove their values. (3) Water is far from abundant, but in many localities during most seasons is probably sufficient. (4) Mineralization, however, is widespread, as is also the gold in the alluvium. (5) Some of the bench deposits are very favorably located for profitable exploitation by hydraulic methods. (6) There is probably some ground which can be dredged, but as yet few facts in regard to it are available.

In the opinion of the writer the Kougarok district may become one of the important gold producers of the peninsula, though it is not to be expected that its output will ever be comparable to that of some of the older districts, such as Nome and Ophir Creek. It is a field where profits can be expected only by a careful counting of costs and conservative business management.

GEOLOGY AND MINERAL RESOURCES OF IRON CREEK.[a]

By Philip S. Smith.

INTRODUCTION.

Iron Creek, one of the largest tributaries of the Kruzgamepa, joins that river near the great bend about 11 miles east of Salmon Lake. Although really continuous, Iron Creek bears three names in different parts of its valley; thus from its mouth to Left Fork, a distance of 7 miles, the stream is called Iron Creek; above Left Fork as far as Eldorado Creek, a distance of 1 mile, it is called Dome Creek, and from Eldorado Creek to the divide it is called Telegram Creek. This confusion of names is due to the interpretation of the mining laws which permits the staking of additional claims on different creeks— i. e., creeks having different names. There are four main tributaries, the three largest being from the south and the fourth and smallest from the north.

Owing to the fact that some errors occur in the reconnaissance map of 1900,[b] the only map prepared by the Geological Survey of this region, it has seemed advisable to correct such inaccuracies as were noted in a hasty trip along the stream in 1906. Much assistance in platting the district was afforded by the transit notes of a ditch survey made by J. M. Love, of the Gold Beach Development Company. A corrected map of the Iron Creek basin is shown in fig. 19. It will be noted that this basin is roughly triangular. The western side of the triangle forms the divide between the drainage of the Kruzgamepa and that of Iron Creek. The southern side of the triangle in the western part separates the Iron Creek drainage from the headwaters of Gassman and Venetia creeks, both tributaries of Eldorado River. The divide from Venetia Creek is low, being only about 800 feet above the mouth of Iron Creek, or 1,000 feet above the sea, so that it affords a good route for a road to Nome. The eastern portion of the southern side of the triangle forms the divide between the Iron

[a] Extract from Bull. U. S. Geol. Survey No. 314, 1907, pp. 157–163.

[b] Reconnaissances in Cape Nome and Norton Bay regions, Alaska, in 1900; a special publication of the U. S. Geol. Survey, 1901, pl. 17.

Creek and Casadepaga drainages. A low pass, with an elevation of about 1,000 feet, permits a fair wagon road to run up Telegram Creek and down Lower Willow Creek to the Council City and Solomon River Railroad, a distance from the junction of Iron and Canyon creeks of 13 miles. Now, however, that the Seward Peninsula Railway from Nome to the Kougarok has been completed beyond the mouth of Iron Creek, it is probable that with reasonable freight rates the use of the wagon roads will decrease, though freight is now

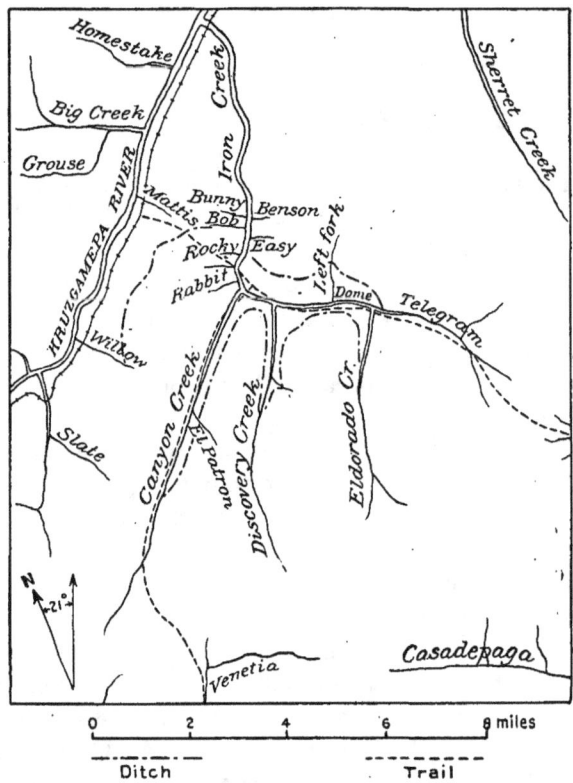

Fig. 19.—Sketch map of Iron Creek basin.

delivered at Iron Creek by winter hauling from Nome for only 2 cents a pound. The northern side of the Iron Creek drainage basin forms the divide from Sherret Creek and several smaller streams which flow northward into the Kruzgamepa.

PHYSIOGRAPHY.

STAGES OF VALLEY CUTTING.

The physical features of the district are complex, and only the more striking facts can be presented here. An older topography, in which the present stream is intrenched, is preserved in the upper slopes of the valley walls. In this portion the bed rock is so covered over that exposures are practically wanting. This is due to a period of erosion and the accumulation of a heavy mantle of waste that reaches nearly to the top of the divide. The higher portions of the divides are generally bare, consisting of fantastically curved pinnacles of rock from which taluses with steep slopes, practically uncovered by vegetation, descend; gradually merging with the smoother moss-covered slopes of the middle portion of the valley walls.

The streams in their lower courses flow in rock-walled canyons. In tracing any one of the streams headward the canyon is found to decrease in height and to merge gradually with the older topography previously mentioned. The history of these features suggests that the former topography of gentle slopes and wide, open valleys was produced by the long-continued erosion of rivers and weathering. Subsequently uplift of the region renewed the down-cutting power of the streams, so that canyons were carved in the floors of the old valleys. This erosion allowed rapid reassortment of the old gravels and waste and thus effected the concentration of any gold or other heavy minerals which may have been contained in the gravels.

The canyon cutting ceased, however, before it had progressed beyond the lower portions of the streams. The interruption was produced either by a movement of the land or, as is more probably the case, by a climatic change which decreased the amount of water transported by the streams. Such a change may have also been responsible for the disappearance of the local glaciers which were formerly present in the Kigluaik Mountains. Whether the climatic change had anything to do with the retreat of the glaciers or not is, however, of slight importance in this discussion. Some change must have occurred, for the streams are no longer down-cutting, but actually building up the floors of their valleys. The reason for believing that the change must have been one which affected the rainfall rather than the elevation of the district with respect to sea level is based on the shape of the rock canyon. The canyon has a broad, swinging course which is so symmetrical that it could not have been produced by the straggling present stream, which occupies only a small portion of the floor between the rock walls. Many other streams in different parts of Seward Peninsula show this same feature. The extensive development of this phenomenon suggests a widespread cause, such as climatic change, rather than a local cause, such as uplift.

EVIDENCES OF GLACIATION.

Another feature of some theoretical interest is the presence of granite bowlders on the divide near the low sag at the head of Mattis Creek. In the rapid reconnaissance it was impossible to examine the district with sufficient care to make a final statement as to the origin of these bowlders. It is known, however, that there is no granite of similar character south of the Kigluaik Mountains. Furthermore, the granite bowlders are unweathered, showing that they have not been in their present position a very long time geologically. Although the question has not been carefully studied in the field, it is suggested that possibly these bowlders have been brought by glaciers from Kigluaik Mountains and carried into their present position by ice blocks floating on a lake formed by glacial obstruction of the drainage. This suggestion is to be regarded only as a working hypothesis, but it fits in with the known facts, which may be summarized as follows: The angular, unweathered form and foreign character of the granite and the presence of shore lines at considerable elevations. Lakes of this type are common in regions that are at present glaciated, and evidences of such lakes have been recognized in many places where glaciers have now disappeared.

GENERAL GEOLOGY.

The bed rock of the district belongs to the Nome series. It consists of a series of much faulted and contorted limestones and schists and some greenstones. The greatest development of limestone occurs in the lower part of Iron Creek, but a great number of thinner beds interlaminated with schists are encountered even up to the headwaters. It is believed that the numerous alternations of schist and limestone are indicative of the source of the Iron Creek gold. Although no extensive proof of mineralization at the limestone-schist contacts has been found in this locality, the fact that such contacts are the loci of mineralization has been very well established in other parts of Seward Peninsula.

The rocks of the Iron Creek district trend northeast and southwest and dip toward the southeast, but there are numerous exceptions to this general direction, as the rocks are complexly folded and faulted. The deformation and consequent shattering that the rocks have undergone has undoubtedly resulted in the formation of zones of pervious rock in which mineralization has taken place. The streams also have taken advantage of the northeast-southwest structure and practically all the tributaries are arranged parallel to this direction.

In lithologic character the rocks are similar to the Nome series as described for other parts of the field. The schists present two main lithologic facies, namely, graphitic and chloritic. No boundary between the two can be drawn at the present time, although it is believed that detailed study would solve their interrelation and structure. The chloritic schists are most extensively developed in Iron Creek below Telegram Creek. They are thinly laminated, with wavy cleavage, and are rusty brown to greenish gray in color. Chlorite and quartz are the only minerals distinguishable in hand specimens. Graphitic schists are most abundant on Telegram Creek above Eldorado Creek. These rocks are in general but slightly schistose and would better be described as dark, nearly black, graphitic quartzites, with here and there schistose phases. Hand specimens show considerable quartz and a little chlorite. The other constituents are not distinguishable by the eye, through the presence of graphite is recognized by its soiling the hands. Here and there some sulphides are found, especially in the places where dislocations occur.

The greenstones which occur in the Iron Creek region have not been studied in detail, but seem to be similar to those found in the adjacent country nearer Nome. If this correlation is correct, they are mainly of intrusive origin. Rumors were heard of an extrusive flow of greenstone south of Iron Creek, but neither was it found in place nor was any float of an extrusive greenstone seen, so that doubt is felt about the occurrence of a surface flow.

MINING DEVELOPMENTS.

GENERAL CONDITIONS.

Mining on Iron Creek has been much retarded by the inaccessibility of the region, but this obstacle is now disappearing with the building of railroads and wagon roads. Freight from Nome can now be delivered by the Seward Peninsula Railway at the mouth of Iron Creek, but the schedule of rates was not learned. It has already been noted that in winter supplies can be brought in by team at a cost of 2 cents a pound. The cost of summer hauling by team to Iron Creek is now, owing to the fair condition of the road to Nome, but little higher than the winter rate.

DITCH CONSTRUCTION.

In 1906 work on Iron Creek and its tributaries had almost ceased at the time of the writer's visit in the later part of September. With one exception the work for the season seemed to have been carried on by small outfits of only one to five men each, and a liberal estimate of the output of the creek and tributaries for the year would not exceed $50,000. The most important work during the last season has been ditch construction, about 13 miles having been built. One ditch taps Eldorado Creek at a point 1 mile above its junction with Telegram Creek and leads the water along the south wall of Iron Creek to a penstock near the junction of Discovery and Iron creeks. A second ditch takes water from Canyon Creek 5 miles above its mouth and leads it along the east wall of Iron Creek, thence following the south slope of the valley to the west side of Discovery Creek, along which it runs southward to a point 2 miles above the mouth of the stream, where it crosses and extends along the east side of the valley to the penstock previously noted. Another ditch on the north side of Iron Creek, which takes its water from the junction of Telegram and Eldorado creeks, is also being constructed by the same company. Between sixty and seventy men at a time have been employed in the construction of these ditches. They were not completed until the latter part of September, so that water for washing the gravels was available for only about two weeks. The ground operated at present by the company is on Iron Creek at the mouth of Discovery Creek. A hydraulic elevator has been installed to handle the flood-plain gravels, and active mining operations will be conducted during the coming year.

MINING ON MAIN STREAM.

Between Discovery Creek and Left Fork on Iron Creek there is a fractional claim which has been worked for the last two years on a small scale. From one to five men have been employed on this claim all summer. The gold is coarse and easily saved. Both rusty and bright gold are found. The values occur in a thin pay streak on limestone and in the cracks and crevices of this bed rock. The small amount of ground in this claim has prevented any large-scale developments.

At the junction of Left Fork and Iron Creek three men have been continuously employed all summer working creek gravels. The method of working these gravels has been by means of a bed-rock drain and sluice boxes. Several nuggets worth $30 or $40 each have been found in this place. The bed rock is a much shattered limestone with thin bands of chloritic schist both above and below it.

A short distance upstream from Left Fork the largest nugget recorded from Iron Creek was found. This nugget weighed over 30 ounces and was valued at $600 on the assumption that the gold was worth $18.50 an ounce. It is a fact of some significance that upstream from this point, which is about half a mile above Left Fork, all the gold is rusty, whereas below both rusty and bright gold occur. The reason for the absence of bright gold above is believed to be that this point marks the place where the older and newer valleys merge. In other words, upstream the creek flows in the nearly unmodified old valley, while downstream it has cut below that level. The result of the down cutting has been to wear some of the gold and expose fresh, shiny surfaces; whereas the gold that has been practically unmoved has a rusty coating.

Between Left Fork and Eldorado Creek only one camp was in operation in 1906. Five or six men have been at work at this place, but as it is understood that this portion of the creek has already been worked over three times it is doubtful whether subsequent work will be remunerative. The gravels are apparently similar to those already noted.

Above Eldorado Creek the main stream, as has been previously stated, is called Telegram Creek. One man only has been at work on this stream during the last year. This claim is located at a point about a mile from the divide. The bed rock is mostly graphitic schist with some thin limestone and schist bands. Several nuggets, worth as much as $100 apiece, have been found on this claim, and it is reported that very coarse gold is found even on the crest of the divide from Willow Creek. The water supply of Telegram Creek is small, especially in a dry year, such as 1906. Often this lack has hindered or in large measure prevented exploration of the gold gravels that have been found by prospectors in this part of the Iron Creek basin.

MINING ON TRIBUTARIES.

On the tributaries of Iron Creek but little work has been done. Bunny Creek, the fifth stream which enters from the west below Canyon Creek, in not over three-fourths of a mile in length. Two men have done a little work on this stream last summer, but it probably produced not more than $100 or $200. On Bobs Creek, the next small stream south of Bunny Creek, the only work done during the last season has been on the upper part. This claim has been worked with water brought over the divide from Willow Creek, the first tributary of the Kruzgamepa east of Rock and Slate creeks. Considerable trouble has been experienced with the ditch, as a large part of it is built on frozen ground, which melts under the water. This ditch carries only about 400 miner's inches of water. Even this small amount is more than is yielded throughout the season by Willow Creek, and it is proposed to extend the ditch next year 3 or 4 miles to Slate Creek.

Easy Creek, which enters Iron Creek from the east opposite Bobs Creek, has shown good values in the lower portion. Three men were at work at this place last summer, but closed down rather early in the season owing to the drought. Little more than assessment work has been done on the other claims along Easy Creek. The next small stream to the south is Lulu or Benson Creek. Four men have been operating on this creek the entire summer. On Rapid, Rocky, and Rabbit creeks, the three other small tributaries of Iron Creek from the west below Canyon Creek, little more than assessment work has been done during the last season, although they are completely staked,

Except on Canyon Creek, no work has been done on any of the larger tributaries of Iron Creek. On a little tributary of Canyon Creek, called El Patron Creek, about 3 miles above the junction with Iron Creek, one man has been at work all summer. However, but very little gold has been produced, owing to the lack of water. It is expected that with the completion of the Canyon Creek ditch, water may be purchased, so that work will be pushed with greater activity in the coming summer.

SUMMARY.

In summarizing the Iron Creek region it may be said that the gold is mostly coarse and easily saved; that it has been derived from a relatively local source; that water for the economic development of the placers is at hand; and that the questions of freighting and transportation are rapidly being effectively and satisfactorily settled.

INDEX.